The Earth's Ionosphere

This is Volume 43 in the
INTERNATIONAL GEOPHYSICS SERIES
A series of monographs and textbooks
Edited by RENATA DMOWSKA and JAMES R. HOLTON

A complete list of the books in this series appears at the end of this volume.

The Earth's Ionosphere
Plasma Physics and Electrodynamics

Michael C. Kelley
SCHOOL OF ELECTRICAL ENGINEERING
CORNELL UNIVERSITY
ITHACA, NEW YORK

with contributions from

Rodney A. Heelis
DEPARTMENT OF SPACE PHYSICS
UNIVERSITY OF TEXAS AT DALLAS
RICHARDSON, TEXAS

ACADEMIC PRESS, INC.
Harcourt Brace Jovanovich, Publishers
San Diego   New York   Berkeley   Boston
London   Sydney   Tokyo   Toronto

ACADEMIC PRESS, INC.
San Diego, California 92101

United Kingdom Edition published by
ACADEMIC PRESS LIMITED
24-28 Oval Road, London NW1 7DX

Library of Congress Cataloging-in-Publication Data

Kelley, Michael C.
　　The earth's ionosphere : plasma physics and electrodynamics /
　Michael C. Kelley, with contributions from Rodney A. Heelis.
　　　p.　cm. — (International geophysics series)
　　Bibliography: p.
　　Includes index.
　　ISBN　0-12-404012-8　(hardcover) (alk. paper)
　　ISBN　0-12-404013-6　(paperback) (alk. paper)
　　1. Space plasmas. 2. Plasma electrodynamics. 3. Ionosphere. .
　I. Heelis, Rodney A. II. Title. III. Series.
　QC809.P5K45　1989
　551.5'145—dc19　　　　　　　　　　　　　　　88-7575
　　　　　　　　　　　　　　　　　　　　　　　　　　CIP

DSZI

K

91　　00787

To my family and especially
to Mom, Dad, and Patricia

Contents

Chapter 4   Equatorial Plasma Instabilities

Chapter 5   The Mid-Latitude Ionosphere

Chapter 6   High-Latitude Electrodynamics

Chapter 7   Effects of Plasma Flow at High Latitudes

Chapter 8   Instabilities and Structure in the High-Latitude Ionosphere

Appendix A   Ionospheric Measurement Techniques

Appendix B   Reference Material and Equations

Preface

In the past two decades, the underlying physical principles on which ionospheric research is based have undergone a drastic change. Prior to this period the main emphasis was on production and loss of electrons and ions by photoionization and particle beams. Ion chemistry, geomagnetic field variations, and other "aeronomical" processes formed primary subfields as well. With the development of incoherent scatter radar techniques as well as scientific satellites, a whole new realm of ionospheric processes became amenable to scientific study. In particular, space scientists were able to reliably probe the dynamics of the ionosphere and to realize the fundamental role played by electric fields. In a parallel development, the physics of plasmas became an important branch of science due to its role in controlled fusion research.

In this book, study of plasma physics and electrodynamics of the ionosphere is given the highest priority, and the many aeronomical influences are given short shrift indeed. This is an unfortunate necessity. The text by Rishbeth and Garrcott, published in the same International Geophysics Series, remains a valuable treatise on these topics, however, and the reader is referred there for information on aeronomy.

The ionosphere is somewhat of a battleground between the earth's neutral atmosphere and the sun's fully ionized atmosphere, in which the earth is imbedded. One of the challenges of ionosphere research is to know enough about these two vast fields of research to make sense out of ionospheric phenomena. A single text cannot treat all three of these disciplines, but we try to give the reader some insights into how these competing sources of mass, momentum, and energy vie for control of the ionosphere.

After some introductory material, the study is begun in earnest with the equatorial ionosphere. Here the earth's magnetic field is horizontal, and many unique and fascinating phenomena occur. It is interesting to note also that the Coriolis force vanishes at the geographic equator, which makes for some unusual ocean and atmospheric dynamics as well. In Chapter 3 the electrodynamics of this zone are studied, while in Chapter 4 the plasma instabilities are discussed.

In Chapter 5 the mid-latitude zone is studied and, in particular, the role of gravity waves and neutral wind dynamics is emphasized. Again both electrodynamics and plasma instabilities are covered. This is the ionospheric footprint of that portion of the near-space region of the earth which is dominated by the rotation of the planet.

In Chapter 6 the interaction of the solar wind with the magnetic field of the earth is reviewed, with particular regard to the imposition of electric fields and field-aligned currents on the ionosphere. At high latitudes the imposed electric field from this source overcomes the field due to the rotation of the earth. The influence of the interplanetary magnetic field is described as are the flow patterns which arise in the ionosphere. The effect of these electric fields on the ionospheric plasma as well as on the neutral atmosphere is the topic of Chapter 7. In the final chapter, a number of processes which create structure in the high-latitude ionosphere are presented, including, but not limited to, the plasma instabilities which occur there.

Throughout the text we treat only the region above 90 km. This ignores the so-called D region entirely. Although unfortunate, this is somewhat justified by the decreasing importance of plasma physics as the guiding discipline in the dense and lower atmosphere.

I am very much indebted to legions of students who have read and constructively criticized the text as it evolved, as well as to my many colleagues at Cornell and throughout the world. The study of space plasma physics is by definition an international discipline and epitomizes the way in which the human race can cooperate perfectly well at the highest technical and interpersonal levels. I have learned a great deal about both space physics and my fellow passengers on spaceship earth while traveling around to various rocket ranges, radar observatories, and scientific laboratories. Special thanks go to Rod Heelis who helped considerably with Chapters 6 and 7. My chief hope is that this text puts a little of that knowledge back into circulation in an understandable way.

Michael C. Kelley

Chapter 1 | Introductory and Background Material

In this introductory chapter we present a qualitative treatment of some peripheral topics that we hope is sufficient to proceed with our study of ionospheric physics. The chapter begins with historical comments and a description of the limitations we have set for the text. In particular, we do not repeat or significantly update the material published by Rishbeth and Garriott (1969) earlier in this same International Geophysics Series. Rather, our emphasis is on electrodynamics and plasma physics and we refer the interested reader to the former text for more information about formation of the ionosphere, its ion chemistry, heat balance, and other aeronomic properties.

1.1 Scope and Goals of the Text

1.1.1 Historical Perspective

The earth's ionosphere is a partially ionized gas that envelops the earth and in some sense forms the interface between the atmosphere and space. Since the gas is ionized it cannot be fully described by the equations of neutral fluid dynamics. In fact, a major revolution in ionospheric physics has occurred in the past decade or so as the language and concepts of plasma physics have played an increasing role in the discipline. On the other hand, the number density of the neutral gas exceeds that of the ionospheric plasma and certainly neutral particles cannot be ignored. A student of the ionosphere must thus be familiar with both classical fluid dynamics and plasma physics. Even a working knowledge of these two "pure" branches of physics is not sufficient, however. Since the ionosphere lies at the interface between two very different and dynamic media, we must understand enough of both atmospheric dynamics and deep space plasma physics to understand how the ionosphere is formed and buffeted by sources from above and below. Added to these two is the requirement for a sufficient knowledge of ion chemistry and photochemistry to deal with production and loss processes.

Ionospheric physics as a discipline grew out of a desire to understand the origin and effects of the ionized upper atmosphere on radio wave propagation. The very discovery of the ionosphere came from radio wave observations and the recognition that only a reflecting layer composed of electrons and positive ions could explain the characteristics of the data. Most of the early work was aimed at explaining the various layers and their variability with local time, latitude, season, etc. The ionosonde, a remote sensing device yielding electron density profiles up to but not above the altitude of the highest concentration of charged particles, was the primary research tool. Such measurements revealed a bewildering variety of ionospheric behavior ranging from quiet, reproducible profiles to totally chaotic ones. Furthermore, a variety of periodic and aperiodic variations was observed with time scales ranging from the order of the solar cycle (11 years) to just a few seconds. As time passed, the emphasis of ionospheric research shifted from questions dealing with formation and loss of plasma toward the dynamics and plasma physics of ionospheric phenomena.

Ionospheric research has greatly benefited from the space program with the associated development of instruments for balloons, rockets, and satellites (see Appendix A). The combination of remote ionospheric sensing and direct *in situ* measurements made from spacecraft has accelerated the pace of ionospheric research. Equally important has been the development of plasma physics as a theoretical framework around which to organize our understanding of ionospheric phenomena. Likewise, the continuing development of magnetospheric and atmospheric science has very much influenced our understanding of the ionosphere.

In parallel with these developments, the incoherent scatter radar technique was devised (see Appendix A) and a number of large facilities built to implement it. The primary advantage of this method was the ability to make quantitative measurements of numerous ionospheric parameters as a function of altitude at heights inaccessible to ground-based ionosondes. Since interpretation of the scattered signal requires detailed understanding of the interaction between the electrodynamic waves and thermal fluctuations in a plasma, a working knowledge of plasma physics became necessary for understanding the diagnostics as well as the science.

Ionospheric science has thus evolved toward the point of view that is encompassed by the term space plasma physics. The central theme of this text is the treatment of ionospheric physics within this context.

1.1.2 Organization and Limitations

As indicated in the previous section, our goal is to treat the electrodynamics and plasma physics of the ionosphere in some detail. We do not, therefore, have the space to start from the first principles of ionospheric science, particularly with regard to photoproduction, ion chemistry, recombination, and related topics. In this first chapter, we attempt a broad general introduction to this area that we hope will provide enough background to make the remainder of the text meaningful.

It is also necessary to introduce elements of neutral atmospheric and magnetospheric physics since the ionosphere is very much affected by processes that originate in these two regions. The ionosphere coexists with the upper portion of the neutral atmosphere and receives considerable energy and momentum from the lower atmosphere as well as from the magnetosphere. The energy and momentum fluxes are carried by particles, electromagnetic fields, and atmospheric waves. We devote some time in this chapter to a qualitative description of lower atmospheric and magnetospheric science. Although useful as a starting point, the descriptive approach is not sufficient in all aspects and we will treat some particularly important processes such as gravity waves in detail as they arise in the text.

Some understanding of the experimental techniques alluded to above is necessary for a student of ionospheric dynamics, if only to understand the sources of the data. Accordingly, an appendix is included describing some of the most important measurement methods. The choice of instruments so dealt with is not exhaustive and certainly reflects the bias and expertise of the author.

Finally, we need to limit the scope of the text. We have somewhat regretfully chosen not to include a detailed discussion of the acceleration of auroral particle beams. Part of the rationale for this lies in the height range where the electron acceleration usually takes place, some 2000 km or more above the earth's surface, which is the limiting height range of the ionosphere as we define it here. A perhaps more accurate reason for omission of particle acceleration is that the physics of this process is not yet at all well understood, certainly not by the author! Thus, even though particle acceleration can and does occur much lower than the 2000-km height, we consider this topic outside the domain of this book, if not always outside the height range under detailed consideration.

At the lower altitudes we ignore entirely the so-called D layer of the ionosphere, the height range below about 90 km. Our rationale is that although the ion chemistry and composition of this region are unique and interesting, the dynamics are mostly dominated by the neutral atmosphere. In some sense, plasma physics is not the proper framework for study of this region and we cannot realistically extend the scope of the text sufficiently to treat it.

In Chapter 2 some basic equations and concepts are developed for use throughout the text. Some of these concepts would be quite useful as background for reading the rest of this first chapter, particularly the magnetospheric section. We thus recommend that a reader unfamiliar with the field read Chapter 1 twice, once for background and once again after reading Chapter 2. Chapters 3 and 4 deal with the electrodynamics and plasma physics of the equatorial ionosphere. This region is singled out since the fact that the earth's magnetic field is horizontal there leads to a number of unique phenomena. An interesting analogy exists with atmospheric and ocean dynamics. Since the Coriolis force vanishes at the equator, there are some unique phenomena common to both meteorology and oceanography that make equatorial dynamics very unusual and interesting. Chapter 5 deals with the tropical and mid-latitude ionosphere, where the earth's magnetic field has a sizable inclination but is not vertical and does not link the ionosphere with the hot, tenuous, flowing plasmas of the magnetosphere or solar wind. In the remainder of the text after Chapter 5 we study the high-latitude region. As mentioned above, several experimental techniques are described in Appendix A. In Appendix B some formulas and tables are gathered that help in describing ionospheric phenomena. In that appendix we also define various magnetic activity indices used in the text, such as K_p and DST.

1.2   Structure of the Neutral Atmosphere and the Ionosphere

Owing to the pervasive influence of gravity, the atmosphere and ionosphere are to first order horizontally stratified. Atmospheric structure can be neatly organized by a representative temperature profile, while the ionosphere is more sensibly organized by the number density of the plasma. Typical mid-latitude profiles of temperature and plasma density are given in Fig. 1.1. The atmospheric temperature initially decreases with altitude from the surface temperature with a "lapse rate" of about 7 K/km in the troposphere. At about 10-km altitude this temperature trend reverses (at the tropopause) and the stratosphere begins. This increase is due primarily to the absorption, by ozone, of part of the ultraviolet portion of the solar radiation. This effect maximizes at 50 km, where the temperature trend again reverses at the stratopause. Radiative cooling creates a very sharp temperature decrease to a minimum in the range 130–190 K at about 80 km. For heights above the altitude of the temperature minimum (the mesopause), the temperature increases dramatically to values that are quite variable but are often well above 1000 K. Not surprisingly, this region is termed the thermosphere. The atmosphere is relatively uniform in composition below about 100 km due to a variety of turbulent mixing phenomena. Above the "turbopause" the constituents begin to separate according to their various masses.

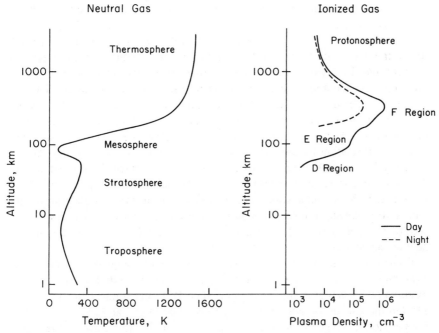

Fig. 1.1. Typical profiles of neutral atmospheric temperature and ionospheric plasma density with the various layers designated.

The temperature increase in the thermosphere is at least partially explained by absorption of another portion of the solar UV spectrum. This same energy source is also responsible for the production of plasma in the sunlit hemisphere since the solar photons have sufficient energy to ionize the neutral atmosphere. Equal numbers of positive ions and electrons are produced in this ionization process.

One requirement for a gas to be termed a plasma is that it very nearly satisfies the requirement of charge neutrality, which in turn implies that the number density of ions, n_i, must be nearly equal to the number density of electrons, n_e. Experimenters who try to measure n_i or n_e usually label their results with the corresponding title, ion or electron density. In this text, we shall usually refer to the *plasma* density n, where $n \simeq n_i \simeq n_e$ is tacitly assumed to hold. Of course, if n_i exactly equaled n_e everywhere, there would be no electrostatic fields at all, which is not the case. A note on units is in order here. Rationalized mks units will be used except in cases where tradition is too firmly entrenched. For example, measuring number density per cubic centimeter is so common that we shall use it rather than per cubic meter. Likewise, the mho and the mho per meter will be used for conductivity. Some further discussion of units and of various parameters of interest to ionospheric physics is included in Appendix B.

Returning to Fig. 1.1, two plasma density profiles are given in the right-hand part of the figure, one typical of daytime mid-latitude conditions and one typical of nighttime. In daytime, the extreme ultraviolet (EUV) radiation in the solar spectrum is incident on a neutral atmosphere that is increasing exponentially in density with decreasing altitude. Since the photons are absorbed in the process of photoionization, the beam itself decreases in intensity as it penetrates. The combination of decreasing beam intensity and increasing neutral density provides a simple explanation for the basic large-scale vertical layer of ionization shown in Fig. 1.1. The peak plasma density occurs in the so-called F layer and attains values as high as 10^6 cm^{-3} near noontime. The factor that limits the peak density value is the recombination rate, the rate at which ions and electrons combine to form a neutral molecule or atom. This in turn very much depends on the type of ion that exists in the plasma and its corresponding interaction with the neutral gas.

Some experimental data on the ion and neutral composition above 100 km are reproduced in Fig. 1.2. Below and near that height, N_2 and O_2 have the same ratio as in the lower atmospheric regions, about $4:1$, and dominate the gas. Near 110 km the amount of atomic oxygen reaches that of O_2, and above about 250 km the atomic oxygen density also exceeds that of N_2. This trend is due to the photodissociation of O_2 by solar UV radiation coupled with the absence of turbulent mixing above the turbopause. The dominance of atomic oxygen in the

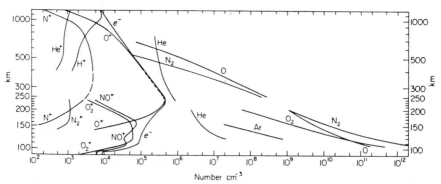

Fig. 1.2. International Quiet Solar Year (IQSY) daytime atmospheric composition, based on mass spectrometer measurements above White Sands, New Mexico (32°N, 106°W). The helium distribution is from a nighttime measurement. Distributions above 250 km are from the Elektron II satellite results of Istomin (1966) and Explorer XVII results of Reber and Nicolet (1965). [C. Y. Johnson, U.S. Naval Research Laboratory, Washington, D.C. Reprinted from Johnson (1969) by permission of the MIT Press, Cambridge, Massachusetts. Copyright 1969 by MIT.]

neutrals is mirrored by the plasma composition. The envelope of the plot is similar to the right-hand side of Fig. 1.1 and represents the electron density (thus labeled with e^-). Near the peak in the plasma density the ions are nearly all O^+, corresponding to the high concentration of atomic oxygen in the neutral gas. The altitude range 150–500 km is termed the F region and the altitude of maximum density there is termed the F peak. Below the peak NO^+ and O_2^+ become more important, dominating the plasma below about 150 km. The altitude range 90–150 km is called the E region and the ionization below 90 km is, not surprisingly, termed the D region. At the highest altitudes shown on the plot, hydrogen becomes the dominant ion in a height regime referred to as the protonosphere.

Turning to the nighttime profile in Fig. 1.1, the plasma density near the F peak is reduced in magnitude but not nearly so drastically as is the density at the lower altitudes. With reference to Fig. 1.2, this difference mirrors the change in composition. That is, in lower altitude regions where molecular ions dominate, the density is sharply curtailed at night. The O^+ plasma density, on the other hand, is sustained through the night. This trend is dramatically illustrated in Fig. 1.3 by the series of plasma density profiles that were obtained by the incoherent scatter radar facility in Arecibo, Puerto Rico (18.5°N, 66.8°W). The sunrise and sunset effects are very dramatic at the lower altitudes but are almost nonexistent in the F region.

This substantial difference in ion behavior is due to the fact that molecular ions have a much higher recombination rate with electrons than do atomic ions. The two reactions that occur in a recombination are of the type

$$NO^+ + e^- \rightarrow N + O$$

and

$$O^+ + e^- \rightarrow O + \text{photon}$$

The former process is called dissociative recombination since the molecule breaks apart, while the latter is termed radiative recombination since emission of a photon is required to conserve energy and momentum. The former process has a reaction rate nearly 1000 times higher than the latter, which results in a much shorter lifetime for molecular ions than for atomic ions. Since the molecular ions are much shorter lived, when their production is curtailed at night, rapid recombination quickly reduces the plasma concentration. The O^+ plasma at higher altitudes often survives the night at concentrations between 10^4 and 10^5 cm^{-3}. This property of atomic ion chemistry also explains the numerous sharp layers of enhanced plasma density seen at low altitudes in Fig. 1.3. These contain heavy atomic ions such as Fe^+ and Mg^+ that are deposited by meteors in this height range. Just why they are gathered into sharp layers is discussed in some detail in Chapter 5.

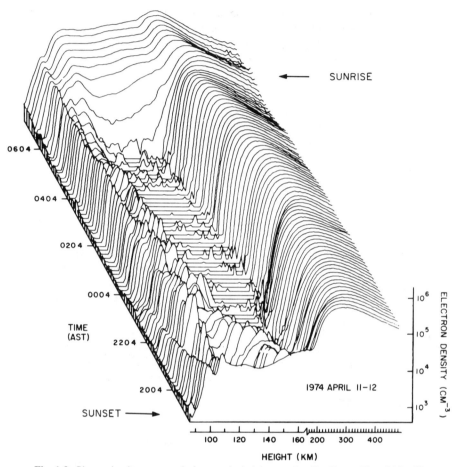

Fig. 1.3. Plasma density contours during a typical night over Arecibo, Puerto Rico. [After Shen *et al.* (1976). Reproduced with permission of the American Geophysical Union.]

Photoionization by solar UV radiation is not the only source of plasma in the ionosphere. Ionization by energetic particle impact on the neutral gas is particularly important at high latitudes. Visible light is also emitted when particles strike the atmosphere. These light emissions create the visible aurora. A view of the aurora looking down on the earth is offered in Fig. 1.4. The picture was taken on the Defense Meteorology Satellite Program (DMSP) satellite from about 800-km altitude and shows some of the complex structure and intricate detail that the auroral emission patterns can have. This complexity is mirrored in the ionization that also results from particle impact. The photograph shows how much this ionization can vary in space. Variability in time is illustrated in Fig. 1.5, where consecutive plasma density profiles taken 20 s apart are displayed. The data were taken at the Chatanika Radar Observatory in Alaska (65.06°N,

Fig. 1.4. Photograph of an active auroral display taken from a high-altitude meteorology satellite over Scandinavia. The center-to-center distance of the two fiducial dots located on the lower edge is 1220 km, and the diameter of each dot is 100 km. [After Kelley and Kintner (1978). Reproduced by permission of the University of Chicago Press.]

147.39°W). Within the 40-s period, the E-region plasma density peak varied by almost one order of magnitude. The particle energies determine their penetration depth and the distribution in their energies determines the resulting ionization profile. The typical energy of electrons in the nighttime aurora is 3–10 keV, and electron impact is the dominant ionization source. This energy range results in large plasma production in the E region as shown in the profiles in Fig. 1.5. The production rates are more than 10 times those provided by noontime photo-ionization in the E region. Lower-energy electrons dominate the auroral precipitation pattern at other local times. These particles deposit their energy at higher altitudes, creating enhanced and highly variable F-region plasma concentrations.

A more global view of the aurora is provided by the Dynamics Explorer 1 satellite imager, which has provided photographs looking down on the earth from

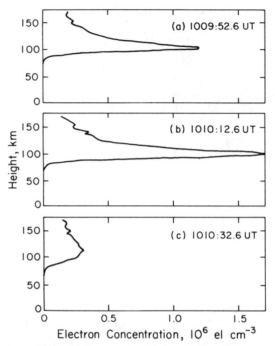

Fig. 1.5. Sequence of E-region electron concentration profiles obtained using the Chatanika, Alaska radar on September 27, 1971 near 1010 UT; azimuth = 209.04°, elevation = 76.58°. [After Baron (1974). Reproduced with permission of the American Geophysical Union.]

an altitude of several thousand kilometers. The photo in Fig. 1.6 encompasses the entire polar region and shows that the band of auroral light extends completely around the polar zone. The auroral emissions wax and wane in a complicated manner related to the rate of energy input from the solar wind as well as to the storage and release rate of this energy in the earth's magnetosphere and its ultimate release into the earth's upper atmosphere and ionosphere. Figure 1.6 shows data from a scan through the ionosphere using an incoherent scatter radar simultaneously with the DE photograph. Contour plots of the electron density in the insert show a striking correlation with the light emissions.

1.3   The Earth's Magnetic Field and Magnetosphere

To first order the earth's magnetic field is that of a dipole whose axis is tilted with respect to the spin axis of the earth by about 11°. This offset, which is common to several planetary magnetic fields, is presently such that the dipole axis in the northern hemisphere is tilted toward the North American continent. The magnetic field **B** points down toward the surface of the earth in the northern

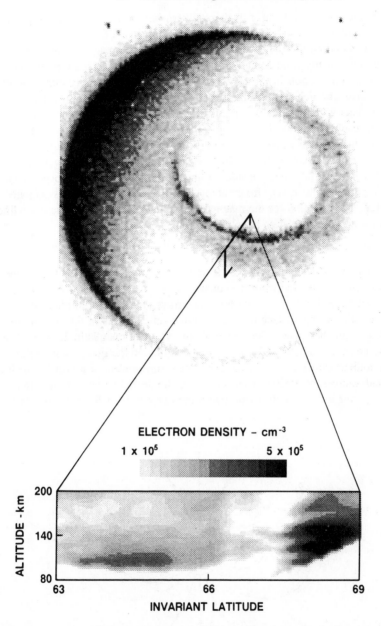

Fig. 1.6. A view of the auroral oval from the DE-1 satellite, along with simultaneous contour plots of electron density for a small portion of the photograph. [After Robinson *et al*. (1989). Reproduced with permission of the American Geophysical Union.]

hemisphere and away from it in the southern hemisphere. The dipole position wanders with time and paleomagnetic studies show that it flips over occasionally, say every hundred thousand years or so. The field is created by currents in the molten, electrically conducting core of the earth, currents that are in turn driven by thermal convection in the core. This convection is certainly quite complex but the magnetic field contributions that are of higher order than the dipole term fall off faster with distance, leaving the dipole term dominant at the surface.

Some useful equations for a dipole field are gathered in Appendix B. The field at the earth's surface varies from about 0.25×10^{-4} tesla (0.25 gauss) near the magnetic equator to about 0.6×10^{-4} tesla near the poles. Sketches of magnetic field lines are useful when considering the ionosphere and magnetosphere since plasma particles move very freely along field lines. A dipole field is sketched in Fig. 1.7. The equation for the magnetic field lines can be written in a modified spherical coordinate system as

$$r = L \cos^2 \theta$$

where r is measured in units of earth radii ($R_e = 6371$ km), θ is the magnetic latitude, and L is the radius of the equatorial crossing point of the field line (also measured in earth radii). A field line that crosses the equatorial plane at $L = 4$ thus exits the earth's surface ($r = 1$) at a magnetic latitude of 60°. Owing to the dipole tilt, this field line is located at about 49° geographic latitude in the North American sector and about 71° geographic latitude in the Euro-Asian region.

As indicated in Fig. 1.7, if the earth were surrounded by a vacuum with no external sources of electrical currents, the dipole field lines would extend in loops of ever increasing dimension with the magnitude of **B** decreasing as $1/r^3$.

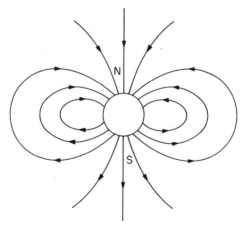

Fig. 1.7. Sketch of dipole magnetic field lines extending into a vacuum.

The earth is, however, immersed in the atmosphere of the sun. Like the earth's thermosphere, the upper atmosphere of the sun is very hot, so hot that the hydrogen and helium can escape gravitational attraction and form a steadily streaming outflow of material called the solar wind. Because of its high temperature and constant illumination by the sun, the solar wind is a fully ionized plasma (unlike the ionosphere, which contains more neutral particles than plasma). One surprising feature of the solar wind is that it is supersonic. Simply heating a gas cannot create supersonic flow but a combination of heating, compression, and subsequent expansion can create this condition, as it does, for example, in a rocket exhaust nozzle. In the solar wind case the solar gravitational field acts analogously to the rocket nozzle and the solar wind becomes supersonic above a few solar radii. The sun has a very complex surface magnetic field created by convective flow of the electrically conducting solar material. Sunspots, in particular, have associated high magnetic fields. The expanding wind drags the solar magnetic field outward, with the result that the earth's magnetic field is continually bathed in a hot, magnetized, supersonic collisionless plasma capable of conducting electrical current and carrying a large amount of kinetic and electrical energy.

Some of the solar wind energy flowing by the earth finds its way into the ionosphere and the upper atmosphere. There it powers the aurora, creates magnetic storms and substorms which affect power grids and communication systems, heats the polar upper atmosphere, drives large neutral atmospheric winds, energizes much of the plasma on the earth's magnetic field lines, and creates a vast circulating system of hot plasma in and around the earth's nearby space environment. We cannot come close to exploring these topics in any detail, but we need a framework about which to organize the ionospheric effects of these phenomena.

Suppose we first ignore the supersonic nature of the solar wind as well as the interplanetary magnetic field (IMF) and surround the earth and its magnetic field with a subsonic, streaming plasma. The magnetic force on a particle of charge q moving at velocity $\mathbf{V}$ is

$$\mathbf{F} = q\mathbf{V} \times \mathbf{B} \qquad (1.1)$$

As illustrated in Fig. 1.8, because of the polarity of the earth's field, this force will deflect solar wind ions to the right and electrons to the left as they approach the earth. Once deflected, they will spiral around the magnetic field lines and drift around the earth. Since the magnetic field strength increases as the plasma approaches the earth, a distance is eventually reached where the force is sufficient to keep the particles from penetrating any farther and they flow around the obstacle. Notice that since ions are deflected toward dusk and electrons toward dawn, a net duskward current exists in a thin sheet (extending also out of the plane of the diagram) where the force balance occurs. The secondary magnetic

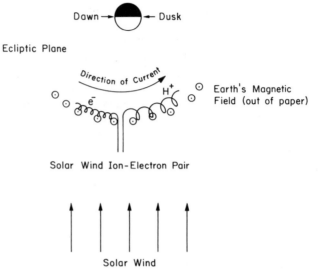

Fig. 1.8. Schematic diagram showing deflection of solar wind particles by the earth's magnetic field. The view is in the ecliptic plane.

field generated by this current sheet is parallel to the earth's field in the region between the earth and the current sheet and is antiparallel to the earth's magnetic field in the solar wind. This magnetic field cancels the earth's field on the sunward side of the boundary and increases the value of the magnetic field inside the current sheet. If this were the entire interaction, the resulting configuration of flow and field would look something like the sketch in Fig. 1.9a. The volume inside the elongated region is termed the magnetosphere, the region dominated by the earth's magnetic field. Experiments show that the magnetosphere is indeed terminated by a very sharp boundary on the sunward side, which is called the magnetopause, but is very much more elongated in the antisunward direction than the sketch shows. (We return to this point later when we include the interplanetary magnetic field in the discussion.) A configuration such as that in Fig. 1.9a is called a closed magnetosphere since there is almost no direct access to particles from the solar wind and all of the magnetic field lines in the magnetosphere have two ends on the earth. The sketch in Fig. 1.9b shows the complication that arises when the solar wind is supersonic. A shock wave must form, across which the solar wind density and temperature rise abruptly. The solar wind velocity decreases, allowing for subsonic flow around the obstacle. The sketch in Fig. 1.9b also shows that a "closed" magnetosphere need not be static even though particles cannot readily enter. A "viscous" interaction may occur (Axford and Hines, 1961) in which plasma inside and near the flanks of the magnetosphere is weakly coupled to the solar wind flow, possibly by waves that

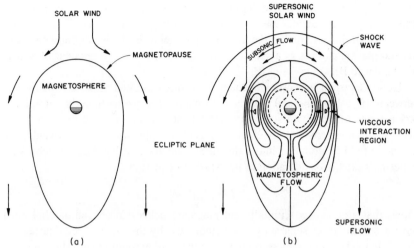

Fig. 1.9. (a) Schematic diagram of a closed magnetosphere. The bow shock wave and internal flow pattern are sketched in (b).

propagate across the boundary. Some magnetospheric plasma is then dragged along in the antisunward direction, and because of the resulting pressure buildup in the nightside of the magnetosphere, there must be a flow back toward the sun in the center of the magnetosphere. The circulation that results (sketched in Fig. 1.9b) is in good qualitative agreement with observations but is now thought to be less important than the magnetospheric flow driven by a process termed "reconnection," which involves the interplanetary magnetic field and results in a partially open magnetosphere. This process is described next.

Before discussing this, we must point out that both the single-particle dynamic and magnetofluid dynamic viewpoints discussed in Chapter 2 are such that the relationships

$$\mathbf{V}_\perp = \mathbf{E} \times \mathbf{B}/B^2 \tag{1.2a}$$

and

$$\mathbf{E}_\perp = -\mathbf{V} \times \mathbf{B} \tag{1.2b}$$

hold in a tenuous, magnetized plasma, where $\mathbf{E}$ is the electric field in the plasma, $\mathbf{V}$ is its flow velocity, and the subscript $\perp$ indicates the component perpendicular to $\mathbf{B}$ (Schmidt, 1966). These two expressions are, in fact, equivalent, since taking the cross product of (1.2b) with $\mathbf{B}$ and dividing by B^2 yields (1.2a). For our present purposes we accept these expressions as accurate relationships between $\mathbf{E}_\perp$ and $\mathbf{V}_\perp$. In the direction parallel to $\mathbf{B}$, charged particles move very freely.

This means that magnetic field lines usually act like perfect electrical conductors, transmitting perpendicular electric fields and voltages across vast distances with no change in the potential in the direction parallel to **B**. Thus, any flowing magnetized plasma can act as a source of voltage if there is a component of **V** perpendicular to **B**.

These electrical properties become very important factors in the interaction between the solar wind and the magnetosphere once we include the interplanetary magnetic field in our solar wind model. Schematic views of the solar magnetic field are given in Figs. 1.10 and 1.11. As the solar wind expands, the magnetic field is stretched into a disklike geometry that has flutes much like a ballerina's skirt. Consideration of the Maxwell equation

$$\nabla \times \mathbf{B} = \mu_0 \mathbf{J}$$

shows that a geometry with adjacent magnetic fields that are antiparallel must have a current sheet separating them (indicated by the crosshatched surface in Fig. 1.10). The entire pattern rotates with the 27-day rotation period of the sun. As the pattern rotates past the earth, the position of the current sheet is alternately above and below the ecliptic plane. As illustrated in Fig. 1.11, the interplanetary magnetic field can be considered approximately as a series of spirals emanating

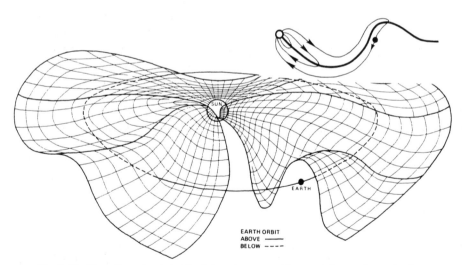

Fig. 1.10. Three-dimensional sketch of the solar equatorial current sheet and associated magnetic field lines. The current sheet is shown as lying near the solar equator with spiraled, outward-pointing magnetic fields lying above it and inward-pointing fields lying below it. The current sheet contains folds or flutes. When the sun rotates, an observer near the ecliptic will alternately lie above and below the current and will see a changing sector pattern. The inset above shows a meridional cross section with the earth below the current sheet. (Figure courtesy of S.-I. Akasofu.)

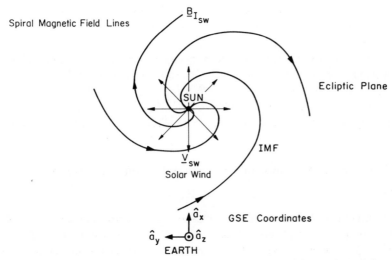

Fig. 1.11. The spiral magnetic field viewed in the ecliptic plane as it is stretched out from the solar surface by the solar wind. The illustrated coordinate system is the GSE system. In this sketch there is one "away" sector and one "toward" sector in the magnetic structure of the solar wind as it sweeps by the earth.

from the sun. Near the earth the magnetic field in the spiral is directed toward or away from the sun depending on whether the current sheet is above or below the plane. In addition, the inclination of the current sheet with respect to the ecliptic plane produces a northward or southward component of the IMF relative to an axis normal to the plane. Considerable small-scale turbulence structure exists in both the solar wind velocity and magnetic fields, and the interplanetary medium is also greatly affected by shock waves and solar flares. Taken together, these properties of the solar wind result in a highly variable buffeting of the earth's magnetosphere.

It is customary to specify the solar wind parameters in terms of three mutually perpendicular components with respect to fixed axes. In these geocentric solar ecliptic (GSE) coordinates, the x axis points from the center of the earth to the sun while the z axis is positive parallel to the earth's spin axis (to the north) and is perpendicular to the ecliptic plane. The y axis makes the third mutually perpendicular right-handed system. Figure 1.11 shows the situation looking in the negative z direction down onto the ecliptic plane. The interplanetary magnetic field is stretched out by the radially flowing solar wind but remains anchored to the rotating sun. The resulting geometry is similar to that of the water stream from a rotating garden sprinkler. At the earth's location this "garden hose angle" is such that the y component and the x component of the interplanetary magnetic field ($\mathbf{B}_{sw}$) are roughly equal and usually have opposite signs. The z component may be positive or negative, and the terms "B_z positive" or "northward" and "B_z negative" or "southward" are used to describe the orientation of the IMF with respect to the z axis.

We now briefly describe the interplanetary electric field. To first order the solar wind blows radially outward from the sun, so that in the ecliptic plane

$$\mathbf{V}_{sw} = -|\mathbf{V}_{sw}|\hat{a}_x$$

In a reference frame fixed to the earth, there will be an electric field given by

$$\mathbf{E}_{sw} = -\mathbf{V}_{sw} \times \mathbf{B}_{sw}$$

This will generate a potential difference across the earth's magnetosphere given by

$$\mathbf{V}_m = -\mathbf{E}_{sw} \cdot L\hat{a}_y$$

where L is the effective width of the magnetosphere perpendicular to $\mathbf{V}_{sw}$. Note that when $\mathbf{B}_{sw}$ has a positive z component, $\mathbf{E}_{sw}$ has a dusk-to-dawn direction (negative y component). When $\mathbf{B}_{sw}$ has a negative z component the interplanetary electric field is in the dawn-to-dusk direction across the earth. For typical values of $V_{sw} = 500$ km/s and $B_{sw_z} = 5 \times 10^{-9}$ tesla, $E_{sw_y} = 2.5$ mV/m. Integrating this electric field across the front of the magnetosphere, which is roughly a distance $L = 20R_e$ across, yields an available potential difference of the order of 300,000 volts. The magnitude of this voltage and its polarity fluctuate along with the parameters V_{sw} and B_{sw_z}. The earth is thus immersed in a magneto-hydrodynamic electrical generator capable of hundreds of kilo-electron-volts of potential.

We now turn to the problem of getting this energy into the magnetosphere. Continuing our analogy with an MHD electrical generator, we need to tap this power by closing the electrical circuit associated with the solar wind–magnetosphere interaction. This is accomplished by the connection of magnetic field lines attached to the earth with those in the interplanetary medium. Current can then flow down into the magnetosphere and close the circuit through the conducting ionosphere. Since the magnetic field lines are equipotentials, the interplanetary electric field maps into the near-space region of the earth, into the ionosphere, and even deep into the stratosphere.

This "connection" can occur most easily when the IMF is southward as illustrated in Fig. 1.12, where the cross-sectional view is now in the noon–midnight plane. This viewpoint was first suggested by Dungey (1961). The numbers correspond to sequential times as the interplanetary magnetic field connects to the earth's field and is then swept antisunward at the solar wind speed. The northern and southern hemispheric field lines are stretched into a magnetic tail and eventually reconnect with each other deep in the antisunward region. We spend considerable time in Chapter 6 discussing the electric field and associated $\mathbf{E} \times \mathbf{B}/B^2$ circulation patterns that arise in the magnetosphere for various solar

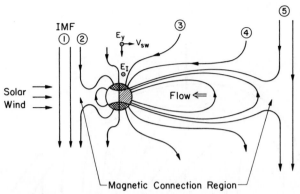

Fig. 1.12. A "connected" and "reconnected" magnetic field topology for B_z south. The view is in the noon–midnight plane. $\mathbf{E}_y$ is the interplanetary electric field and $\mathbf{E}_I$ is the electric field mapped into the ionosphere down the magnetic field lines.

wind configurations. In the "B_z south" case shown here, the interplanetary dawn–dusk electric field shown at the top of the sketch maps directly into the polar ionosphere along the connected magnetic field lines. There, the electric field $\mathbf{E}_I$ is "out of the paper" and the $\mathbf{E} \times \mathbf{B}$ drift is antisunward. Once reconnection occurs in the tail, the forces associated with the stretched magnetic field lines brings the plasma rapidly back toward the earth. This is called the region of sunward convection.

A cutaway view of the magnetosphere is given in Fig. 1.13, with the location of various regions and phenomena indicated. When the interplanetary magnetic field connects to the earth's magnetic field it produces open field lines with only one foot on the earth. These field lines extend far into the magnetotail and form the northern and southern tail lobes. These open field lines also cross the magnetopause, where they connect to the interplanetary field in the magnetosheath. In so doing, they come into contact with plasma populations just inside the magnetopause which are called the boundary layer and the mantle. The field lines that reach down the magnetotail have opposite directions on the northern and southern sides of the equatorial plane and a current sheet must exist to separate the northern and southern tail lobes. This so-called neutral sheet where $|\mathbf{B}| \simeq 0$ is also the region in which the open field lines may reconnect to form closed field lines that convect back to the earth. As the hot plasma sheet flows toward the earth into an increasing magnetic field, the differential motion of the ions and electrons produces a ring current somewhat analogous to the interaction between the solar wind and the magnetosphere itself on the sunward side of the earth. Some of these hot plasma sheet particles move along the magnetic field lines and precipitate into the atmosphere in a ring around the polar regions called the auroral oval. This ring forms the light pattern seen in Fig. 1.6. Most of the solar

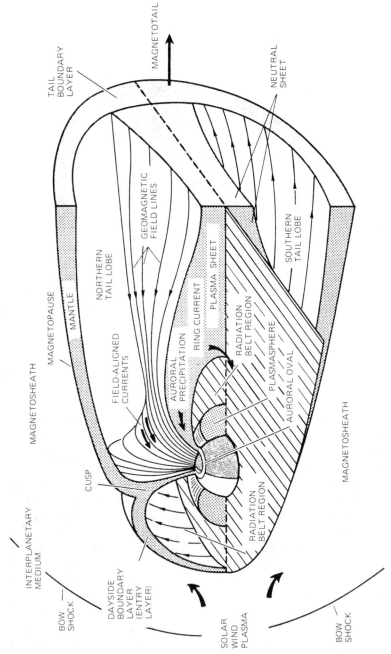

Fig. 1.13. Schematic representation of the magnetosphere. (Courtesy of J. Roederer.)

wind and magnetospheric effects occur at high latitudes, although many interesting phenomena do extend to lower latitudes during active times. The active region is skewed in the antisunward direction, extending to 70° magnetic latitude on the sunward side and 60° on the nightside. In the course of this text, some of the phenomena indicated in Fig. 1.13 will be treated in detail and others in a more sketchy fashion. The interested reader is referred to the text by Lyons and Williams (1984) for a more complete treatise on magnetospheric processes.

Particle energization processes will not be discussed quantitatively at all in this text but a brief sketch is in order here. Solar wind electrons have thermal energies of only a few electron volts, while the bulk flow energy of the ions is about a kilovolt. The earth's bow shock wave, shown in Fig. 1.13, energizes the electrons to many tens or even hundreds of electron volts in the magnetosheath. Some of these hot electrons and ions directly penetrate into the ionosphere in the connection process or through the cusp in the magnetic field geometry. This precipitation of energetic particles contributes to the dayside part of the auroral oval. As the magnetic field lines reconnect in the tail and jet back toward the earth, the average electron energy is boosted to about a kilovolt in the region termed the plasma sheet. The plasma in this region provides a source of steady precipitation that corresponds visually to the widespread emissions called diffuse aurora. This precipitation also acts as an ionization source for the ionosphere. The hot flowing plasma has an associated electric field [see (1.2b)] that is impressed on the ionosphere. Electrical currents flow in response to this electric field. For nighttime conditions the precipitating plasma itself provides most of the conductivity, so an electrical feedback can develop between the plasma sheet and the ionosphere as currents flow back and forth between them.

This feedback is most complex in the region of the discrete aurora which contains the very bright, active, auroral displays. Here, at some point along the magnetic field (usually above 3000 km), electric fields develop parallel to the magnetic field lines that accelerate electrons down into the atmosphere and ions out into the magnetosphere. Typical field-aligned potential drops associated with this parallel electric field range from 100 V to 10 kV. The electron density in the ionospheric E region can exceed 10^6 cm^{-3} below a strong auroral display (see Fig. 1.5) and the horizontal electrical current in the ionosphere can create perturbations in the surface magnetic field of up to 3000 nanotesla, nearly 5% of the earth's surface field at high latitudes. Many other acceleration processes besides parallel electric fields operate in the magnetosphere and are beyond our capability to list, let alone discuss in detail.

References

Axford, W. I., and Hines, C. O. (1961). A unifying theory of high-latitude geophysical phenomena and geomagnetic storms. *Can. J. Phys.* **39**, 1433.

Baron, M. J. (1974). Electron densities within auroras and other auroral E-region characteristics. *Radio Sci.* **9**, 341.

Dungey, J. W. (1961). Interplanetary magnetic field and the auroral zones. *Phys. Rev. Lett.* **6**, 47.

Istomin, V. G. (1966). Observational results on atmospheric ions in the region of the outer ionosphere. *Ann. Geophys.* **22**, 255.

Johnson, C. Y. (1969). Ion and neutral composition of the ionosphere. *Ann. IQSY* **5**.

Kelley, M. C., and Kintner, P. M. (1978). Evidence for two-dimensional inertial turbulence in a cosmic-scale low β-plasma. *Astrophys. J.* **220**, 339.

Lyons, L. R., and Williams, D. J. (1984). "Quantitative Aspects of Magnetospheric Physics." Reidel, Boston, Massachusetts.

Reber, C. A., and Nicolet M. (1965). Investigation of the major constituents of the April–May 1963 heterosphere by the Explorer XVII satellite. *Planet. Space Sci.* **13**, 617.

Rishbeth, H., and Garriott, O. K. (1969). "Introduction to Ionospheric Physics," Int. Geophys. Ser., Vol. 14. Academic Press, New York.

Robinson, R. M., Vondrak, R., Craven, J., Frank L., and Miller, K. (1989). Comparison of Dynamics Explorer auroral imaging data with Chatanika radar electron density measurements. *JGR, J. Geophys. Res.,* in press.

Schmidt, G. (1966). "Physics of High Temperature Plasmas: An Introduction." Academic Press, New York.

Shen, J. S., Swartz, W. E., Farley D. T., and Harper, R. M. (1976). Ionization layers in the night-time E-region valley above Arecibo. *JGR, J. Geophys. Res.* **81**, 15517.

Chapter 2 | Fundamentals of Ionospheric Plasma Dynamics

In this chapter we model the ionospheric plasma as three interpenetrating fluids, with the electron and ion fluids immersed in the neutral gas. At all heights of interest in the study of ionospheric phenomena, the neutral gas density exceeds that of the plasma. In fact, the plasma density does not become comparable to that of the neutrals until several thousand kilometers altitude. The primary difference between ionospheric plasma dynamics and thermospheric neutral gas dynamics is the effect of electromagnetic forces. The various forces acting on charged particles drive electric currents, which, in turn, create electric fields that modify the plasma dynamics. The electrical conductivity of the medium is thus extremely important and is derived in this chapter. We discuss briefly the generation of electric fields in the ionosphere and the transmission of electric fields along magnetic field lines between the ionosphere and the magnetosphere. At middle and low latitudes the electric field is generated primarily by the neutral wind field. In later chapters electric fields impressed on the ionosphere by solar wind and magnetospheric processes will be taken into account, as will their occasional penetration into the middle- and low-latitude sectors. In the analysis that follows, we first obtain the equations for a neutral fluid and then extend them to ionized gases. Finally, we develop the equations needed to describe collisionless plasmas in the absence of a neutral fluid, which is the appropriate approximation for the magnetosphere. Although not the primary topic of this text, such a development is necessary since we must be able to describe certain key magnetospheric phenomena in some detail.

2.1   The Basic Fluid Equations

The ions, electrons, and neutrals can be considered as three interpenetrating fluids coupled by collisions. The gas of ions and electrons, taken together, is often referred to as a plasma in the text. As usual in a fluid description, we assume that an element small enough to be treated as a differential volume in the mathematical sense still contains a sufficient number of molecules that statistical techniques can be used to define such quantities as the temperature, density, and mean flow velocity. Such a volume element is macroscopically small even though microscopically it contains many particles colliding in a random fashion. We use this approach almost exclusively in the text.

2.1.1   Conservation of Mass

Conservation of mass dictates that the flux of material into or out of a volume through its surface be equal to the rate of increase or decrease of mass inside the volume. The mathematical statement in integral form relates the mass density ρ and the fluid velocity $\mathbf{U}$ through

$$\iiint_V \partial\rho/\partial t \, dV \; = \; -\iint_\Sigma \rho\mathbf{U}\cdot d\mathbf{a} \tag{2.1}$$

where the surface Σ encloses the volume V and $d\mathbf{a}$ is normal to the surface and directed outward. The differential form of (2.1) can be derived from Gauss's theorem, which relates the surface integral on the right-hand side to a volume integral of the divergence of $\rho\mathbf{U}$, that is,

$$\iiint_V \partial\rho/\partial t \, dV \; = \; -\iint_\Sigma \rho\mathbf{U}\cdot d\mathbf{a} \; = \; -\iiint_V \nabla\cdot(\rho\mathbf{U}) \, dV$$

and hence

$$\partial\rho/\partial t \; = \; -\nabla\cdot(\rho\mathbf{U}) \tag{2.2}$$

This constitutes the continuity equation for the neutral atmosphere with mass density ρ and velocity $\mathbf{U}$.

Expanding the divergence operator, (2.2) can also be written

$$\partial\rho/\partial t \, + \, \mathbf{U}\cdot\nabla\rho \, + \, \rho(\nabla\cdot\mathbf{U}) = 0 \tag{2.3}$$

or, equivalently,

$$d\rho/dt \, + \, \rho(\nabla\cdot\mathbf{U}) = 0 \tag{2.4}$$

where the total time derivative

$$d/dt \; = \; \partial/\partial t \, + \, \mathbf{U}\cdot\nabla \tag{2.5}$$

has been used. This operation yields the time rate of change of a quantity, ρ in this case, moving with the flow. Equation (2.4) states that the rate of change of ρ moving with the flow is determined only by the divergence of the velocity field. For an incompressible fluid the mass density of a parcel cannot change as it moves ($d\rho/dt = 0$) and hence from (2.4) it follows that $\nabla \cdot \mathbf{U} = 0$. In an incompressible fluid the velocity field is divergence free. The advective derivative ($\mathbf{U} \cdot \nabla$) in (2.5) is important since it describes the temporal variation of a quantity, at a point in space, due to the transport of that quantity into the region. For example, in the case of incompressible flow, the mass density can only change with time at a fixed point in space via this term since

$$\partial\rho/\partial t = -(\mathbf{U} \cdot \nabla\rho) \qquad (2.6)$$

Figure 2.1 illustrates that if a parcel of material with a high mass density is advected into a region down a density gradient, the local mass density will increase. This result is verified by (2.6) since $\mathbf{U}$ and $\nabla\rho$ are oppositely directed in such a case.

In the case of a partially ionized medium, ion and electron pairs may be produced by impact of a photon or energetic particle or lost through recombination between positively and negatively charged particles. These processes are very important for the ionospheric plasma. If P_j denotes the rate of production of ions (and electrons) per cubic meter per second and L_j the rate of loss, then the mass conservation equation for each of the ionized species is

$$\partial\rho_j/\partial t + \nabla \cdot (\rho_j \mathbf{V}_j) = (P_j - L_j)M_j \qquad (2.7)$$

where M_j is the mass of each species. Notice that we use the notation $\mathbf{V}_j$ to represent the velocity of the charged species, reserving $\mathbf{U}$ for the velocity of the neutral gas. Since electric charge is a conserved quantity, it must be the case that the total number of electrons gained or lost equals the sum of all the different types of ions gained or lost, that is,

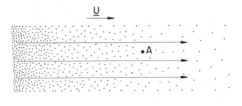

Fig. 2.1. Advection of mass by the velocity $\mathbf{U}$ in the downgradient direction increases the local mass density at A.

$$\sum_{j=1}^{N}(P_j - L_j) = P_e - L_e$$

We ignore negative ions since their formation is unimportant above about 80-km altitude. Because the number density of the neutral gas far exceeds that of the ion and electron gas for heights below several thousand kilometers, we can ignore the loss of neutral particles when ion–electron pairs are formed. Furthermore, if we are not interested in the possible composition changes of the neutrals (e.g., formation of atomic oxygen by photodissociation) we may use (2.2) for the neutral atmosphere while for the ionized particles we need (2.7).

2.1.2   Equation of State

For an ideal gas the mass density and pressure, p_j, are related by

$$p_j = \rho_j k_B T_j / M_j = n_j k_B T_j \qquad (2.8)$$

This is the equation of state for each of the fluids we consider (ions, electrons, neutrals), and we relate the mass density ρ_j to the number density n_j through $\rho_j = n_j M_j$. In this text we use k_B to represent Boltzmann's constant. The letter k used alone or with any other subscript represents a wave number.

2.1.3   Conservation of Momentum

The continuity equations and the equation of state must be supplemented by a dynamical equation relating the fluid velocity to the forces acting on the fluid. This is derived from the principle of conservation of momentum, which requires that the change of momentum per unit time within a volume be equal to the pressure gradient force and the total external force field **F** acting on the material inside the volume plus the momentum flux carried across the surface bounding the volume by a viscosity, advection, or wave flux. We first treat the neutral gas. Here, as in most ionospheric and upper atmospheric studies, the wind direction is indicated by where the wind is going; that is, an eastward wind is a wind toward the east. (In meteorology an easterly wind comes from the east.)

Advection of a vector quantity such as momentum across a boundary by flow is conceptually no more difficult than discussed above for the scalar mass density. However, the mathematical description is more complex and is most readily accomplished by the use of tensor notation as employed here. The equation equivalent to (2.1) for the time rate of momentum change in a volume is

$$\iiint_V \frac{\partial}{\partial t}(\rho \mathbf{U}) \, dV = \iiint_V (-\nabla p) \, dV + \iiint_V \mathbf{F} \, dV$$
$$- \iint_\Sigma \boldsymbol{\pi}_m \cdot d\mathbf{a} - \iint_\Sigma \boldsymbol{\pi}_w \cdot d\mathbf{a} \qquad (2.9)$$

where **F** is the external force, p the pressure, $\boldsymbol{\pi}_m$ the momentum flux density tensor due to material motions, and $\boldsymbol{\pi}_w$ the momentum flux density tensor due to waves in the medium. Applying Gauss's theorem again, (2.9) becomes

$$\partial(\rho\mathbf{U})/\partial t = -\nabla p + \mathbf{F} - \nabla \cdot \boldsymbol{\pi}_m - \nabla \cdot \boldsymbol{\pi}_w \qquad (2.10)$$

The pressure gradient term should be familiar. The external force **F** can be of many different types which will be treated as they arise in the text.

To understand the material momentum tensor, consider a single particle of mass m moving at velocity **v**. The momentum, $m\mathbf{v}$, is carried along by the velocity **v**, resulting in a momentum flux tensor given by

$$\boldsymbol{\pi}_m = \mathbf{v}m\mathbf{v}$$

Written as a 3×3 matrix, this becomes

$$(\boldsymbol{\pi}_m)_{jk} = mv_j v_k$$

An analogous form for a fluid of mass density ρ characterized by a mean flow velocity **U** is given by

$$(\boldsymbol{\pi}_m)_{jk} = \rho U_j U_k \qquad (2.11)$$

This part of the momentum flux tensor describes how momentum is transferred within a fluid by the motion of the fluid. If there is any net divergence of this momentum flux, a net force will occur as indicated in (2.10). The divergence of a tensor is a vector and may be written

$$(\nabla \cdot \boldsymbol{\pi})_j = \sum_k \frac{\partial \pi_{jk}}{\partial x_k}$$

where x_k is the kth Cartesian coordinate. Applying this to the tensor in (2.11),

$$(\nabla \cdot \boldsymbol{\pi}_m)_j = \sum_k \frac{\partial}{\partial x_k} (\rho U_j U_k)$$

$$= \sum_k \left[U_k \frac{\partial}{\partial x_k} (\rho U_j) + \rho U_j \frac{\partial}{\partial x_k} (U_k) \right]$$

which in vector form is

$$\nabla \cdot \boldsymbol{\pi}_m = \mathbf{U} \cdot \nabla(\rho\mathbf{U}) + \rho\mathbf{U}(\nabla \cdot \mathbf{U})$$

The first term is the advective time derivative of the momentum.

We are not quite finished since the term given in (2.11) is not the only contribution to the momentum flux tensor $\boldsymbol{\pi}_m$. Consider the situation illustrated in Fig. 2.2, where the x component of the mean fluid velocity increases in the z direction. Particles randomly crossing the plane $z = z_0$ from above and colliding with particles below will, on average, contribute more x momentum to the fluid below $z = z_0$ than particles crossing in the other direction will contribute above.

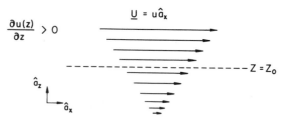

Fig. 2.2. If the wind increases with increasing z, the viscous force will tend to reduce the velocity for $z > z_0$ and to increase it for $z < z_0$.

This momentum transfer results in forces on the fluid such that the velocity gradient is reduced. Unlike the momentum flux described by (2.11), this "viscous" force depends on collisions and is a dissipative process. Energy in the mean flow is converted into heat when viscosity is important. The exact form of the viscous momentum flux density tensor is quite complicated and we refer the reader to a discussion by Landau and Lifshitz (1959) for details. In its simplest form the momentum flux density depends on the first derivatives of the velocity with two associated proportionality constants η and η'. When the divergence of the momentum flux tensor is taken, as required by (2.10) the viscous force $\mathbf{F}_v$ is given by

$$\mathbf{F}_v = \eta \nabla^2 \mathbf{U} + \eta' \nabla(\nabla \cdot \mathbf{U}) \tag{2.12}$$

For incompressible flow $(\nabla \cdot \mathbf{U}) = 0$ and (2.12) reduces to the more familiar form

$$\mathbf{F}_v = \eta \nabla^2 \mathbf{U} \tag{2.13}$$

where η is called the dynamic viscosity coefficient. Equation (2.10) becomes

$$\partial(\rho \mathbf{U})/\partial t + \mathbf{U} \cdot \nabla(\rho \mathbf{U}) + \rho \mathbf{U}(\nabla \cdot \mathbf{U}) = -\nabla p + \mathbf{F}$$

$$- \nabla \cdot \boldsymbol{\pi}_w + \text{viscous terms} \tag{2.14}$$

Using the continuity equation, the left-hand side is just $\rho d\mathbf{U}/dt$. Finally, using (2.13)

$$\rho \, d\mathbf{U}/dt = -\nabla p + \mathbf{F} - \nabla \cdot \boldsymbol{\pi}_w + \eta \nabla^2 \mathbf{U} \tag{2.15}$$

where as before d/dt is the total time derivative. Notice that the Laplacian operator in (2.15) acts on each velocity component. That is, the x component of the viscous force is given by

$$(\mathbf{F}_v)_x = \eta \nabla^2 U_x$$

For large-scale flow patterns $(\nabla \cdot \mathbf{U}) = 0$ is a good approximation and (2.13) is used almost exclusively for viscous forces. We must note, however, that in the lower atmosphere the coefficient η is determined not by molecular collisions but by the interaction between eddies in the flow. In other words, the effective viscosity coefficient is much larger than the molecular coefficient calculated from kinetic theory. Above the turbopause, where turbulent mixing ceases, the classical molecular viscosity coefficient is appropriate.

We now consider the forces that can contribute to the external force $\mathbf{F}$ in (2.15). The body force $\mathbf{F}$ includes the important gravitational term $\rho \mathbf{g}$. In the thermosphere $\mathbf{F}$ also includes a frictional force $\mathbf{F}_p$ on the neutrals due to the existence of the plasma in the ionosphere. This force is transmitted by collisions between particles when there is a net mean relative motion between the neutral gas and the plasma. Electrons have such little mass that most of this friction is due to differential motion between the neutral gas and the ions. At first glance one might suppose that the ions would always move with the neutrals since the latter are much more numerous. As we shall see, this argument is not valid since the ions are constrained to gyrate around the magnetic field lines and, unlike the neutrals, cannot move freely across them. We return to this point in more detail below. The force on the neutrals due to the plasma is thus often referred to as "ion drag" and has the mathematical form

$$\mathbf{F}_p = -\rho \nu_{ni}(\mathbf{U} - \mathbf{V}_i) \tag{2.16}$$

where ν_{ni} is the neutral–ion collision frequency for momentum transfer and $\mathbf{V}_i$ is the mean ion velocity. For simplicity, we have assumed that only one ion species is present.

The upward flux of momentum $\boldsymbol{\pi}_w$ due to waves can be very important in the upper atmosphere. Waves from the dense lower regions tend to grow in amplitude as they propagate upward since the mass density decreases exponentially with altitude. If the waves are absorbed at any given height then $(\nabla \cdot \boldsymbol{\pi}_w) \neq 0$ and the local atmosphere will be accelerated. Ocean swimmers often experience a phenomenon of this type when water waves break on the beach at an angle that is not exactly perpendicular to the shoreline. Some of the wave momentum is absorbed in the surf region and a "long-shore" current results. This current transports sand (and people) along the coast and is an important factor in the formation, structure, and location of beaches.

This would complete the momentum equation for the neutrals except for the fact that the earth is rotating. Newton's laws as expressed in (2.15) refer to a frame of reference at rest or one moving with constant linear velocity. In a reference frame (R) rotating with constant angular velocity $\boldsymbol{\Omega}$, the time rate of change of a vector is related to the time derivative in an inertial frame (I) by

$$(d\mathbf{A}/dt)_I = (d\mathbf{A}/dt)_R + \boldsymbol{\Omega} \times \mathbf{A} \tag{2.17}$$

An excellent discussion of this result is given by Goldstein (1950). In the rotating frame where we live and take measurements, the time derivative of the velocity in (2.15) must thus be replaced by

$$(d\mathbf{U}_I/dt)_I = (d\mathbf{U}_I dt)_R + \boldsymbol{\Omega} \times \mathbf{U}_I \tag{2.18}$$

We have included a subscript on the velocity vector to show specifically that even though we now have the proper time derivative in the rotating frame, the vector $\mathbf{U}_I$ is still the velocity in inertial space. Since the goal is to describe dynamics totally in the rotating coordinates, $\mathbf{U}_I$ must be expressed in that frame also. Viewed from inertial space, an object moving on the earth's surface with velocity $\mathbf{U}_R$ has an additional velocity $\boldsymbol{\Omega} \times \mathbf{r}$ where $\mathbf{r}$ is the position vector drawn from the center of the earth; that is,

$$\mathbf{U}_I = \mathbf{U}_R + \boldsymbol{\Omega} \times \mathbf{r} \tag{2.19}$$

If we assume that $\boldsymbol{\Omega}$ is constant, which is a valid approximation for all but extremely long-period dynamics, the time derivative of (2.19) in the inertial frame is

$$(d\mathbf{U}_I/dt)_I = (d\mathbf{U}_R/dt)_I + \boldsymbol{\Omega} \times (d\mathbf{r}/dt)_I$$

Now each of the derivatives on the right-hand side must be replaced by the operation (2.18) in order to have the expression given entirely by quantities measured in the rotating coordinates, and hence

$$(d\mathbf{U}_I/dt)_I = (d\mathbf{U}_R/dt)_R + \boldsymbol{\Omega} \times \mathbf{U}_R + \boldsymbol{\Omega} \times [(d\mathbf{r}/dt)_R + \boldsymbol{\Omega} \times \mathbf{r}]$$

For a parcel of fluid moving across the surface, $(d\mathbf{r}/dt)_R = \mathbf{U}_R$, so we have

$$(d\mathbf{U}_I/dt)_I = (d\mathbf{U}_R/dt)_R + 2\boldsymbol{\Omega} \times \mathbf{U}_R + \boldsymbol{\Omega} \times (\boldsymbol{\Omega} \times \mathbf{r})$$

The second term is known as the Coriolis force. The last term is equal to $r\Omega^2 \cos\theta$ where θ is the latitude and has components both radially inward and equatorward. This term may be combined with $\mathbf{g}$ to describe an effective gravitational field. We will continue to use the symbol $\mathbf{g}$ for these combined terms and move the Coriolis term to the right-hand side of (2.15), which yields the following equation of motion of the neutral atmosphere in a rotating frame:

$$d\mathbf{U}/dt = -\nabla p + \rho\mathbf{g} + \eta\nabla^2\mathbf{U} - \nabla\cdot\boldsymbol{\pi}_w$$

$$- 2\rho(\boldsymbol{\Omega} \times \mathbf{U}) - \rho\nu_{ni}(\mathbf{U} - \mathbf{V}_i) \tag{2.20}$$

Turning now to the plasma, each species will have its own momentum equation. Although neutral fluid dynamics on a rotating object such as the earth is strongly affected by the Coriolis force, this velocity-dependent force has little importance in geophysical plasma analysis since the magnetic force (which is also velocity dependent) is much greater. The important body forces which *do* act on the ionospheric plasma and which we include in the force F are as follows:

Gravitational: $\rho_j\mathbf{g}$

Electric: $n_j q_j\mathbf{E}$

Magnetic: $n_j q_j (\mathbf{V}_j \times \mathbf{B})$

where q_j is the charge of the jth species and $\mathbf{E}$ and $\mathbf{B}$ are the electric and magnetic fields. In addition, there is a frictional force exerted on each species by collisions with all the other species. For example, electrons will collide with neutrals as well as with the various ions. The force is proportional to the respective collision frequency and to the differential velocity between the particular fluid and the other fluids. The frictional force on each species may be written

$$\mathbf{F}_j = -\sum_{\substack{k \\ j \neq k}} \rho_j \nu_{jk} (\mathbf{V}_j - \mathbf{V}_k)$$

where again we leave any detailed discussion of the ν_{jk} coefficients for other texts. The momentum equation we shall primarily use for each ionized species is then

$$\rho_j \frac{d\mathbf{V}_j}{dt} = -\nabla p_j + \rho_j \mathbf{g} + n_j q_j (\mathbf{E} + \mathbf{V}_j \times \mathbf{B}) - \sum_{\substack{k \\ j \neq k}} \rho_j \nu_{jk} (\mathbf{V}_j - \mathbf{V}_k)$$

Viscosity and momentum transfer by waves are ignored in this equation and will be discussed only briefly in the text where appropriate.

2.1.4 The Complete Equation Sets

The equations we have thus far obtained for the fluids making up the ionosphere may be summarized as follows. The neutral atmospheric equations of continuity, momentum, and state are

$$\partial \rho / \partial t = -\nabla \cdot (\rho \mathbf{U}) \tag{2.21a}$$

$$\rho \, d\mathbf{U}/dt = -\nabla p + \rho \mathbf{g} + \eta \nabla^2 \mathbf{U} - \nabla \cdot \boldsymbol{\pi}_w$$
$$- 2\rho \boldsymbol{\Omega} \times \mathbf{U} - \rho \nu_{ni} (\mathbf{U} - \mathbf{V}_i) \tag{2.21b}$$

$$p = \rho k_B T_n / m_n = n_n k_B T_n \tag{2.21c}$$

where the subscript "n" stands for neutrals.

The corresponding equations for the ionized species are

$$\partial \rho_j / \partial t + \nabla \cdot (\rho_j \mathbf{V}_j) = (P_j - L_j) M_j \tag{2.22a}$$

$$\rho_j \, d\mathbf{V}_j/dt = -\nabla p_j + \rho_j \mathbf{g} + n_j q_j (\mathbf{E} + \mathbf{V}_j \times \mathbf{B})$$
$$- \sum_{\substack{k \\ j \neq k}} \rho_j \nu_{jk} (\mathbf{V}_j - \mathbf{V}_k) \tag{2.22b}$$

$$p_j = \rho_j k_B T_j / M_j = n_j k_B T_j \tag{2.22c}$$

Owing to the complexity of the equation sets (to which Maxwell's equations must yet be added), we will not consider independent heat equations. This is equivalent to treating temperature profiles as given quantities. In this book we primarily study variations in the plasma density and velocity fields due to dynamical rather than aeronomic processes and do not attempt to include a thermal analysis. This simplification would be very poor if our main interest was thermospheric neutral gas dynamics. For the plasma, however, many interesting phenomena may be studied without including temperature changes self-consistently.

To treat the electric and magnetic fields, Maxwell's electrodynamic equations are needed, which, in their full form, are given by

$$\nabla \times \mathbf{E} = -\partial \mathbf{B}/\partial t \qquad (2.23a)$$

$$\nabla \cdot \mathbf{E} = \rho_c/\varepsilon_0 \qquad (2.23b)$$

$$\nabla \times \mathbf{B} = \mu_0(\mathbf{J} + \varepsilon_0 \partial \mathbf{E}/\partial t) \qquad (2.23c)$$

$$\nabla \cdot \mathbf{B} = 0 \qquad (2.23d)$$

where ρ_c is the charge density ($\rho_c = \Sigma_j\, n_j q_j$) and $\mathbf{J}$ the current density ($\mathbf{J} = \Sigma_j\, n_j q_j \mathbf{V}_j$). To this must be added the principle of conservation of charge

$$\frac{\partial}{\partial t} \iiint_V \rho_c\, dV = -\iint_\Sigma \mathbf{J} \cdot \mathbf{da}$$

This equation states that the buildup or decay of charge inside a volume V is determined by the net electric conduction current across the surface Σ. Note the similarity to the equation (2.1) for the conservation of mass. In differential form this may be written

$$\nabla \cdot \mathbf{J} = -\partial \rho_c/\partial t \qquad (2.24)$$

Neither production nor recombination affect this equation since the net charge does not change in either process. In the ionosphere the conduction current is larger than the vacuum displacement current $\varepsilon_0 \partial \mathbf{E}/\partial t$ in (2.23c) for all wave frequencies of interest here and the displacement current is therefore dropped. Furthermore, since the largest changes in the magnetic field at ionospheric heights are such that $\delta B/B \leq 5\%$, we take (2.23c) as a diagnostic equation only. That is, given some current $\mathbf{J}$, we can find $\delta \mathbf{B}$, the perturbation to the geomagnetic field, from (2.23c), but the $\mathbf{B}$ we use in the rest of the dynamical equations is the earth's main field and hence we may take $\partial \mathbf{B}/\partial t = 0$. We are thus left with the set

$$\nabla \times \mathbf{E} = 0 \qquad (2.25a)$$

$$\nabla \cdot \mathbf{E} = \rho_c/\varepsilon_0 \qquad (2.25b)$$

$$\nabla \times \mathbf{B} = \mu_0 \mathbf{J} \tag{2.25c}$$

$$\nabla \cdot \mathbf{B} = 0 \tag{2.25d}$$

$$\nabla \cdot \mathbf{J} = -\partial \rho_c / \partial t \tag{2.25e}$$

Equation (2.25a) shows that the electric field is derivable from a potential function $\phi(\mathbf{r}, t)$ through $\mathbf{E} = -\nabla\phi$. Finally, one further important simplification is possible. In an ionized medium, very small charge differences create large electric fields. Thus, a plasma must nearly exhibit charge neutrality. This implies

$$\nabla \cdot \mathbf{J} = -\partial \rho_c / \partial t \simeq 0 \tag{2.26}$$

That is, the divergence of the electric current on any macroscopic time scale must be zero. As we shall see, in most cases we use $\nabla \cdot \mathbf{J} = 0$ to calculate the electric field rather than (2.25b). Poisson's equation [i.e., (2.25b)] is, of course, still valid, but it is not very useful since the charge difference associated with geophysically important electric fields is very small. This situation is similar to the problem of calculating the vertical velocity in atmospheric dynamics. The vertical component of the momentum equation is not very useful since the vertical pressure gradient and the gravitational terms nearly cancel. In practice, the vertical velocity is often found from the divergence of the horizontal wind, much as we use $\nabla \cdot \mathbf{J} = 0$ to find the electric field. The following combined set of dynamic and electrodynamic equations for the plasma (plus the equation of state) remains after these simplifications:

$$\rho_j \, d\mathbf{V}_j / dt = -\nabla p_j + \rho_j \mathbf{g} + \frac{q_j \rho_j}{M_j} (\mathbf{E} + \mathbf{V}_j \times \mathbf{B}) \tag{2.27a}$$
$$- \sum_{\substack{k \\ j \neq k}} \rho_j \nu_{jk} (\mathbf{V}_j - \mathbf{V}_k)$$

$$\frac{\partial \rho_j}{\partial t} + \nabla \cdot (\rho_j \mathbf{V}_j) = (P_j - L_j) M_j \tag{2.27b}$$

$$p_j = \rho_j k_B T_j / M_j = n_j k_B T_j \tag{2.27c}$$

$$\mathbf{E} = -\nabla\phi \tag{2.27d}$$

$$\nabla \cdot \mathbf{J} = 0 = \nabla \cdot \mathbf{B} \tag{2.27e}$$

$$\nabla \times \mathbf{B} = \mu_0 \mathbf{J} \tag{2.27f}$$

Note that from (2.26) we have the important result that the number of electrons per unit volume must be almost equal to the number of positive ions of all types

$$n_e \simeq \sum_{\text{ions}} n_j$$

This means we can define a plasma density n which is equal to both n_e and $\Sigma \, n_j$.

2.2 Steady-State Ionospheric Plasma Motions Due to Applied Forces

In this section we start with (2.27a) and derive the equations that determine the electrodynamic response of a partially ionized plasma to applied steady forces. For the present we specify the distribution of plasma density and the wind field. Considering (2.27a), with the plasma pressure distribution specified, we can argue that the response of the plasma constituents to changing forces occurs very quickly (i.e., $d\mathbf{V}_j/dt \approx 0$). This can be seen by comparing the terms in (2.27a) that include the velocity. The acceleration terms on the left-hand side are of order V_j/τ and V_j^2/L, where τ is the response time to a new set of forces and L is a distance scale for velocity change; the Lorentz term (the third term on the right-hand side) is of order $\mathbf{V}_j\Omega_j$, where Ω_j is the gyrofrequency ($q_j\mathbf{B}/M_j$); and the frictional term is of order $\mathbf{V}_j\nu_j$, where ν_j is the collision frequency. As long as $\tau \gg \Omega_j^{-1}$ or ν_j^{-1} the acceleration term can be neglected. Collision frequencies and gyrofrequencies are sufficiently high that in most problems of interest to macroscopic dynamics the acceleration term can be neglected. Also, since fluid velocities are usually subsonic, if L is greater than the gyro radius and the mean free path, the advective term is small. The plasma constituents are thus assumed to be in velocity equilibrium with the existing force fields.

The steady-state fluid velocity of each species may now be found from (2.27a) by setting the time derivative equal to zero and specifying the force fields and pressure distributions

$$0 = -\nabla(n_j k_B T_j) + n_j M_j \mathbf{g} + q_j n_j (\mathbf{E} + \mathbf{V}_j \times \mathbf{B})$$
$$- \sum_{\substack{k \\ k \neq j}} n_j M_j \nu_{jk} (\mathbf{V}_j - \mathbf{V}_k) \tag{2.28}$$

The collision frequencies ν_{jk} play a crucial role in a partially ionized plasma. Texts by Rishbeth and Garriott (1969) and Banks and Kockarts (1973) discuss the physics in some detail. Here it suffices to plot and discuss representative collision profiles such as the ones given in Fig. 2.3, which are valid for high sunspot conditions (e.g., see Johnson, 1961). The numerical values of various ionospheric and atmospheric parameters used in making this plot are given in Appendix B. Useful approximate formulas for these collision frequencies are

$$\nu_{\text{in}} = 2.6 \times 10^{-9} (n_n + n_i) A^{-1/2} \tag{2.29a}$$

where A denotes the mean neutral molecular mass in atomic mass units, and

$$\nu_e \equiv \nu_{en} + \nu_{ei} = 5.4 \times 10^{-10} n_n T_e^{1/2}$$
$$+ [34 + 4.18 \ln(T_e^3/n_e)] n_e T_e^{-3/2} \tag{2.29b}$$

where T_e is measured in kelvin and all the densities are expressed per cubic centimeter. The ion–electron collision frequency is not plotted since it is negli-

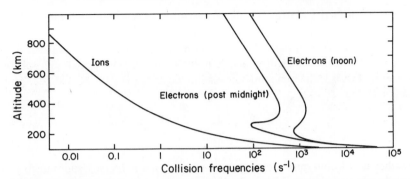

Fig. 2.3. Typical electron and ion–neutral collision frequencies at high sunspot number. [From "Satellite Environment Handbook," edited by F. Johnson (1961). Reproduced by permission of Stanford Univ. Press. Copyright © 1961 by the Board of Trustees of the Leland Stanford Junior University.]

gible over the altitude range of interest. In the postmidnight period the electron–neutral collision frequency equals the electron–ion collision frequency near about 280 km. As we shall see, the equatorial F-region plasma is often driven to very high altitudes in the evening hours, and in such a case the transition height can be much higher. During the daytime, transition between ν_{en} and ν_{ei} occurs at about 200 km. For the moment we consider the case that electron–neutral collisions are more common than electron–ion collisions, which simplifies the algebra considerably.

It is instructive and useful to solve (2.28) for the ion and electron velocities in terms of the driving forces. For simplicity we consider a single ion species of mass M, using the symbol m for the electron mass and the symbol e for the elemental charge, which is taken to be positive. For spatially uniform ion and electron temperatures we have the two equations

$$0 = -k_B T_i \nabla n + nM\mathbf{g} + ne(\mathbf{E} + \mathbf{V}_i \times \mathbf{B}) - nM\nu_{in}(\mathbf{V}_i - \mathbf{U}) \quad (2.30a)$$

$$0 = -k_B T_e \nabla n + nm\mathbf{g} - ne(\mathbf{E} + \mathbf{V}_e \times \mathbf{B}) - nm\nu_{en}(\mathbf{V}_e - \mathbf{U}) \quad (2.30b)$$

where we have used the fact that $n_i = n_e = n$, the plasma density. The electric field in this equation is the one which would be measured in an earth-fixed coordinate system. This is usually the electric field which is measured in ionospheric experiments. It is nevertheless instructive to express these equations in a reference frame moving with the neutral flow velocity $\mathbf{U}$. Transformation between two coordinate systems moving at a relative velocity $\mathbf{U}$ does not leave the electric field invariant even if $|\mathbf{U}| \ll c$, where c is the speed of light. Jackson (1975) discusses transformation of the electromagnetic fields $\mathbf{E}$ and $\mathbf{B}$ between two coordinate systems and shows that in the moving frame,

$$\mathbf{E}' = (\mathbf{E} + \mathbf{U} \times \mathbf{B})/(1 - U^2/c^2)^{1/2} \quad (2.31a)$$

$$\mathbf{B}' = (\mathbf{B} - \mathbf{U} \times \mathbf{E}/c^2)/(1 - U^2/c^2)^{1/2} \quad (2.31b)$$

where the primed variables are those measured in the moving frame and the unprimed variables are measured in the earth-fixed frame.

It is easy to show that the $\mathbf{U} \times \mathbf{E}/c^2$ term in (2.31b) is small compared to $\mathbf{B}$ for any reasonable values of $\mathbf{U}$ and $\mathbf{E}$ in the earth's atmosphere or ionosphere. However, $\mathbf{U} \times \mathbf{B}$ is the same order of magnitude as $\mathbf{E}$ and must be retained. Thus for $|\mathbf{U}| \ll c$

$$\mathbf{E}' = \mathbf{E} + \mathbf{U} \times \mathbf{B} \tag{2.32a}$$

$$\mathbf{B}' = \mathbf{B} \tag{2.32b}$$

Another way to interpret these equations is that in a nonrelativistic transformation the current density is not significantly changed, $\mathbf{J}' \simeq \mathbf{J}$, but the charge density is, $\rho_c' \neq \rho_c$. The explanation for this "asymmetry" between the electric and magnetic fields lies in the fact that very small charge densities can produce significant electric fields in a plasma.

Following the notation of Haerendel (personal communication, 1973) we may now transform the terms in (2.28) to a reference frame moving with the neutral wind $\mathbf{U}$. We use the subscript notation again for brevity where j stands for i (the single-ion gas) or e (the electron gas). Since $\mathbf{V}_j' = \mathbf{V}_j - \mathbf{U}$, (2.28) becomes

$$0 = -k_B T_j \nabla n + nM_j \mathbf{g} + nq_j \mathbf{E}' + nq_j(\mathbf{V}_j' \times \mathbf{B}) - nM_j \nu_{jn} \mathbf{V}_j' \tag{2.33}$$

where everything is expressed in the moving reference frame (note that ∇n and $\mathbf{g}$ are unchanged in a nonrelativistic transformation). If we divide through by $nM_j \nu_{jn}$ and gather terms, this can be written

$$\mathbf{V}_j' - \kappa_j(\mathbf{V}_j' \times \hat{B}) = -D_j \nabla n/n + b_j \mathbf{E}' + (D_j/H_j)\hat{g} \equiv \mathbf{W}_j' \tag{2.34}$$

where $\hat{B}$ is a unit vector in the $\mathbf{B}$ direction, $\hat{g}$ is a unit vector in the $\mathbf{g}$ direction, κ_j is the ratio of gyrofrequency to collision frequency ($q_j B/M_j \nu_{jn}$), which has the same sign as the particle charge, D_j is the diffusion coefficient ($k_B T_j/M_j \nu_{jn}$), b_j is the mobility ($q_j/M_j \nu_{jn}$), which also has the algebraic sign of q_j, and H_j is the scale height ($k_B T_j/M_j g$). Notice that the velocity $\mathbf{W}_j'$ is the fluid velocity which would arise in an unmagnetized plasma subject to the same forces.

The absolute value of κ_j determines whether a particle does or does not make a cycle about the magnetic field before a collision takes place. For a small absolute value of κ_j, many collisions occur and the particle basically moves parallel to the applied forces as if there were no magnetic field. This is illustrated in Fig. 2.4a for ions and electrons subject to an electric field. The collisionless case (κ infinite) is shown in Fig. 2.4b for particles initially at rest. After about one gyroperiod the particles are moving at right angles to the electric field. In this important case the final velocity is identical for ions and electrons and equal to $\mathbf{E} \times \mathbf{B}/B^2$. For an absolute value of $\kappa = 1$ (Fig. 2.4c), the net motion is at a 45° angle to the electric field.

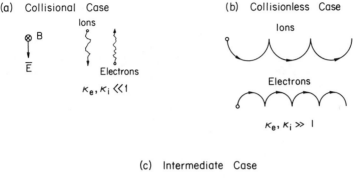

(a) Collisional Case

(b) Collisionless Case

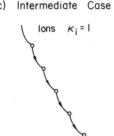

(c) Intermediate Case

Fig. 2.4. Ion and electron trajectories for various values of κ.

These results can be seen analytically from (2.34). Consider first the case of a very high collision frequency ($\kappa_j \ll 1$). Then the first term on the left-hand side dominates and

$$\mathbf{V}'_j = \mathbf{W}'_j = b_j \mathbf{E}' - D_j \nabla n/n + (D_j/H_j)\hat{g} \qquad (2.35)$$

which is the same as the fluid velocity that would arise in the case of an unmagnetized plasma. The velocity is parallel to the forces. For κ_j very large (a "collisionless" plasma) the component of (2.34) parallel to **B** is unchanged:

$$(V'_j)_{\|} = [b_j \mathbf{E}' - D_j \nabla n/n + (D_j/H_j)\hat{g}] \cdot \hat{B} \qquad (2.36a)$$

In the perpendicular direction the second term on the left-hand side of (2.34) dominates and the components perpendicular to **B** become

$$(\mathbf{V}'_j)_{\perp} = \kappa_j^{-1} [b_j \mathbf{E}' - D_j \nabla n/n + (D_j/H_j)\hat{g}] \times \hat{B} \qquad (2.36b)$$

or, equivalently, when κ_j is evaluated we have

$$(\mathbf{V}'_j)_{\perp} = (1/B^2)[\mathbf{E}' - (k_B T_j/q_j)\nabla n/n + (M_j/q_j)\mathbf{g}] \times \mathbf{B} \qquad (2.36c)$$

The individual terms on the right-hand side of (2.36c) are all perpendicular to the forces that drive them. Furthermore, since the first term does not depend on the charge, it is identical for ions and electrons. The ions and electrons move together at the "$E \times B$" velocity in a collisionless plasma and no net current flows in response to an applied electric field. Thus Ohm's law in its usual form is of little use in a collisionless magnetized plasma. There is another interesting subtlety in (2.36c). Ignoring ∇n and $\mathbf{g}$, we express all the remaining variables in earth-fixed coordinates:

$$(V_j - U)_\perp = [(E + U \times B) \times B]/B^2$$

Carrying out the triple cross product, $[(U \times B) \times B]/B^2 = -U_\perp$ and hence

$$(V_j)_\perp = E \times B/B^2$$

This equation shows that in the collisionless case the plasma moves at the $E \times B$ velocity in any reference frame, provided the electric field and velocity are expressed in that reference frame. In the earth-fixed frame then

$$(V_j)_\perp = (1/B^2)[E - (k_B T_j/q_j)\nabla n/n + (M_j/q_j)g] \times B \qquad (2.36d)$$

which is identical in form to (2.36c). In (2.36a) we have left the prime on E' but it should be noted that the transformation (2.32a) leaves the component of E parallel to B invariant.

The solutions of (2.34) for intermediate values of κ are given by

$$(V_j')_\parallel = (W_j')_\parallel \qquad (2.37a)$$

and

$$V_{j\perp}' = \frac{W_{j\perp}'}{1 + \kappa_j^2} + \frac{\kappa_j W_{j\perp}'}{1 + \kappa_j^2} \times \hat{B} \qquad (2.37b)$$

where we have again expressed the result in terms of the steady-state unmagnetized velocity solution W_j'. These expressions show explicitly that for small κ, V_j' tends toward W_j', while for large κ the motions tend to be perpendicular to the forces. The absolute values of κ_e and κ_i are plotted in Fig. 2.5 for an equatorial ionosphere with $B = 2.5 \times 10^{-5}$ tesla (0.25 gauss). The transition from a molecular ion plasma (NO^+ and O_2^+) to an atomic ion plasma (O^+) has been included in the calculation of κ. The absolute value of κ_e passes through unity near 75 km, while κ_i does so at 130 km. In making the plot of κ_e we have used the total electron collision frequency $\nu_e = \nu_{en} + \nu_{ei}$. This is not entirely consistent with the discussion above, which assumes $\nu_e = \nu_{en}$. However, the modification is of little importance since the absolute value of κ_e is very large above 100 km.

The relationship between J' and E' may now be determined from the definition $J' = ne(V_i' - V_e')$ with V_j' given by (2.37) and (2.35). The result may be

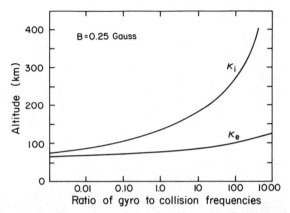

Fig. 2.5. Typical values for κ_e and κ_i in the equatorial ionosphere for a magnetic field of 2.5×10^{-5} tesla.

expressed through a tensor relationship $\mathbf{J}' = \boldsymbol{\sigma} \cdot \mathbf{E}'$, where

$$\boldsymbol{\sigma} = \begin{pmatrix} \sigma_P & -\sigma_H & 0 \\ \sigma_H & \sigma_P & 0 \\ 0 & 0 & \sigma_0 \end{pmatrix} \tag{2.38}$$

To obtain this form, $\mathbf{B}$ has been taken to be parallel to the z axis and we have defined

$$\sigma_0 = ne(b_i - b_e) \tag{2.39a}$$

$$\sigma_P = ne[b_i/(1 + \kappa_i^2) - b_e/(1 + \kappa_e^2)] \tag{2.39b}$$

$$\sigma_H = (ne/B)[\kappa_e^2/(1 + \kappa_e^2) - \kappa_i^2/(1 + \kappa_i^2)] \tag{2.39c}$$

The three conductivity parameters, σ_0, σ_P, and σ_H, are called the specific, Pedersen, and Hall conductivities, respectively.

Plots of σ_0, σ_P, and σ_H for a typical daytime mid-latitude ionosphere are given in Fig. 2.6. These plots correspond to the daytime collision frequencies in Fig. 2.3 and a magnetic field of 5×10^{-5} tesla (0.5 gauss). The specific or parallel conductivity σ_0 is dominated by the high electron mobility and is equal to ne^2/mv_e to a good approximation. At high altitudes when electron–neutral collisions become rare, the plasma density factor in v_{ei} cancels the same factor in the numerator of σ_0 and therefore σ_0 is independent of density above 400 km. The variation above that height displayed in Fig. 2.6 is related to the electron temperature since, according to (2.29b), v_{ei} is very nearly proportional to $(T_e)^{-3/2}$. The parallel conductivity is so high that the ratio σ_0/σ_P is greater than 1×10^4 above 130 km. Above about 75 km, κ_e is very large and the electrons only move

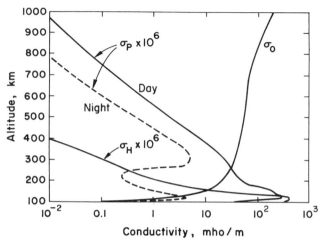

Fig. 2.6. Typical conductivity values for the mid-latitude daytime ionosphere. Notice the change of scale for σ_P and σ_H. The dashed curve is a typical nighttime profile of σ_P, also multiplied by 10^6. [From "Satellite Environment Handbook," edited by F. Johnson (1961). Reproduced by permission of Stanford Univ. Press. Copyright © 1961 by the Board of Trustees of the Leland Stanford Junior University.]

perpendicular to the forces that act on them in the plane perpendicular to **B**. Above this height the Pedersen conductivity may be written in the form

$$\sigma_P \;=\; ne^2/M\nu_{in}(1 \;+\; \kappa_i^2) \tag{2.40a}$$

For $\kappa_i \gg 1$ (above 130 km) this expression becomes even simpler,

$$\sigma_P \;=\; ne^2\nu_{in}/M\Omega_i^2 \;=\; nM\nu_{in}/B^2 \tag{2.40b}$$

The Hall conductivity σ_H falls off more rapidly with height than does σ_P and is important only in a narrow height range where three conditions are met: $\kappa_e \gg 1$, $\kappa_i \approx 1$, and n is large. A typical nighttime curve for σ_P is also given.

Finally, we remind the reader that the calculations have thus far been performed in the neutral reference frame where $\mathbf{J}' = \boldsymbol{\sigma} \cdot \mathbf{E}'$. More usually we measure the neutral wind **U** and electric field **E** in the earth-fixed frame. However, since $\mathbf{E}' = \mathbf{E} + \mathbf{U} \times \mathbf{B}$ and $\mathbf{J} = \mathbf{J}'$ for nonrelativistic transformations, we have the important and most usual form of the current equation

$$\mathbf{J} \;=\; \boldsymbol{\sigma} \cdot (\mathbf{E} \;+\; \mathbf{U} \times \mathbf{B}) \tag{2.41}$$

where all parameters are measured in the earth-fixed coordinates. Earth-fixed measurements of electric fields, of course, can determine only the **E** in (2.41), not the entire quantity $\mathbf{E}' = \mathbf{E} + \mathbf{U} \times \mathbf{B}$.

To summarize, we note that the ionospheric plasma is subject to electromagnetic forces in addition to those felt by the neutral atmosphere. The dipole nature of the magnetic field is not greatly affected by ionospheric currents. The result is that the magnetic field creates geometric constraints on the plasma behavior; constraints that are quite different at different magnetic latitudes. Electric fields, on the other hand, come and go quite rapidly and put the plasma constituents in motion perpendicular to **B**. Electric fields thus play a dominant role in the dynamics of the ionosphere. In the next sections we briefly discuss the generation and mapping of electric fields. Specific electric field sources are discussed as they arise in subsequent chapters.

2.3   Generation of Electric Fields

Although we have allowed for the possibility of an electric field in the earth's upper atmosphere, we have yet to show that such fields exist. The other forces in (2.28) are, for our present purpose, given quantities—that is, the forces related to the plasma pressure gradients associated with aeronomic processes, to the magnetic and gravitational fields, and to the atmospheric winds. Electric fields arise as a result of these forces when the ions and electrons respond differently to them. This is expressed quantitatively via the current divergence equation

$$\nabla \cdot \mathbf{J} = -\partial \rho_c / \partial t \qquad (2.42)$$

Any charge density, of course, must create electric fields through Poisson's equation

$$\nabla \cdot \mathbf{E} = \rho_c / \varepsilon_0 \qquad (2.43)$$

Given the complexity of the forces in (2.28), it is not surprising that the electric current associated with the difference between the ion and electron velocities has a finite divergence. However, this divergence creates a charge density via (2.42) that, via (2.43), creates an electric field which forces the divergence to zero. In other words, if the complex forces acting on the ion and electron fluids create a divergence in **J**, an electric field builds up quickly to modify the fluid velocities so that once again $\nabla \cdot \mathbf{J} = 0$.

For example, consider an electric field of 10 mV/m, which is large by equatorial and mid-latitude standards. Assuming a scale length of 1 km in (2.43) we find $\rho_c = 8.85 \times 10^{-17}$ C/m³, which amounts to an excess of ions or electrons of a few thousand per cubic meter compared to a total of at least 10^9 m⁻³. The time scale for buildup of such a charge density can be estimated from (2.42) and (2.43). From (2.42)

$$\tau \sim \rho_c / \nabla \cdot \mathbf{J} = \varepsilon_0 \nabla \cdot \mathbf{E} / \nabla \cdot \mathbf{J}$$

Assuming for the moment that σ is uniform and isotropic, then $\mathbf{J} = \sigma \mathbf{E}$ and

$$\tau \simeq \varepsilon_0 / \sigma \qquad (2.44)$$

Using the lowest value of any component of σ in the ionosphere we find the largest value for $\tau = 10^{-6}$ s. Electric fields thus build up very quickly indeed in response to any divergence of $\mathbf{J}$. Such divergences arise whenever there are spatially varying forces on the plasma or when the conductivity changes in space. In practice, it is not possible to calculate ρ_c from (2.42) and thence $\mathbf{E}$ from (2.43). Rather, the electric field is treated as a free parameter that adjusts in magnitude and direction to fit the requirement that $\nabla \cdot \mathbf{J} = 0$. In the next chapters we show in detail how electric fields arise in the earth's ionosphere.

When an electric field is created by a wind, the process is often called a dynamo in analogy to a motor-driven electric generator in which a conductor is moved across a magnetic field. In this case, or any case in which electrical energy is created, the quantity $(\mathbf{J} \cdot \mathbf{E})$ should be negative and, in addition, the electrical forces must act in opposition to the source of the charge separation.

When electric fields are applied from an external source, such as occurs at high latitudes due to the solar wind–magnetosphere interaction, it is usually the case that $\mathbf{J} \cdot \mathbf{E} > 0$ in the ionosphere. In this case electrical energy is converted into mechanical energy in the ionosphere and is released in the form of heat. Such Joule heating is a very important process at high latitudes and may greatly affect the thermospheric winds. In this case the ionosphere acts like an electrical load on some external generator. Likewise, momentum may be transferred to the thermospheric gas through the "ion drag" term if the ions are driven very strongly by an externally applied electric field. In such a case the ionosphere–magnetosphere system acts like a motor with electrical energy converted into mechanical energy.

2.4 Electric Field Mapping

The high conductivity parallel to the earth's magnetic field, σ_0, has important implications concerning the transmission of electric fields for long distances along $\mathbf{B}$. In fact, if σ_0 were infinite, there would be zero potential drop along the magnetic field and the potential difference between any two field lines would be constant. In such a case any electric field generated at ionospheric heights would be transmitted along the magnetic field lines to very high altitudes. For example, an electric field generated at $60°$ magnetic latitude would be communicated to the equatorial plane at an altitude over 25,000 km. Likewise, electric fields of solar wind or magnetospheric origin could be transmitted to ionospheric heights.

This phenomenon can be studied quantitatively as follows (following Farley, 1959, 1960). Suppose first the conductivity is anisotropic but uniform and that the neutral wind is absent. If the electric field perpendicular to $\mathbf{B}$ is $\mathbf{E}_\perp$ and the field parallel to $\mathbf{B}$ is $\mathbf{E}_\parallel$ the total current is

$$\mathbf{J} = \sigma_P \mathbf{E}_\perp - \sigma_H (\mathbf{E}_\perp \times \hat{B}) + \sigma_0 \mathbf{E}_\parallel \qquad (2.45)$$

For an electrostatic field $\mathbf{E} = -\nabla\phi$. Substituting this expression for $\mathbf{E}$ into (2.45), taking the divergence, setting $\nabla\cdot\mathbf{J} = 0$, and taking $\mathbf{B}$ to be in the z direction yields

$$-(\sigma_P)\,\partial^2\phi/\partial x^2 - (\sigma_H)\,\partial^2\phi/\partial x\partial y - (\sigma_P)\,\partial^2\phi/\partial y^2$$
$$+ (\sigma_H)\,\partial^2\phi/\partial y\partial x - (\sigma_0)\,\partial^2\phi/\partial z^2 = 0$$

The terms containing σ_H cancel, leaving

$$\partial^2\phi/\partial x^2 + \partial^2\phi/\partial y^2 + (\sigma_0/\sigma_P)\,\partial^2\phi/\partial z^2 = 0 \qquad (2.46)$$

Making the change of variables

$$dz' = (\sigma_P/\sigma_0)^{1/2}\,dz$$
$$dx' = dx$$
$$dy' = dy$$

converts (2.46) to

$$(\nabla')^2\phi = 0 \qquad (2.47)$$

which is Laplace's equation in the "reduced" coordinate system. That is, the substitution has transformed the real medium into an equivalent isotropic medium with a greatly reduced depth parallel to the magnetic field (the z direction in the calculation). The ratio $(\sigma_0/\sigma_P)^{1/2}$ is plotted in Fig. 2.7 for a typical ionospheric profile. Above 130 km the ratio exceeds 100, reaching 1000 at 300 km. At high altitudes, σ_0 becomes independent of density. The ratio σ_0/σ_P continues to increase as the ion–neutral collision frequency and the plasma density which determine σ_P continue to decrease. One of the basic approximations of magnetohydrodynamics (MHD) is that if the conductivity parallel to the magnetic field becomes very large then the parallel electric field component vanishes. For many

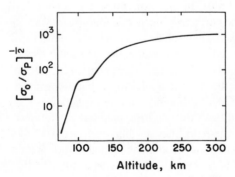

Fig. 2.7. The mapping ratio $(\sigma_0/\sigma_P)^{1/2}$ plotted as a function of height for a typical mid- to high-latitude ionosphere.

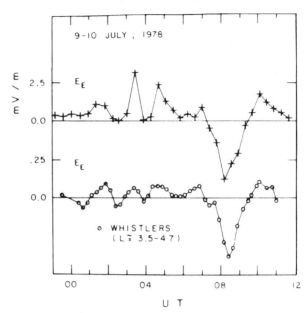

Fig. 2.8. Electric field components perpendicular to the magnetic field. The top panel is a measurement of the zonal electric field in the ionosphere. The lower panel is a measurement of the zonal electric field at the equatorial plane on the same magnetic field line. [After Gonzales *et al.* (1980). Reproduced with permission of the American Geophysical Union.]

applications, MHD theory applies on the high-altitude portions of the field lines which contact the ionosphere. This theory is discussed in the next section in the context of the "frozen-in" condition.

The implication of these calculations is that electrical features perpendicular to **B** map for long distances along the earth's magnetic field lines. This has been verified experimentally via the simultaneous measurements shown in Fig. 2.8 (Gonzales *et al.*, 1980). In this experiment the zonal electric field component was measured in the northern hemispheric ionosphere with a radar (see Appendix A) and in the inner magnetosphere at a point very close to where the same magnetic field line crossed the equatorial plane. The latter measurement was accomplished using the whistler technique (Carpenter *et al.*, 1972). The two measurements clearly have the same temporal form, but the magnetospheric component is 10 times smaller. This difference may be explained as a geometric effect arising from the spreading of the magnetic field lines as follows. First we take σ_0 to be infinite so that the field lines act like perfect conductors. This implies that there is no potential difference along them and in turn that the voltage difference between two lines is conserved. The magnetic flux density (measured in tesla) decreases along the field line as a function of distance from the mid- or high-latitude ionosphere to the equatorial plane in the magnetosphere. Since the voltage between adjacent field lines is constant, the perpendicular electric field must also decrease along the field lines. For a dipole field Mozer (1970) has shown that the two electric field components (meridional E_{MI} and zonal E_{ZI}) map from the ionosphere to the magnetosphere in the equatorial plane as

$$E_{MI} = 2L\left(L - \frac{3}{4}\right)^{1/2} E_{RM} \qquad (2.48a)$$

$$E_{ZI} = L^{3/2} E_{ZM} \qquad (2.48b)$$

where the L value is the distance from the center of the earth to the equatorial crossing point measured in earth radii (R_e), E_{RM} is the radial magnetospheric component at the equatorial plane, and E_{ZM} is the zonal magnetospheric component there. The equation for the zonal component (2.48b) is in excellent agreement with the corresponding data in Fig. 2.8. Notice that the zonal ionospheric electric field component maps to a zonal field in the equatorial plane but that the meridional component in the ionosphere, E_{MI}, becomes radial in the equatorial plane (E_{RM}). In particular, a poleward ionospheric electric field points radially outward at the equatorial plane. Thus, large-scale electric fields generated in the E region and F region of the ionosphere can map upward to the magnetosphere and create motions there. Likewise, electric fields of magnetospheric and solar wind origin can map from deep space to ionospheric heights and have even been detected by balloons at stratospheric heights.

Farley (1959) studied the upward mapping process realistically by including the z dependence of the conductivities in his analysis. The basic equations, $\nabla \cdot \mathbf{J} = 0$, $\mathbf{E} = -\nabla\phi$, and $\mathbf{J} = \boldsymbol{\sigma} \cdot \mathbf{E}$, are the same used in deriving (2.47) and, with the assumption that variations of $\boldsymbol{\sigma}$ occur only in the z direction, they yield

$$\partial^2\phi/\partial x^2 + \partial^2\phi/\partial y^2 + (1/\sigma_P)\,\partial/\partial z(\sigma_0\,\partial\phi/\partial z) = 0$$

The same change of variables now yields

$$(\nabla')^2\phi + (\partial\phi/\partial z')(\partial/\partial z')[\ln(\sigma_0\sigma_P)^{1/2}] = 0 \qquad (2.49a)$$

where $(\sigma_0\sigma_P)^{1/2} = \sigma_m$ is termed the geometric mean conductivity. Furthermore, if σ_m can be modeled in the form $\sigma_m = c\,\exp(c_0 z')$, then the equation simplifies to

$$(\nabla')^2\phi + c_0(\partial\phi/\partial z') = 0 \qquad (2.49b)$$

This differential equation has a straightforward analytical solution. By considering the solutions with different Fourier wave numbers in the source field, Farley showed that (a) larger-scale features map more efficiently to the F region from the E region than small-scale features, (b) the height of the source field is very important, with upper E-region structures very much favored as F-region sources, and (c) roughly speaking, perpendicular structures with scale sizes greater than a few kilometers map unattenuated to F-region heights. The implication here is that if very large scale electric fields are generated in the E region, the potential differences thereby created map up into the F-region ionosphere and beyond along the magnetic field lines deep into space. As we shall see, these low-altitude electric fields dominate motions of the plasma throughout the dense plasma region around the earth termed the plasmasphere.

Considering sources at F-region and even higher altitudes, for example, in the magnetosphere and solar wind, the previous analysis can be used to show that the mapping efficiency to the E region is even greater. In fact, within the magnetospheric and solar wind plasmas the parallel conductivity is often taken to be infinite and hence the parallel electric field vanishes even when finite field-aligned currents flow. As we shall see later, this assumption that $E_\parallel = 0$ is a powerful analytical device, allowing great conceptual simplifications in the understanding of magnetospheric electric field and flow patterns. On the other hand, it is exactly in the regions where the assumption of infinite conductivity breaks down that very interesting phenomena occur. The generation of the aurora is an example.

Since large-scale electric fields map along the magnetic field lines, we may consider them to be independent of z in the magnetic coordinate system used above (**B** in the z direction). This has some interesting consequences for ionospheric electrodynamics. Consider the current divergence equation separated into its perpendicular and parallel parts:

$$\nabla_\perp \cdot \mathbf{J}_\perp = -\partial J_z / \partial z$$

where $\mathbf{J}_\perp = \boldsymbol{\sigma}_\perp \cdot (\mathbf{E}_\perp + \mathbf{U} \times \mathbf{B})$ and $\boldsymbol{\sigma}_\perp$ is the 2×2 perpendicular conductivity matrix. Ignoring the neutral wind for the moment,

$$\nabla_\perp \cdot (\boldsymbol{\sigma}_\perp \cdot \mathbf{E}_\perp) = -\partial J_z / \partial z$$

Integrating from the top of the northern hemisphere ionosphere ($z = 0$) to a value z_0 below which no significant perpendicular currents flow yields

$$J_z(0) - J_z(z_0) = \int_0^{z_0} [\nabla \cdot (\boldsymbol{\sigma}_\perp \cdot \mathbf{E}_\perp)] \, dz$$

Although small currents do exit the base of the ionosphere and link up with atmospheric electrical currents to complete the global atmospheric electrical circuit, we ignore this interesting phenomenon by setting $J_z(z_0) = 0$. $J_z(0)$ is then the field-aligned current entering the ionosphere from above and is related to the divergence of the perpendicular current density. Now, since $\mathbf{E}_\perp$ is independent of z, we may move it and the divergence operator through the integral to yield

$$J_z = \nabla \cdot (\mathbf{\Sigma} \cdot \mathbf{E}_\perp) \tag{2.50a}$$

where $\mathbf{\Sigma}$ is the perpendicular height-integrated conductivity tensor

$$\mathbf{\Sigma} = \begin{pmatrix} \Sigma_P & -\Sigma_H \\ \Sigma_H & \Sigma_P \end{pmatrix}$$

and we have dropped the agreement of J_z and assume that we measure J_z at a sufficiently high altitude that all perpendicular ionospheric currents are below it. Since conductivity has units of mhos per meter, $\mathbf{\Sigma}$ has units of mhos. The E-region values vary from a few tenths of a mho in the nighttime mid-latitude region to several tens of mhos during a strong auroral precipitation event.

Some insight can be achieved by letting $\mathbf{\Sigma}$ be uniform in the horizontal directions. Then the perpendicular divergence operator does not operate on it and

$$J_z = \Sigma_P(\partial E_x/\partial x) - \Sigma_H(\partial E_y/\partial x) + \Sigma_H(\partial E_x/\partial y) + \Sigma_P(\partial E_y/\partial y)$$

which can be written

$$-J_z = \Sigma_P(\nabla_\perp \cdot \mathbf{E}_\perp) + \Sigma_H[(\partial E_x/\partial y - \partial E_y/\partial x)]$$

The last term is the z component of $-\nabla \times \mathbf{E}$, which vanishes for steady magnetic fields, and we have

$$J_z = \Sigma_P(\nabla_\perp \cdot \mathbf{E}_\perp) \tag{2.50b}$$

In the auroral zone $\Sigma_P \sim 10$ mho, $\mathbf{E}_\perp \sim 50$ mV/m, and $\nabla_\perp \sim 1/L \sim 10^{-5}$ m, which yields a typical large-scale J_z of order 5 μA/m². Small-scale currents can be more than an order of magnitude higher. Field-aligned currents at mid- and low latitudes are much smaller but still play a very important role in the physics. Equation (2.50b) plus (2.25b) show that current flows downward in the northern hemisphere when the net charge density in the ionosphere is positive. This seems curious but corresponds to the fact that the ionosphere is a load in this model.

Finally, we turn to the question of downward mapping of magnetospheric and ionospheric electric fields into the region below the ionosphere, that is, into the lower atmosphere. In the late 1960s it was realized that these fields could be detected at balloon altitudes (~30 km) since most of the atmospheric resistivity is in the last atmospheric scale height, which has a value of about 8 km near the surface of the earth (Mozer and Serlin, 1969). At the surface of the earth, of course, horizontal fields must vanish since the earth is a good conductor compared with the atmosphere. Researchers have only recently begun to treat unified models of atmospheric and space electricity which include such effects, and the future is bright for such studies.

2.5   Elements of Magnetospheric Physics

2.5.1   Collisionless Plasmas

The neutral atmosphere is so tenuous at magnetospheric heights that, unlike the situation in the ionosphere, the plasma can be considered fully ionized. Furthermore, the plasma itself is so rarefied that $\nu_{ei} \ll \Omega_i$ and the medium is treated as a collisionless magnetoplasma. The fluid description now has only two constituents. The production and loss terms in (2.22a) can be dropped since there are no neutrals, as can the collision term in (2.22b). Following Spitzer (1962), we define the plasma velocity in terms of the ion and electron velocities as

$$\mathbf{V} = (1/\rho)(n_i M \mathbf{V}_i + n_e m \mathbf{V}_e)$$

where $\rho = n_i M + n_e m$. Then adding the equations of motion for the two species yields

$$\rho(d\mathbf{V}/dt) \doteq \mathbf{J} \times \mathbf{B} - \nabla p + \rho \mathbf{g} \tag{2.51}$$

with $p = p_i + p_e$ and $\mathbf{J} = n_i e \mathbf{V}_i - n_e e \mathbf{V}_e$. We have dropped the volume electrical force term $(n_i - n_e)e\mathbf{E}$ in this expression since $n_i \simeq n_e$. The plasma flow velocity is dominated by the ions since $M \gg m$ implies $\mathbf{V} \simeq \mathbf{V}_i$. For conditions that are slowly varying in time and space, $d/dt \simeq 0$, and using $n_i = n_e = n$ and $\rho \approx \rho_i$, (2.22b) can be solved for the perpendicular component of the ion velocity, which is equivalent to the perpendicular plasma velocity

$$\mathbf{V}_\perp = (\mathbf{E} \times \mathbf{B})/B^2 - (\nabla p_i \times \mathbf{B})/neB^2 + \rho(\mathbf{g} \times \mathbf{B})/neB^2 \tag{2.52}$$

The parallel dynamics are described by an equation analogous to (2.36a), where the prime can be dropped since there is no neutral atmosphere to contend with and where only ion–electron collisions are considered. Often ∇p_i and $\mathbf{g}$ can be ignored, yielding the result

$$\mathbf{V}_\perp = \mathbf{E} \times \mathbf{B}/B^2 \tag{2.53a}$$

Equation (2.51) can be solved for $\mathbf{J}_\perp$ to yield the macroscopic perpendicular current density

$$\mathbf{J}_\perp = \mathbf{B} \times \nabla p/B^2 + \rho \mathbf{g} \times \mathbf{B}/B^2 - \rho(d\mathbf{V}/dt) \times \mathbf{B}/B^2 \tag{2.53b}$$

These two equations, (2.53a) and (2.53b), show that a steady electric field determines the perpendicular plasma velocity but has nothing to do with the perpendicular current. That is, the electrons and ions undergo an $\mathbf{E} \times \mathbf{B}$ drift at the same velocity and no net current results. Perpendicular currents are caused by pressure gradients, inertial effects, and gravitational fields.

Current can flow quite easily parallel to $\mathbf{B}$ according to the difference between the ion and electron velocities given by (2.36a). As we have noted above, very small electric fields parallel to $\mathbf{B}$ can usually drive whatever current is necessary in that direction to keep the voltage drop small along **B**. Closing such field-aligned currents perpendicular to **B** in a collisionless plasma is another matter, since it requires either pressure gradients or time/space-varying velocity fields. The field-aligned currents which flow into and out of the ionosphere at high latitudes must close across the field lines somewhere in the magnetosphere.

There are (at least) two philosophies one may adopt in describing a collision-less magnetized plasma. In the magnetohydrodynamic approach the plasma is considered as a *single* conducting fluid with a certain bulk flow velocity deter-mined by the forces acting on the fluid. These forces must include the electro-magnetic force as well as the gravitational, pressure, and viscous forces acting on a hydrodynamic system. Another viewpoint considers the microscopic mo-tions of particles subject to the various forces. Both descriptions are used in space plasma physics and are discussed in the next two sections.

2.5.2 The Guiding Center Equations and the Adiabatic Invariants

One way to view particle dynamics is to consider the motion of the so-called guiding centers. In this picture the near-circular gyromotion about the magnetic field is considered to be the dominant motion with a time scale (Ω^{-1}) much shorter than any other dynamical time scale τ $(\Omega^{-1} \ll \tau)$.

To illustrate this, consider the single-particle motion associated with a general force **F** and the magnetic field **B**,

$$(M \, d\mathbf{V}/dt) = q(\mathbf{V} \times \mathbf{B}) + \mathbf{F}$$

Taking the component parallel to **B** gives

$$M(d\mathbf{V}_\parallel/dt) = \mathbf{F}_\parallel$$

which is straightforward. Perpendicular to **B**, this leaves

$$(M \, d\mathbf{V}_\perp/dt) = q(\mathbf{V}_\perp \times \mathbf{B}) + \mathbf{F}_\perp \qquad (2.54)$$

Now let

$$\mathbf{V}_\perp = \mathbf{W}_D + \mathbf{u}$$

where

$$\mathbf{W}_D = (\mathbf{F} \times \mathbf{B}/qB^2) \qquad (2.55)$$

This substitution is motivated by the result we already have that $\mathbf{W}_D = (\mathbf{E} \times \mathbf{B})/B^2$ when $\mathbf{F} = q\mathbf{E}$. Substituting (2.55) into (2.54) yields

$$M \, d\mathbf{W}_D/dt + M \, d\mathbf{u}/dt = (\mathbf{F}_\perp \times \mathbf{B}/B^2) \times \mathbf{B} + q\mathbf{u} \times \mathbf{B} + \mathbf{F}_\perp$$

The first term on the left vanishes, while the first term on the right-hand side equals $-\mathbf{F}_\perp$ and hence cancels the other $\mathbf{F}_\perp$ term. This leaves

$$M \, d\mathbf{u}/dt = q(\mathbf{u} \times \mathbf{B})$$

The solution to this of course is just the gyromotion at frequency

$$\Omega = qB/M$$

The interpretation we make is that in a frame moving at $\mathbf{W}_D$, the particle motion is pure gyration. This yields the concept of a guiding center motion since $\mathbf{W}_D$ gives the velocity of the center of gyromotion. Some examples of guiding center drifts due to various forces are as follows. For an electric field, $\mathbf{F} = q\mathbf{E}$ and

$$\mathbf{W}_D = q\mathbf{E} \times \mathbf{B}/qB^2 = \mathbf{E} \times \mathbf{B}/B^2$$

For the gravitational field, $\mathbf{F} = M\mathbf{g}$ and

$$\mathbf{W}_D = M\mathbf{g} \times \mathbf{B}/qB^2$$

For the inertial force $\mathbf{F} = -M\, d\mathbf{W}_D/dt$ and, letting $\mathbf{W}_D = \mathbf{E} \times \mathbf{B}/B^2$,

$$\mathbf{W}_D = (\mathbf{F} \times \mathbf{B})/qB^2 = (-1/qB^2)[M\, d/dt(\mathbf{E} \times \mathbf{B}/B^2) \times \mathbf{B}]$$

$$\mathbf{W}_D = -(M/qB^2)(d\mathbf{E}/dt \times \mathbf{B}) \times \mathbf{B}$$

$$\mathbf{W}_D = (M/qB^2)(d\mathbf{E}/dt)$$

Notice that this expression can be related to a displacement current in the plasma since using $\mathbf{J} = (ne\mathbf{W}_{Di} - ne\mathbf{W}_{De})$ and $M \gg m$ yields

$$\mathbf{J} = ne(M/eB^2)\, d\mathbf{E}/dt = (nM/B^2)\, d\mathbf{E}/dt$$

Substituting this into Maxwell's equations gives

$$\nabla \times \mathbf{B} = \mu_0\mathbf{J} + \mu_0\varepsilon_0\, \partial\mathbf{E}/\partial t$$

$$= \mu_0(nM/B^2 + \varepsilon_0)\, \partial\mathbf{E}/\partial t$$

so we can define the low-frequency plasma dielectric constant via the expression

$$\varepsilon = (nM/B^2) + \varepsilon_0 \simeq (nM/B^2)$$

Now an assumption necessary for use of the guiding center approximation is that the time scale $\tau > \Omega_i^{-1}$. Hence, this dielectric constant should be valid for electromagnetic waves with frequencies $f < \Omega_i$. Indeed, the expression for the phase velocity of an electromagnetic wave is

$$V_{ph} = (1/\mu_0\varepsilon)^{1/2}$$

Substituting the dielectric constant derived above yields

$$V_{ph} = B/(\mu_0\rho)^{1/2}$$

This velocity is the Alfvén speed, which is the velocity of an electromagnetic wave propagating parallel to $\mathbf{B}$ in a magnetized plasma when its frequency satisfies $f \ll \Omega_i$ (e.g., Spitzer, 1962, or any elementary plasma text).

If the magnetic field is curved as in a dipole field, particles moving along **B** will feel a force given by

$$\mathbf{F} = (MV_{\parallel}^2/R)\hat{n}$$

where $\hat{n}$ is a unit vector pointed inward, $V_{\parallel}$ is the particle velocity parallel to **B**, and R is the radius of curvature (see Fig. 2.13 for a sketch of the coordinate system). Substituting this force into (2.55), we find that the particles drift perpendicular to the field with velocity

$$\mathbf{W}_D = (MV_{\parallel}^2/R)(\hat{n} \times \mathbf{B}/qB^2)$$

Similarly, if there is a gradient in the magnetic field with a scale large compared to a gyroradius, we can consider the force on a magnetic dipole of moment μ, which is given by

$$\mathbf{F} = -\mu \nabla B$$

A gyrating particle has a magnetic moment since it carries a current I and surrounds an area πr_g^2, where r_g is the gyroradius, so

$$\mu = IA = I\pi r_g^2$$

The current equals the charge multiplied by the gyrofrequency Ω, and using

$$\Omega^{-1} = (2\pi r_g/V_{\perp}) \quad \text{and} \quad r_g = (MV_{\perp}/qB)$$

yields the magnetic moment

$$\mu = (MV_{\perp}^2/2B) = K_{\perp}/B$$

where $K_{\perp}$ is the perpendicular kinetic energy of the particle. Thus, the force due to the magnetic field gradient is

$$\mathbf{F} = -(K_{\perp}/B)\nabla B$$

and the corresponding "gradient drift" is

$$\mathbf{W}_D = -\mu(\nabla B \times \mathbf{B}/qB^2) = (MV_{\perp}^2/2B)(\mathbf{B} \times \nabla B/qB^2)$$

Notice that the gradient and curvature drifts are proportional to the particle perpendicular and parallel energies, respectively. The gradient-driven motion can be visualized easily with reference to Fig. 2.9. If ∇B is downward as in the figure, a gyrating particle will have a slightly smaller radius of curvature in one portion of its cycle than in the other. A net drift results to the right for positively charged particles and to the left for negatively charged particles.

To summarize, for time scales greater than Ω^{-1} and length scales greater than r_g the guiding center perpendicular drift equation for particle motion in a magnetic field is given by a combination of all the forces, starting with the electric

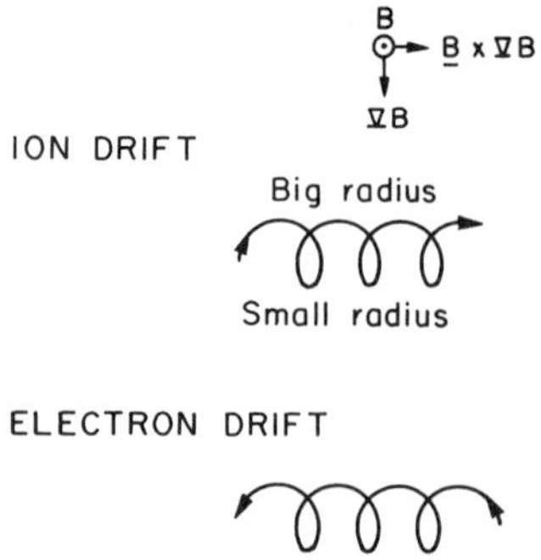

ION DRIFT

Big radius

Small radius

ELECTRON DRIFT

Fig. 2.9. Gyromotion in a magnetic field with a gradient pointing downward.

field, that is,

$$\mathbf{W}_D = (\mathbf{E} \times \mathbf{B}/B^2) + (M/qB^2) \, d\mathbf{E}/dt + (M/qB^2)(\mathbf{g} \times \mathbf{B})$$
$$+ (1/qB^2)(MV_\perp^2/2B)(\mathbf{B} \times \nabla B) - (MV_\parallel^2/qB^4)[\mathbf{B} \times (\mathbf{B} \cdot \nabla)\mathbf{B}] \tag{2.56}$$

Since the particle motions described above are quite complex, it has proved useful to develop an intuition based on parameters that are nearly conserved in the motion. The most "rugged" of these adiabatic invariants is the particle magnetic moment μ. Here, we prove that if the time scale for field changes is larger than $\tau_g = \Omega^{-1}$ (the gyroperiod), then μ is conserved. We refer the reader to Schmidt (1966) for proof that conservation also holds if the length scale for changes in $\mathbf{B}$ is much greater than r_{gi} (the ion gyroradius).

If $\mathbf{B}$ is uniform in space but time varying,

$$\partial \mu / \partial t = \partial / \partial t (K_\perp / B) = (\partial K_\perp / \partial t)/B - (\partial B / \partial t)(K_\perp / B^2)$$

The rate of change of perpendicular energy in the guiding center approximation can be approximated by the energy gained in one gyration divided by the time the gyration takes. Thus,

$$\Delta K_\perp / \Delta t = (\Delta K_g / \tau_g)$$

and

$$\Delta K_g = \int \mathbf{F} \cdot d\mathbf{l} = q \int (\mathbf{E} + \mathbf{V} \times \mathbf{B}) \cdot d\mathbf{l}$$

where the integral is around one gyroloop of the particle motion. We note that $d\mathbf{l} = \mathbf{V} \, dt$ so

$$(\mathbf{V} \times \mathbf{B}) \cdot d\mathbf{l} = (\mathbf{V} \times \mathbf{B}) \cdot \mathbf{V} \, dt = 0$$

This is an example of the fact that magnetic forces do not change particle energy.

The energy change in one gyration is thus

$$\Delta K_g = \int q\mathbf{E} \cdot d\mathbf{l}$$

We can transform this to a surface integral

$$\Delta K_g = q \iint (\nabla \times \mathbf{E}) \cdot d\mathbf{a}$$

and using (2.23a)

$$\Delta K_g = -q \iint (\partial \mathbf{B}/\partial t) \cdot d\mathbf{a}$$

Note that the vectors $q\mathbf{B}$ and $d\mathbf{a}$ are in opposite directions. Now since $\mathbf{B}$ is uniform in space we can move it outside the integral and

$$\Delta K_\perp/\Delta t = \Delta K_g/\tau_g = (q\pi r_g^2/\tau_g)(\partial B/\partial t) = (K_\perp/B)(\partial B/\partial t)$$

and finally

$$\partial \mu/\partial t = (K_\perp/B)(\partial B/\partial t)/B - (\partial B/\partial t)(K_\perp/B^2) = 0$$

and hence μ is conserved.

For illustrative purposes, we can apply these findings to a "magnetic bottle" geometry similar to the earth's magnetic field as illustrated in Fig. 2.10. Suppose

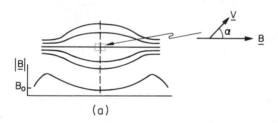

(a)

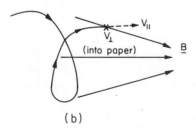

(b)

Fig. 2.10. (a) Sketch of the magnetic field lines and the corresponding magnetic field magnitude along the axis of a magnetic bottle. (b) The gyromotion associated with a converging magnetic field.

a particle starts on the symmetry axis at the place where the field strength is B_0 and the velocity vector of the particle makes an angle α (the pitch angle) with respect to the magnetic field line. At the magnetic equator, then,

$$V_\parallel = V \cos \alpha, \qquad V_\perp = V \sin \alpha$$

The velocity $V_\parallel$ carries the particle into a region of larger field strength as shown in Fig. 2.10b. However, if μ is conserved, as $|\mathbf{B}|$ increases $MV_\perp^2/2$ must also increase. Since the particle energy must also be conserved, this can only come at the expense of $V_\parallel$ and the particle slows down in its parallel motion. That the particle eventually bounces off the increasing magnetic field and returns toward the equatorial plane and the other hemisphere can be seen as follows. The magnitude of the Lorentz force is $|\mathbf{F}_L| = qV_\perp B$. From Fig. 2.10b it is clear that the direction of the force is to the left, which is opposite to the particle parallel velocity. Even at the right-hand side of the figure where the parallel velocity has gone to zero we see that $\mathbf{F}_L = q\mathbf{V}_\perp \times \mathbf{B}$ still has a small component of force toward the magnetic equator due to the convergence of the magnetic field lines. The parallel velocity of the particle will increase until the equatorial plane is crossed. Then the axial component of the Lorentz force reverses sign and the particle again slows down and is eventually reflected. If the initial pitch angle is too small, the particle will penetrate so deeply into the atmosphere at the end of the "bottle" that it will be lost by collisions. A "loss cone" then develops in the distribution function such that particles with pitch angles less than a certain value α_0 escape and particles with $\alpha > \alpha_0$ are trapped for many bounce cycles.

This bounce motion leads to another time scale with a time constant τ_{bounce} which can be considered in the particle dynamics. A second adiabatic invariant is related to this dynamical time scale. The second invariant holds only for time variations much longer than τ_{bounce}. Finally, the gradient and curvature drifts take particles entirely around the earth, creating another time scale τ_{drift} which is related to a third adiabatic invariant. This has an even longer time scale and hence is the easiest condition to break down. In general, we have

$$\tau_{\text{drift}} \gg \tau_{\text{bounce}} \gg \tau_{\text{gyration}}$$

and the particles gyrate many times between mirror points and bounce between the mirrors many times while the gradient and curvature drifts move them around the earth. Sketches of these three oscillatory motions are presented in Fig. 2.11. Neglecting scattering and plasma instabilities, the particles could be trapped in the earth's magnetic field forever. In practice, some species with particular energies (e.g., protons with energies ~ 100 MeV) are trapped for ~ 100 years. These particles travel $\sim 3 \times 10^{10}$ cm/s $\times$ 3×10^9 s $\sim 10^{20}$ cm. Since the scale size of the system is $\sim 10^9$ cm, such particles traverse the system $\sim 10^{11}$ times before escaping.

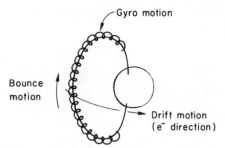

Fig. 2.11. The three oscillatory motions in the earth's magnetic field which are associated with the three adiabatic invariants.

We now raise the interesting question, can the current in some complicated plasma problem be computed using only the guiding center equations? As pointed out by Spitzer, the answer is no, since the guiding center current is only part of the total current. The particle gyromotion produces an additional current if there is a population gradient, exactly analogous to the magnetization current which arises from edge effects in solids. When this contribution is calculated exactly, it contains a term that can even cancel the gradient and curvature drift terms. The actual currents have nothing to do with the values of **B** and ∇B and we must use the macroscopic equations given in Section 2.5.1. It turns out that in actual problems the sign of currents is generally given correctly by the sign of the gradient and curvature drifts but the magnitudes must be derived from other equations. We return to this problem later in the text.

2.5.3 Magnetohydrodynamics

We now take up a second viewpoint on the dynamics of a collisionless magnetized plasma. In magnetohydrodynamics it is conjectured that the electrical conductivity is so high parallel to **B** that in a reference frame moving with velocity

$$\mathbf{V} = \mathbf{E} \times \mathbf{B}/B^2$$

the electric field vanishes both parallel and perpendicular to **B**. Since this is the plasma reference frame, the plasma can be considered as a single fluid having infinite conductivity. Study of such a one-fluid model is referred to as magnetohydrodynamics. Its range of applicability is particularly great in astrophysics due to the large distance scales involved. Thus plasmas in the solar wind, solar flares, sunspots, the earth's magnetosphere, and interstellar regions can all be treated to first order with a magnetohydrodynamic approach.

A close analogy exists with hydrodynamics, in which, as we have noted, a single fluid is subject to a variety of forces: pressure, viscosity, gravity, etc. The presence of an electrical conductivity requires the inclusion of Maxwell's equations and a volume force on the fluid given by $\mathbf{J} \times \mathbf{B}$. We thus have the continuity equation,

$$\partial \rho / \partial t + \nabla \cdot (\rho \mathbf{V}) = 0$$

the equation of motion,

$$\rho (d\mathbf{V} \, dt) = -\nabla p + \rho \mathbf{g} + \eta \nabla^2 \mathbf{V} + \mathbf{J} \times \mathbf{B}$$

and Maxwell's equations,

$$\nabla \times \mathbf{B} = \mu_0 \mathbf{J}$$

$$\nabla \times \mathbf{E} = -\partial \mathbf{B} / \partial t$$

$$\nabla \cdot \mathbf{B} = 0$$

$$\nabla \cdot \mathbf{E} = \rho_c / \varepsilon_0$$

where the vacuum displacement current has been ignored due to the high conductivity. These equations constitute the single-fluid description of a magnetized plasma or of conducting fluids such as the element mercury and the earth's molten core.

If we neglect the displacement current, we can also set $\nabla \cdot \mathbf{E} = 0$ since $\rho_c = 0$. Whatever electric fields are present, they change in space only when $\partial \mathbf{B} / \partial t$ is nonzero or the reference frame is changed. These equations constitute an MHD approximation and to them must be added a relationship between the fields and currents. For a true conducting fluid this is just

$$\mathbf{J}' = \sigma \mathbf{E}'$$

since the conductivity is isotropic. As before, $\mathbf{E}'$ is the electric field in the frame moving with the fluid. In some other frame in which the plasma velocity is $\mathbf{V}$ and the electric field is $\mathbf{E}$ we have $\mathbf{E} = \mathbf{E}' - \mathbf{V} \times \mathbf{B}$ and $\mathbf{J} = \mathbf{J}', \mathbf{B} = \mathbf{B}'$ for $V \ll c$. Thus, $\mathbf{J} = \mathbf{J}' = \sigma \mathbf{E}'$ and

$$\mathbf{J} = \sigma (\mathbf{E} + \mathbf{V} \times \mathbf{B})$$

with $\mathbf{J}, \mathbf{E}, \mathbf{V},$ and $\mathbf{B}$ all measured in the second frame. The conductivity of a plasma is not isotropic, so it is not clear that the MHD fluid approach should work. However, for a collisionless plasma σ_0 is so high parallel to $\mathbf{B}$ that in the frame of reference moving with the plasma, the electric field is zero. This is also true in an infinitely conducting isotropic fluid. Thus, in the limit that σ_0 goes to infinity, the plasma will behave like an infinitely conducting fluid, even though it is an anisotropic material.

Finally, we need an equation of state. Several possibilities exist depending on the properties of the fluid. For a true conducting metallic fluid like mercury or the earth's core we could use the incompressibility condition

$$\nabla \cdot \mathbf{V} = 0$$

For a fluid with a high heat conductivity we could use isothermal conditions

$$dT/dt = d(p/\rho)/dt = 0$$

or for an adiabatic fluid

$$d(p^\gamma/\rho^\gamma)/dt = 0$$

The particular equation of state to be used depends on the application.

To reduce the number of equations and gain physical insight, the current density and electric field can be eliminated among them. Returning to the equation of motion and ignoring viscosity and gravity, we have

$$\rho(d\mathbf{V}/dt) = -\nabla p + \mathbf{J} \times \mathbf{B} = -\nabla p + 1/\mu_0 (\nabla \times \mathbf{B}) \times \mathbf{B} \qquad (2.57)$$

Using the vector identity

$$\nabla(\mathbf{X} \cdot \mathbf{Y}) = (\mathbf{X} \cdot \nabla)\mathbf{Y} + (\mathbf{Y} \cdot \nabla)\mathbf{X} + \mathbf{X} \times (\nabla \times \mathbf{Y}) + \mathbf{Y} \times (\nabla \times \mathbf{X})$$

with $\mathbf{X} = \mathbf{Y} = \mathbf{B}$ yields

$$\nabla(B^2) = 2(\mathbf{B} \cdot \nabla)\mathbf{B} + 2\mathbf{B} \times (\nabla \times \mathbf{B})$$

and hence the $\mathbf{J} \times \mathbf{B}$ term in (2.57) becomes

$$\mathbf{f}_B = 1/\mu_0 (\nabla \times \mathbf{B}) \times \mathbf{B} = -\nabla(B^2/2\mu_0) + (\mathbf{B} \cdot \nabla)\mathbf{B}/\mu_0$$

and the equation of motion becomes

$$\rho(d\mathbf{V}/dt) = -\nabla(p + B^2/2\mu_0) + (\mathbf{B} \cdot \nabla)\mathbf{B}/\mu_0 \qquad (2.58)$$

We study this equation in more detail below. In a similar fashion we may eliminate $\mathbf{E}$ from Maxwell's magnetic field equation,

$$\partial \mathbf{B}/\partial t = -(\nabla \times \mathbf{E}) = -\nabla \times (\mathbf{J}/\sigma - \mathbf{V} \times \mathbf{B})$$

$$= \nabla \times (\mathbf{V} \times \mathbf{B}) - (1/\sigma)(\nabla \times \mathbf{J})$$

$$= \nabla \times (\mathbf{V} \times \mathbf{B}) - \nabla \times (\nabla \times \mathbf{B})/\sigma\mu_0$$

But using $\nabla \times (\nabla \times \mathbf{B}) = \nabla(\nabla \cdot \mathbf{B}) - \nabla^2\mathbf{B} = -\nabla^2\mathbf{B}$ yields finally

$$\partial \mathbf{B}/\partial t = \nabla \times (\mathbf{V} \times \mathbf{B}) + (1/\sigma\mu_0)\nabla^2\mathbf{B} \qquad (2.59)$$

The MHD equations can now be written

$$\partial \rho/\partial t + \nabla \cdot (\rho\mathbf{V}) = 0 \qquad (2.60a)$$

$$\nabla \cdot \mathbf{B} = 0 = \nabla \cdot \mathbf{E} \qquad (2.60b)$$

$$\rho(d\mathbf{V}/dt) = -\nabla(p + B^2/2\mu_0) + (\mathbf{B} \cdot \nabla)\mathbf{B}/\mu_0 \qquad (2.60c)$$

$$\partial \mathbf{B}/\partial t = \nabla \times (\mathbf{V} \times \mathbf{B}) + (1/\sigma\mu_0)\nabla^2\mathbf{B} \qquad (2.60d)$$

One of the equations of state completes the system.

The concepts of magnetic pressure and tension are derivable from (2.60c). One should remember that the terms involving the magnetic field **B** all stem from the **J** × **B** force but that **J** and **B** are inextricably related through Maxwell's equations. Thus, any deviation of **B** from a force-free configuration must be balanced by fluid pressure when the fluid is in equilibrium. Fluid acceleration occurs in the case of a nonequilibrium condition. For example, it is straightforward to show that the dipole magnetic field in the upper portion of Fig. 2.12 is such that the magnetic terms in (2.60c) cancel everywhere outside the core of the earth. The distorted dipole field shown below it, however, has forces that will cause the plasma to flow back toward the earth. Since $\nabla \times \mathbf{B} \neq 0$ in this distorted field, there must be a current in the region, as indicated. Such a current is associated with a **J** × **B** force on the plasma in the direction toward the earth since there is a small component of **B** upward while **J** is out of the page. This **J** × **B** force is equivalent to the magnetic pressure and magnetic tension forces. In a steady state then, there must be a pressure gradient pointing toward the earth.

The meaning of the magnetic terms can be understood in a local coordinate system (see Fig. 2.13) defined by a unit vector $\hat{s}$ parallel to **B**, a unit vector $\hat{n}$

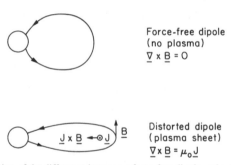

Force-free dipole
(no plasma)
$\underline{\nabla} \times \underline{B} = 0$

Distorted dipole
(plasma sheet)
$\underline{\nabla} \times \underline{B} = \mu_0 \underline{J}$

Fig. 2.12. Illustration of the difference between a force-free dipole and a distorted field configuration. The magnetic tension force is related to the **J** × **B** force shown.

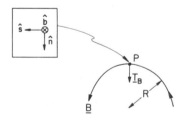

Fig. 2.13. Local coordinate system used to describe a curved magnetic field. The inset gives the directions of the unit vectors at the point P. In a stretched magnetic field the tension force $\mathbf{T}_B$ is parallel to $\hat{n}$ and inversely proportional to R.

normal to **B** (and antiparallel to the radius of curvature), and a unit vector $\hat{b}$ given by $\hat{b} = \hat{n} \times \hat{s}$. We have shown above that $\mathbf{J} \times \mathbf{B}$ may be written in the form of the magnetic force $\mathbf{f}_B = \mathbf{J} \times \mathbf{B}$, or

$$\mathbf{f}_B = (1/\mu_0)(\mathbf{B} \cdot \nabla)\mathbf{B} - \nabla(B^2/2\mu_0) \tag{2.61}$$

where

$$\mathbf{B} = B\hat{s}$$

In terms of $\hat{s}$, $\hat{n}$, and $\hat{b}$,

$$(\mathbf{B} \cdot \nabla)\mathbf{B} = B(\partial/\partial s)(B\hat{s}) = B(\partial B/\partial s)\hat{s} + B^2(\partial\hat{s}/\partial s)$$

Now

$$\partial\hat{s}/\partial s = \hat{n}d\theta/Rd\theta = \hat{n}/R$$

so

$$(\mathbf{B} \cdot \nabla)\mathbf{B} = B(\partial B/\partial s)\hat{s} + (B^2/R)\hat{n}$$

Substituting into (2.61),

$$\mathbf{f}_B = (B^2/\mu_0 R)\hat{n} + (B/\mu_0)(\partial B/\partial s)\hat{s} - \nabla(B^2/2\mu_0)$$

But

$$(1/\mu_0)(B\partial B/\partial s) = \partial(B^2/2\mu_0)/\partial s$$

so this term cancels the $\hat{s}$ component of $\nabla B^2/2\mu_0$ and we have finally

$$\mathbf{f}_B = (B^2/\mu_0 R)\hat{n} - [\partial(B^2/2\mu_0)/\partial n]\hat{n} - [\partial(B^2/2\mu_0)/\partial b]\hat{b}$$

The magnetic $\mathbf{J} \times \mathbf{B}$ forces are therefore equivalent to a pressure $B^2/2\mu_0$ acting isotropically in the plane perpendicular to **B** plus a force $\mathbf{T}_B$ also normal to **B,** which acts in the plane of curvature of the magnetic field. The sense of $\mathbf{T}_B$ is similar to that of forces due to tension in a stretched string, which are parallel to $\hat{n}$ and proportional to $1/R$ (see Fig. 2.13). As noted above for a dipole field the tension and pressure forces in the $\hat{n}$ direction exactly cancel (as they must when $\nabla \times \mathbf{B} = 0$).

The case of a steady-state equilibrium requires $d\mathbf{V}/dt = 0$, which in turn implies from (2.57)

$$0 = -\nabla p + \mathbf{J} \times \mathbf{B}$$

which is equivalent to

$$\nabla(p + B^2/2\mu_0) = (\mathbf{B} \cdot \nabla)\mathbf{B}/\mu_0$$

The magnetic pressure $(B^2/2\mu_0)$ enters just like the particle pressure, while the curvature (tension) term on the right-hand side of this force balance equation. These concepts are quite useful in discussing magnetospheric dynamics and equilibria.

Equation (2.60d) also has an interesting and useful interpretation. Assuming first that σ is infinite,

$$\partial \mathbf{B}/\partial t - \nabla \times (\mathbf{V} \times \mathbf{B}) = 0$$

Consider the magnetic flux ϕ across an arbitrary surface Σ moving with velocity $\mathbf{V}$

$$d\phi/dt = \iint_{\Sigma} (\partial \mathbf{B}/\partial t) \cdot d\mathbf{a} + \int \mathbf{B} \cdot (\mathbf{V} \times d\mathbf{l})$$

The first integral yields the change in ϕ due to the time variation of $\mathbf{B}$ and the second integral yields the change due to the motion of the surface. Working on this second term,

$$\int \mathbf{B} \cdot (\mathbf{V} \times d\mathbf{l}) = -\int (\mathbf{V} \times \mathbf{B}) \cdot d\mathbf{l} = -\iint_{\Sigma} \nabla \times (\mathbf{V} \times \mathbf{B}) \cdot d\mathbf{a}$$

so

$$d\phi/dt = \iint_{\Sigma} [\partial \mathbf{B}/\partial t - \nabla \times (\mathbf{V} \times \mathbf{B})] \cdot d\mathbf{a}$$

But for σ infinite we have shown that the bracket vanishes so

$$d\phi/dt = 0$$

Thus the flux is constant through the surface. We say that the magnetic fluid is "frozen in" and can be considered to move with the fluid. In this sense, if the magnetic field line is labeled by particles on the line at time t, they still label the line at any other time. The concept of frozen-in field lines is very useful and allows visualization of complex flow if we know the magnetic field geometry. Conversely, if the motion is known, the field geometry can be deduced.

If $\sigma \neq \infty$ the field can slip through the fluid and we have

$$\partial \mathbf{B}/\partial t - \nabla \times (\mathbf{V} \times \mathbf{B}) = (1/\mu_0 \sigma)\nabla^2 \mathbf{B}$$

For a stationary case ($\mathbf{V} = 0$),

$$\partial \mathbf{B}/\partial t = (1/\mu_0 \sigma)\nabla^2 \mathbf{B}$$

which is a diffusion equation. If the diffusion scale length is L, the diffusion time constant is given by

$$\tau = \mu_0 \sigma L^2$$

For typical laboratory dimensions L is small, and so even for good conductors τ

is very short. In cosmic plasmas or conducting fluids, however, τ is large and the concept of frozen-in fields is correspondingly important.

Before leaving this section, it is useful to derive an energy relationship based on MHD principles. The particle pressure $p = nk_BT$ is equivalent to the particle energy density, and this is also true for magnetic pressure and magnetic energy density. That is, $B^2/2\mu_0$ yields the energy stored in a magnetic field per unit volume. The total stored magnetic energy in a system is then

$$W_B = (1/2\mu_0) \int B^2 \, dV$$

where dV is the volume element and we have used a single integral sign to designate a triple integral. Changes of this quantity with time can be written

$$\partial W_B/\partial t = (1/\mu_0) \int (\mathbf{B} \cdot \partial \mathbf{B}/\partial t) \, dV$$

Then using $\partial \mathbf{B}/\partial t = -\nabla \times \mathbf{E}$, $\mathbf{B} = \mu_0 \mathbf{H}$, and the vector identity $\nabla \cdot (\mathbf{E} \times \mathbf{B}) = \mathbf{B} \cdot (\nabla \times \mathbf{E}) - \mathbf{E} \cdot (\nabla \times \mathbf{B})$ we have

$$\partial W_B/\partial t = -\int \nabla \cdot (\mathbf{E} \times \mathbf{H}) \, dV - \int \mathbf{E} \cdot (\nabla \times \mathbf{H}) \, dV$$

Using $\mathbf{J} = \sigma(\mathbf{E} + \mathbf{V} \times \mathbf{B})$ and $\mathbf{J} = \nabla \times \mathbf{H}$ and applying the divergence theorem to convert the first volume integral to a surface integral gives

$$\partial W_B/\partial t = -\iint_\Sigma (\mathbf{E} \times \mathbf{H}) \cdot d\mathbf{a} - \int ([\mathbf{J}/\sigma - \mathbf{V} \times \mathbf{B}] \cdot \mathbf{J}) \, dV$$

where the area element $d\mathbf{a}$ points outward from the surface of the volume. Finally, rearranging this equation yields

$$\partial W_B/\partial t = -\iint_\Sigma (\mathbf{E} \times \mathbf{H}) \cdot d\mathbf{a} - \int (J^2/\sigma) \, dV - \int [\mathbf{V} \cdot (\mathbf{J} \times \mathbf{B})] \, dV \quad (2.62)$$

In words, the change in stored magnetic energy in a volume equals the energy flux across the surface into the volume in the form of the Poynting flux ($\mathbf{E} \times \mathbf{H}$), minus the resistive energy loss inside the volume, minus the mechanical work done against the $\mathbf{J} \times \mathbf{B}$ force inside the volume.

In a truly closed magnetosphere with the surface an equipotential, no energy crosses the surface ($\mathbf{E}$ is everywhere normal to an equipotential) and there could be no internal circulation (convection), no dissipation by ionospheric currents and no storage of magnetic energy for later release (e.g., in substorms). Two sources for generating a component of $\mathbf{E}$ parallel to the magnetopause, and hence a net Poynting flux inward, are viscous interaction and reconnection. Both of these processes thus result in a net flow of energy into the magnetosphere. We return to this matter in Chapter 6.

2.6 Coordinate Systems

In standard meteorological practice a local coordinate system has the x axis eastward, the y axis to the north, and the z axis vertically upward. The three components of the neutral wind vector U are usually denoted by (u, v, w) in those coordinates. We will use this notation here as well.

Some complication arises in ionospheric plasma studies due to the importance of the magnetic field direction and the fact that it varies from horizontal at the magnetic equator to vertical at the poles. The reader should be alert to this and to the fact that the magnetic coordinate systems vary somewhat in the text. For example, in Chapter 3 and most of Chapter 4 we use a coordinate system in which the magnetic field is in the direction of the y axis. The conductivity tensor does not then have the form given in (2.38). In deriving (2.38), in Chapters 5–8, and even in the short-wavelength wave analysis in Chapter 4 we have taken B parallel to the z axis. The reader should be aware of the possible confusion caused by the use of different coordinate systems. Furthermore, in the northern hemisphere the "z axis" associated with meteorology is nearly antiparallel to the magnetic field aligned z axis.

References

Banks, P. M., and Kockarts, G. (1973). "Aeronomy," Parts A and B. Academic Press, New York.

Carpenter, D. L., Stone, K., Siren, J. C., and Crystal, T. L. (1972). Magnetospheric electric fields deduced from drifting whistler paths. *J. Geophys. Res.* **77**, 2819.

Farley, D. T. (1959). A theory of electrostatic fields in a horizontally stratified ionosphere subject to a vertical magnetic field. *J. Geophys. Res.* **64**, 1225.

Farley, D. T. (1960). A theory of electrostatic fields in the ionosphere at nonpolar geomagnetic latitudes. *J. Geophys. Res.* **65**, 869.

Goldstein, H. (1950). "Classical Mechanics." Addison-Wesley, Reading, Massachusetts.

Gonzales, C. A., Kelley, M. C., Carpenter, D. L., Miller, T. R., and Wand, R. H. (1980). Simultaneous measurements of ionospheric and magnetospheric electric fields in the outer plasmasphere. *Geophys. Res. Lett.* **7**, 517.

Jackson, J. D. (1975). "Classical Electrodynamics." Wiley, New York.

Johnson, F., ed. (1961). "Satellite Environment Handbook." Stanford Univ. Press, Stanford, California.

Landau, L. D., and Lifshitz, E. M. (1959). "Fluid Mechanics." Pergamon, Oxford.

Mozer, F. S. (1970). Electric field mapping from the ionosphere to the equatorial plane. *Planet. Space Sci.* **18**, 259.

Mozer, F. S., and Serlin, R. (1969). Magnetospheric electric field measurements with balloons. *J. Geophys. Res.* **74**, 4739.

Rishbeth, H., and Garriott, O. K. (1969). "Introduction to Ionospheric Physics," Int. Geophys. Ser., Vol. 14. Academic Press, New York.

Schmidt, G. (1966). "Physics of High Temperature Plasmas: An Introduction." Academic Press, New York.

Spitzer, L. (1962). "Physics of Fully Ionized Gases." Wiley (Interscience), New York.

Chapter 3 | Electrodynamics of the Equatorial Zone

In this chapter we study the electrodynamics of the magnetic equatorial zone. To a great extent our knowledge of the electrical structure of this region comes from measurements made at the Jicamarca Radio Observatory located just east of Lima, Peru. This incoherent scatter facility was designed to optimize measurements of plasma flow perpendicular to the earth's magnetic field, which is nearly horizontal over the site. In the F region of the ionosphere, κ_i and κ_e are very large and hence the ion and electron velocities perpendicular to $\mathbf{B}$ are very nearly equal to each other. This means that a plasma flow velocity can be uniquely defined and related to the electric field. We deal first with the generation of electric fields by thermospheric winds in the F region and follow by an analysis of the E-region dynamo and the equatorial electrojet. The latter is an intense current jet which flows in the E region at the magnetic equator. These dynamos are the primary sources of the low-latitude electric field, but high-latitude processes also contribute and are discussed as well.

3.1 Motions of the Equatorial F Region: The Data Base

We choose first to study the equatorial F-layer dynamo, since it seems conceptually simpler, although it was not discovered first. Most of the data concerning equatorial electrodynamics come from incoherent scatter radar observations near the magnetic equator over Jicamarca, Peru. Details concerning Jicamarca and several other observatories are given in Appendix A, along with a discussion of the measurement method. The Jicamarca system is capable of determining the plasma temperature, density, and ion drift velocity as functions of altitude and

65

time from the backscatter due to thermal fluctuations in the plasma. The radar can be directed perpendicular to the magnetic field, where the frequency width of the backscatter spectrum is very narrow. This means that even small mean Doppler shifts can be detected and converted to very accurate ionospheric drift velocities. In practice, the radar is split into two beams, one oriented 3° to the geomagnetic east and one 3° to the geomagnetic west. The difference of the Doppler shifts detected by the two beams yields the zonal eastward drift speed of the ion gas, while their average yields the vertical drift speed of the ions. In the F region, κ_i is very large and the ion velocity perpendicular to **B** is given by (2.36d), that is,

$$(\mathbf{V}_i)_\perp = [\mathbf{E} - (k_B T_i/e_i)\nabla n/n + (M/e_i)\mathbf{g}] \times [\mathbf{B}/B^2] \qquad (3.1)$$

where all quantities are measured in the earth-fixed frame. Notice that the neutral wind velocity does not appear in this equation. This means that the radar measurements cannot be used directly to measure any F-region wind components perpendicular to **B**. At typical measurement heights, we can estimate the drift velocities due to the pressure and gravity terms. The former is the order of $k_B T_i/q_i LB$, where L is the large-scale density gradient scale length. Taking $L = 10$ km and $T_i = 1000$ K yields a drift velocity of roughly 0.4 m/s. The gravitational drift is even smaller, less than one-tenth of this value. Since the observed drift velocities due to the pressure and gravity terms. The former is the order of $k_B T_i/e_i LB$, where L is the large-scale density gradient scale length. Taking $L =$

These results hold for both daytime and nighttime conditions; that is, incoherent scatter radar measurements in the F region can in principle, and often in practice, yield the ion drift, and hence the electric field, at all local times. There are some limitations of the technique, however. Measurements in the equatorial E region using the incoherent scatter technique are rarely possible due to the common occurrence of plasma instabilities that raise the fluctuation level at 3-m wavelengths to well above thermal levels. The resulting backscatter signal is very strong and yields considerable information on these plasma processes, some of which are discussed in the next chapter. Similarly, a plasma instability termed equatorial spread F often occurs in the evening and, when present, precludes incoherent scatter measurements even in the upper F region. Finally, we note that the incoherent scatter method requires a minimum plasma density in the scattering volume determined by the system noise, antenna size, transmitter power, integration time, etc. For the Jicamarca Radar Observatory this minimum is about 10^4 cm^{-3}. This limitation usually precludes measurements at night in the altitude range below the F peak. In this important height range, we must depend on methods other than incoherent backscatter. Barium ion cloud releases, rocket probes, and radar interferometric methods have been used successfully in this height range.

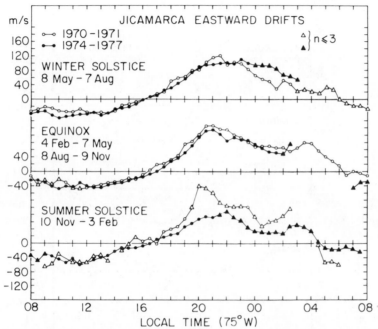

Fig. 3.1. Measured zonal (positive for motions toward the east) plasma drifts in the equatorial F region for various seasons and solar conditions. Open symbols correspond to solar maximum and filled symbols to solar minimum activity. Eastward drifts at 100 m/s correspond to a downward electric field at 2.5 mV/m. [After Fejer *et al.* (1981). Reproduced by permission of the American Geophysical Union.]

A compilation of the Jicamarca data set spanning nearly an entire solar cycle is presented in Figs. 3.1 and 3.2. In Fig. 3.1 the zonal eastward drift measured near the F peak is plotted, while in Fig. 3.2 the vertical drift component is shown. In each case two sets of averages are presented; the open circles and triangles correspond to conditions near solar maximum and the filled symbols to solar minimum conditions. Equinoctial and solstice periods are plotted separately. Both components have a strong diurnal modulation. To first order, the drifts are up and to the west during the day, and down and to the east at night.

Some aspects we wish to emphasize are as follows:

1. The peak eastward drift at night is twice as great as the peak westward drift during the day.

2. The zonal drifts are much larger than the vertical velocities.

3. The vertical drift is often strongly enhanced just after sunset but shows no comparable feature near sunrise.

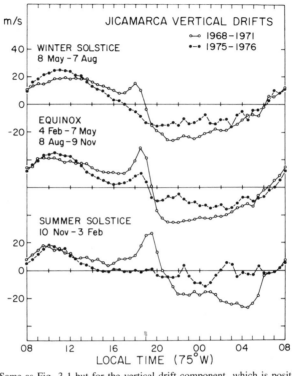

Fig. 3.2. Same as Fig. 3.1 but for the vertical drift component, which is positive for motions upward. [After Fejer *et al.* (1979). Reproduced by permission of the American Geophysical Union.]

4. There are strong solar cycle effects in the vertical drifts and moderate seasonal effects in both data sets.

As discussed above, these drifts may be directly related to the ambient perpendicular electric field (where as usual we mean the component perpendicular to **B**) through

$$\mathbf{E}_\perp = -\mathbf{V}_i \times \mathbf{B} \qquad (3.2)$$

The magnetic field over Jicamarca is about 2.5×10^{-5} tesla. Thus, a zonal eastward drift is due to a vertically downward electric field component, with a 100-m/s drift corresponding to a 2.5-mV/m electric field. Likewise, an upward drift of 40 m/s corresponds to an eastward electric field of about 1 mV/m.

Evidence for the repeatability of the average drifts plotted above is given in Fig. 3.3, where a large number of individual 24-h measurement sets are superimposed. Except for the July 1968 event, which occurred during a strong magnetic substorm, the data are remarkably well behaved. Such a plot gives us con-

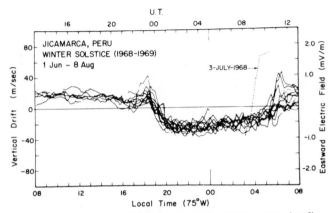

Fig. 3.3. Superimposed vertical drifts (eastward electric fields) measured at Jicamarca, Peru. [After Woodman (1970). Reproduced with permission of the American Geophysical Union.]

fidence that the temporal sequences plotted in these figures can be interpreted equally well in terms of Universal Time measurements, at least in an average sense. That is, if we take a snapshot of the instantaneous electric field pattern around the earth's equatorial zone, it should look very much like the Jicamarca equatorial measurements as a function of local time plotted here. One test of this hypothesis uses the fact that for electrostatic fields, the Maxwell equation $\nabla \times \mathbf{E} = 0$ implies

$$\oint \mathbf{E} \cdot d\mathbf{l} = 0 \tag{3.3}$$

That is, the line integral of the zonal electric field entirely around the earth must vanish at any one time. An equivalent statement is that the voltage difference between the dusk and dawn terminators (see Fig. 3.5) when the electric field is westward (at night) should equal the voltage drop when the field is eastward (during the day). Fejer *et al.* (1979) have performed this calculation for a number of days using the Jicamarca time series and the results are plotted in Fig. 3.4. To first order, the voltage drop across the dayside is close to the drop across the night, both equaling about 8 kV. The agreement is not perfect, however, and the ratio $V_{\text{NIGHT}}/V_{\text{DAY}}$ drops as low as 0.5 in October 1975–1976. This indicates that local conditions may affect the Jicamarca measurements and that the 24-h plots may not represent an instantaneous pattern in some seasons.

From Poisson's equation

$$\nabla \cdot \mathbf{E} = \rho_c/\varepsilon_0$$

we expect the observed dayside and nightside fields to be the result of charge buildup at the terminators, with the dusk terminator charged negatively and the

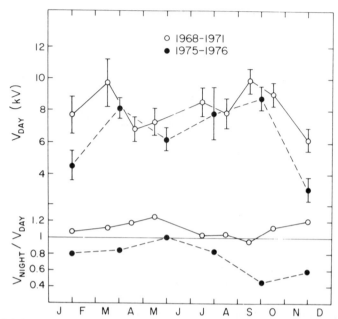

Fig. 3.4. Monthly variation of the local daytime voltage drop between the reversals and of the nighttime-to-daytime ratio. Standard deviation of the latter is about 15–20%. [After Fejer *et al.* (1979). Reproduced with permission of the American Geophysical Union.]

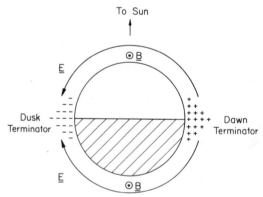

Fig. 3.5. Schematic diagram showing the zonal electric field component and its relationship to the charge densities at the terminators.

dawn terminator positively. Assuming a change of the electric field equal to 2 mV/m over a 2-h local time interval corresponding to a distance of about 3×10^6 m at the equator, $\rho_c \approx 7 \times 10^{-21}$ C/m^3. This is less than one excess elemental charge per cubic meter, again illustrating how very small net charge densities can yield significant electric fields in a plasma. A schematic diagram based on the "charged terminator model" is given in Fig. 3.5 from a viewpoint looking down on the earth from above the northern hemisphere.

3.2 The Equatorial F-Region Dynamo

Before discussing the F-region dynamo, we need to understand a little about the thermospheric winds in the equatorial region since they provide the source of energy that maintains the electric field. Early observational data on thermospheric winds came from studies of the drag exerted on artificial satellites by the neutral atmosphere (King-Hele, 1970). Analysis of such data shows that the change in satellite inclination is directly related to the angular velocity of the atmosphere with respect to the rotating earth. The surprising conclusion was that on the average the low-latitude thermosphere superrotates; that is, there is a net eastward average zonal flow of about 150 m/s near 350-km altitude and about 50 m/s at 200-km altitude. The effect is most pronounced in the 2100–2400 local time period. Direct measurements of thermospheric winds are now available at night using the Fabry–Perot technique to determine the Doppler shift of airglow emissions (Sipler and Biondi, 1978). Some of these data (which were taken over Kwajalein in the South Pacific) are reproduced in Fig. 3.6a in the form of azimuth plots of the tip of a vector in the direction toward which the wind is blowing (note that at Kwajalein 8 UT is equal to 2000 LT). Although the winds display a high degree of variability from day to day, to a first approximation the winds are eastward and quite strong ($\sim$150 m/s) in the postsunset period, decaying in amplitude to less than 50 m/s after midnight. More recent airglow observations in the Peruvian sector are in good agreement with these data (Meriwether et al., 1986).

Airglow observations are restricted to nighttime, but recent satellite observations have yielded full 25-h coverage of the thermospheric winds. Some of these data are presented in Fig. 3.6b, superimposed on the Jicamarca zonal plasma drift pattern. Both are clearly diurnal, with larger winds and plasma drifts occurring at night. There is a lot of scatter in the winds measured by the satellite but the pattern is unmistakable, as is its correlation with the plasma drifts.

Setting aside the superrotation phenomenon for the moment, we can ask whether forcing due to solar heating of the thermosphere can explain the high winds observed in the postsunset period. Since the Coriolis force vanishes at the

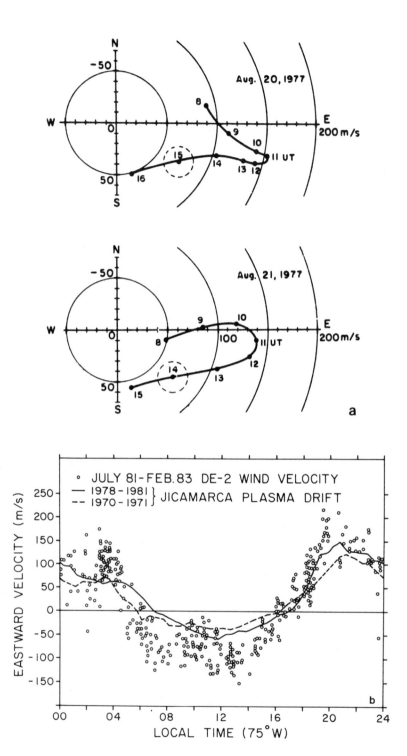

equator, in a steady state the winds should blow in the $-\nabla p$ direction from west to east across the sunset terminator, that is, away from the expected (and observed) subsolar neutral atmospheric temperature bulge which occurs on the sunward side of the earth. The problem is not so much with the direction of the wind, which agrees with the data, as with the magnitude of the theoretically calculated winds—the predictions are too high.

Without a Coriolis term, the primary limitation on the wind speed comes from the various drag terms. If the plasma is ignored, as the early modelers naturally did, viscosity is the only frictional factor. As discussed in Chapter 2, in the lower atmosphere viscosity is provided by the interaction of eddies in the atmospheric flow. Above the turbopause, classical molecular viscosity becomes the dominant factor. An expression for the viscosity coefficient is given by

$$\eta = 4.5 \times 10^{-5} \, (T/1000)^{-0.71} \quad \text{kg/m} \cdot \text{s} \tag{3.4}$$

where atomic oxygen has been assumed to be the dominant neutral (Dalgarno and Smith, 1962). This is a good approximation in the F region. The vertical temperature gradient is also small in the upper thermosphere and η is therefore virtually independent of height. However, when we are comparing viscosity with the acceleration terms such as $\mathbf{F} = -(1/\rho)\nabla p$, it is the kinematic viscosity (η/ρ) that matters. Unlike η, this ratio increases exponentially with height as ρ decreases. This drastic increase of the ratio η/ρ is responsible for the importance of viscosity in the thermosphere.

A quantitative comparison between the various terms in the neutral momentum equation (2.21b) has been provided by Rishbeth (1972) and is reproduced here in Fig. 3.7. The graph was constructed for 45° latitude, but most of the features are applicable to the equatorial case. For now, we are interested in the curves labeled F and $(\eta/\rho H^2)$. The former is the magnitude of the driving horizontal pressure gradient acceleration term and is deduced from neutral atmospheric measurements. The upper scale corresponds to F and is given in meters per second squared. Since F is inversely proportional to ρ, it increases with height. Surprisingly, however, F increases only roughly linearly with height. To plot the viscous term we need to scale it appropriately according to the form it takes in the momentum equation, $(\eta/\rho)\nabla^2 U$. We would not expect the winds to vary more quickly with altitude than the scale height H. An upper limit on the

---

Fig. 3.6. (a) Position of the tip of the neutral wind vector measured over Kwajalein as a function of universal time on August 20 and 21, 1977. The dashed circle indicates the 15-m/s uncertainty in the determination. [After Sipler and Biondi (1978). Reproduced with permission of the American Geophysical Union.] (b) Comparison of the Jicamarca average drifts with thermospheric wind data from DE-2. [After Wharton et al. (1984). Reproduced with permission of the American Geophysical Union.]

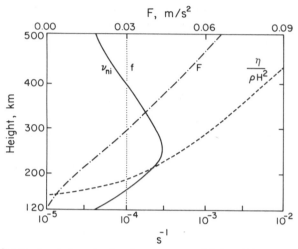

Fig. 3.7. Graphs of neutral acceleration time constants and F for midday at sunspot minimum. The upper scale applies to the acceleration due to the pressure gradient F (-·-·-·). The lower scale refers to the ion drag parameter ν_{ni} (———); the Coriolis parameter for a latitude of 45°, f (·····); and the normalized kinematic viscosity parameter $\eta/\rho H^2$ (----). [After Rishbeth (1972). Reproduced with permission of Pergamon Press.]

viscous effect is then $(\eta/\rho H^2)$, which has units of reciprocal seconds, as plotted along the lower scale. Since the viscous term increases rapidly with height, it will eventually become large enough to suppress all variations of **U** with altitude. Therefore most numerical models have as an upper boundary condition the requirement that $\partial \mathbf{U}/\partial z = 0$.

In an early study of thermospheric winds, Lindzen (1966) took the observed diurnal component of the pressure gradient given experimentally by Harris and Priester (1965) and solved the simplest form of the resulting zonal wind equation at equatorial latitudes:

$$\partial u/\partial t - (\eta/\rho)\partial^2 u/\partial z^2 = -1/\rho(\partial p/\partial x) \tag{3.5}$$

where u is the zonal component of the wind and η is the viscosity. He found an analytic solution to this equation by generating reasonable models for the functional variations of p and ρ with x and z. Even with a low estimate of the pressure term, however, zonal velocities of about 300 m/s were found in this analysis. In later work, Lindzen (1967) used a numerical solution with a more realistic pressure variation and showed that, ignoring the plasma, winds as high as 550 m/s are predicted! We leave this dilemma for the moment to determine the electrodynamic effect of the observed and predicted postsunset zonally eastward thermospheric wind. Later we shall show that including the frictional drag on the neutrals by the ionospheric plasma explains the low wind values that are actually observed.

In the remainder of this chapter and in most of the next, we shall slightly redefine our coordinate system to preserve the conventional notation that the $\hat{a}_z$ axis is upward. At the equator, therefore, we take $\mathbf{B} = B\hat{a}_y$, which is horizontal and northward, and take $\hat{a}_x$ toward the east. The conductivity tensor is then

$$\boldsymbol{\sigma} = \begin{pmatrix} \sigma_P & 0 & \sigma_H \\ 0 & \sigma_0 & 0 \\ -\sigma_H & 0 & \sigma_P \end{pmatrix} \tag{3.6}$$

In the F region $\sigma_P \gg \sigma_H$ and the conductivity tensor is diagonal, although it still holds that $\sigma_P \ll \sigma_0$. Assume first that the horizontal magnetic field lines extend forever or equivalently that they terminate at both ends in an insulating layer. In fact, the field lines bend and enter the E region, which has a finite conductivity that varies with local time, season, and solar activity. We return to this point later. The vertical component of the large-scale neutral wind field in the atmosphere is always small, so we consider a simple model in which the thermospheric wind is eastward and has a magnitude u that is uniform with height. In the text we use the meteorological notation in which $\mathbf{U} = (u, v, w)$, where u, v, and w correspond to the zonal (positive eastward), meridional (positive northward), and upward components, respectively. From (2.41), an electric current will flow with magnitude and direction given by

$$\mathbf{J} \simeq \boldsymbol{\sigma} \cdot (u\hat{a}_x \times \mathbf{B})$$

The wind-driven current is therefore vertically upward with magnitude $J_z = \sigma_P u B$. The current J_z is quite small with a peak value of the order of 0.01 $\mu\mathrm{A}/\mathrm{m}^2$. However, σ_P varies considerably with altitude due to its dependence on the product $n\nu_{\mathrm{in}}$. The zonal wind component, u, may also vary with height, but we assume for now that viscosity keeps this variation small. At any rate, $d(\sigma_P u B)/dz \neq 0$, and an electric field must build up in the z direction to produce a divergence-free current. However the "insulating plates" we have assumed at the ends of the magnetic field lines do not allow a magnetic field-aligned current to flow at all (i.e., $J_y = 0$), so in this first approximation a stronger condition on $\mathbf{J}$ holds than the expression $\nabla \cdot \mathbf{J} = 0$, namely $\mathbf{J} = 0$.

A typical plasma density profile for the postsunset equatorial F layer is shown in Fig. 3.8a. As we noted in Section 2.2, gravitational forces do not cause the plasma (with large κ_i, κ_e) to fall at the magnetic equator since the velocity due to gravity is perpendicular to the gravitational force. Recombination "eats away" at the molecular ions on the bottomside, forming a steep upward density gradient. The result is a dense O^+ plasma with a well-defined lower boundary. To study the electrodynamics of this region in a little more detail we approximate the actual situation shown in Fig. 3.8a with the configuration illustrated in Fig. 3.8b, which shows a slab geometry with σ_P constant inside the slab and zero elsewhere and with the zonal wind u constant everywhere. Since the current is upward inside the layer and zero outside, charges pile up at the two boundaries

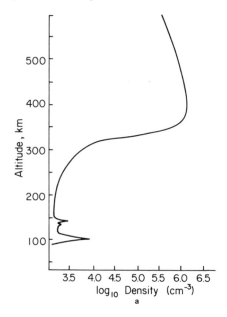

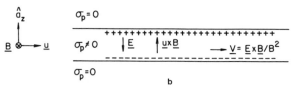

Fig. 3.8. (a) Typical equatorial plasma density profile in the evening local time period. (b) Electrodynamics of the equatorial F region, in which the density and conductivity profile is modeled with a slab geometry, subject to a constant zonal eastward neutral wind.

as shown in the figure. The magnitude of the electric field which builds up as a result of these charges is such that

$$J_z = \sigma_P E_z + \sigma_P u B = 0$$

which yields

$$E_z = -uB \tag{3.7}$$

Note that the plasma inside the slab will drift with an $\mathbf{E} \times \mathbf{B}/B^2$ velocity equal in magnitude and direction to the zonal wind speed. Furthermore, the electric field in the frame of reference of the neutral wind, $\mathbf{E}' = \mathbf{E} + \mathbf{U} \times \mathbf{B}$, vanishes. This must be true since the current is independent of reference frame and we have set the current equal to zero; that is, since $\mathbf{J} = \mathbf{J}' = \sigma_P \mathbf{E}' = 0$, $\mathbf{E}'$ must be zero.

In this simple model there is a very strong shear in the plasma flow velocity at the two interfaces that is not shared by the driving neutral wind; that is, at the interfaces the plasma velocity changes abruptly from u to 0. This "prediction" of a sheared plasma flow is intriguing since such shears have been observed in the equatorial F region and are discussed below. However, in this model the shear is created somewhat artificially by the slab conductivity assumption. As shown below, a more realistic density profile subject to a zonal wind under the requirement that $J_z = 0$ yields $E_z(z) = -u(z)B$ at all heights.

The insulating end plate assumption we have made here is most nearly valid at night when the E-region molecular ion and electron pairs rapidly recombine with no offsetting production by sunlight (see, for example, Rishbeth and Garriott, 1969). The dominant O^+ (atomic) ions in the F layer are much longer lived and support the F-layer dynamo. As shown in Fig. 3.1, the maximum nighttime zonal drift and maximum vertical electric field (downward) are about 150 m/s and 4.5 mV/m, respectively. This observed nighttime plasma velocity is consistent in magnitude and direction with the eastward zonal neutral thermospheric wind found experimentally and discussed above. The vertical pattern of the nighttime electric field is thus consistent with the simple F-region dynamo model we have presented.

To understand the diurnal variation in more detail, however, we need to consider the role of the "end" plates in the E layer, which, contrary to the approximation used above, are good conductors during the day. First consider again an idealized slab geometry that ignores the magnetic field line curvature and dip angle but includes conductivity variations with distance (y) along the magnetic field direction. The actual geometry is shown schematically in Fig. 3.9a with the slab model shown in Fig. 3.9b. There are no variations in the x direction. The wind is a function of z in the F layer but goes to zero in the E layer. The finite density in the E layer and its attendant conductivity act as an electric load on the dynamo. (In this three-level slab model we take the E-layer conductivity to be constant as a function of y.) The current parallel to $\mathbf{B}$ is nonzero and must be supplied by the dynamo and completed in the E layer. The full divergence equation must be used. First, as discussed in Chapter 2, we note that for large-scale dynamo sources, the magnetic field lines are nearly equipotentials due to the high ratio of σ_0 to σ_P ($\geq 10^5$). The electric field is thus mapped down to the E-region altitudes, where we have assumed that the neutral wind vanishes. The $\nabla \cdot \mathbf{J} = 0$ condition in the F region now yields

$$\frac{d[\sigma_P(E_z + uB)]}{dz} = \frac{dJ_y}{dy}$$

We now integrate this equation along the y direction from $y = 0$ to $y = y_1$, which corresponds to integrating from the equatorial plane to the base of the F layer in the northern hemisphere. By symmetry the contribution to the integral

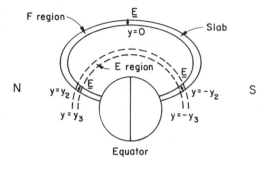

ACTUAL GEOMETRY

a

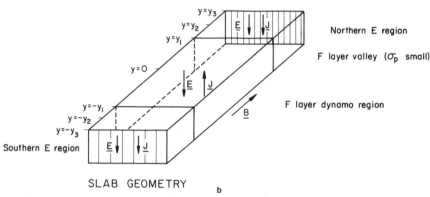

SLAB GEOMETRY

b

Fig. 3.9. (a) Side view of the dipole magnetic field geometry near the magnetic equator. The curves are exaggerated to show the coupling geometry between the F region at the equator and the off-equatorial E region. (b) F-layer slab geometry including conducting end plates in the northern and southern hemispheres.

from the southern hemisphere is identical. Symmetry also requires that the field-aligned current vanish directly at the equator [$J_y(0) = 0$] and the integral yields

$$\int_0^{y_1} \frac{d}{dz}[\sigma_P(E_z + uB)] \, dy = -J_y(y_1) \tag{3.8}$$

Due to the low perpendicular conductivity in the "valley" between the E and F regions, the current we have just calculated, which leaves the F-region dynamo at $y = y_1$, must be equal to the field-aligned current that enters the E region at y

$= y_2$. In the E layer we have only the (mapped) electric field E_z to contend with since we have set the wind speed $u = 0$ there for now. Applying $\nabla \cdot \mathbf{J} = 0$ yields

$$d(\sigma_P E_z)/dz = -dJ_y/dy$$

Integrating from y_2 to y_3 and requiring $J_y(y_3) = 0$, that is, an insulating atmosphere below y_3, we have

$$\int_{y_2}^{y_3} \frac{d}{dz} (\sigma_P E_z) \, dy = J_y(y_2) = J_y(y_1) \tag{3.9}$$

and finally using (3.8) and (3.9)

$$\int_0^{y_1} \frac{d}{dz} [\sigma_P^F(E_z + uB)] \, dy = -\int_{y_2}^{y_3} \frac{d}{dz} (\sigma_P^E E_z) \, dy \tag{3.10}$$

Now E_z is not a function of y (equipotential magnetic field lines), so if we take uB constant with y until the field line enters the F-layer valley, we can perform the y integral in (3.10), giving

$$\frac{d}{dz} [(\Sigma_P^F + \Sigma_P^E)E_z + \Sigma_P^F uB] = 0$$

where Σ_P^F and Σ_P^E are the field-line integrated conductivities discussed in Chapter 2. The solution for $E_z(z)$ is thus

$$E_z(z) = -[u(z)B\Sigma_P^F(z)]/[\Sigma_P^F(z) + \Sigma_P^E(z)] \tag{3.11}$$

In (3.11) we have dropped a constant of integration since the slab geometry is not physically accurate and would be replaced by a smooth transition to a two-dimensional (J_x and J_z) equatorial electrojet current in a more accurate model (see Section 3.4).

Equation (3.11) clearly shows that if the off-equatorial E region is insulating ($\Sigma_P^E = 0$), $E(z)$ will be equal to $-u(z)B$ everywhere. We can now use (3.11) to explain some aspects of the diurnal variation in the vertical electric field. During the nighttime, Σ_P^E becomes quite small and hence $E_z \approx -uB$ in regions where Σ_P^F is large, for example, near the peak in the F-region plasma density. During the daytime, however, the E-region conductivity is comparable to or larger than the magnetic field-line integrated F-region conductivity. For large Σ_P^E the electrodynamic control of the ionosphere is vested in the E region. As discussed below, a dynamo operates there as well but it is driven by tidal winds which are smaller than the thermospheric winds in the F region. This explains the diurnal variation in the F-region zonal plasma drifts plotted in Fig. 3.1. At night the E-region conductivity is low and the high zonal winds at several hundred kilometers altitude determine the vertical electric field and hence the horizontal plasma flow.

For $\Sigma_P^F \gg \Sigma_P^E$ the electric field will be almost equal to uB. The eastward plasma velocity thus nearly matches the neutral wind speed. During the day, however, the F-region dynamo loses control of the electrodynamics and the resulting electric fields are determined by winds in the E region. Since these winds tend to be weaker, the plasma drift is smaller during the day.

An analogous electric circuit is shown schematically in Fig. 3.10. The three batteries correspond to the two (north and south) E-region dynamos and the F-region dynamo. Each battery has a finite internal resistance given by $R = (\Sigma_P)^{-1}$. The voltage measured by the meter corresponds to the electric field times the distance between the two magnetic field lines (the conducting wires). The battery with the lowest internal resistance determines the voltage and hence the electric field. In the F layer during the nighttime Σ_P^F is larger and the F region dominates. During the day, E-region sources determine the electric field and the F layer acts as a load.

A model calculation of the field-line integrated Pedersen conductivity and of the vertical electric field is presented in Fig. 3.11 as a function of height measured at the magnetic equator. The local time is 1900 and the driving eastward neutral wind used in the calculation is taken to be constant, with a value of 160 m/s. The model used for the ionospheric plasma density is similar to the data plotted in Fig. 3.8a except that the E-region density below 150 km was taken to be constant at a value of 2×10^4 cm^{-3}. The parenthetic expressions yield the percentage of the total Σ_P found below 300 km along that particular magnetic field line. At first glance the Σ_P^F values seem high since the value of σ_P at 300

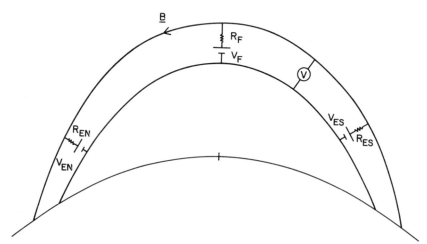

Fig. 3.10. Electric circuit analogy to the voltage sources in the equatorial F region. Off-equatorial E-region wind dynamo competes with the F-region dynamo at the equator to determine the voltage differences between magnetic field lines.

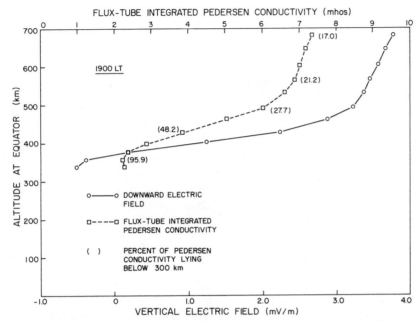

Fig. 3.11. Equatorial F-region electric field and conductivity calculations using realistic iono-spheric parameters, a 160-m/s zonally eastward neutral wind in the F layer, and a diurnal tidal mode in the off-equatorial E region. [After Anderson and Mendillo (1983). Reproduced with permission of the American Geophysical Union.]

km in Fig. 2.6 is only about 5×10^{-5} mho/m. However, near the equator the magnetic field lines are nearly horizontal and are the order of several thousand kilometers long. Large values of Σ_P^F then result. Notice that Σ_P continues to increase with height above the F peak at about 350 km, even though measured at the magnetic equator σ_P decreases with height above 350 km. This is due to the length of the field line and to the fact that the field lines crossing the equator "reenter" the F layer at off-equatorial latitudes where the equatorial anomaly occurs (see Chapter 5). Above 400-km altitude most of the conductivity is located in the F layer and the local thermospheric winds drive the electrical system. However, as the altitude decreases, the E layer begins to short out the electric field and it decreases rapidly. A considerable shear thus occurs in the plasma flow due to the rapid change in the driving electric field even though the thermospheric wind was taken to be uniform with height. This detailed calculation is totally consistent with the simple model presented above. (Note that in Fig. 3.11 the electric field reverses sign since a tidal wind field was included in the E region.)

Such a shear is very difficult to measure with an incoherent scatter radar since

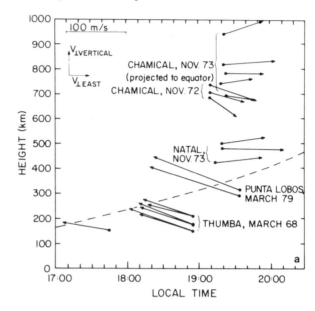

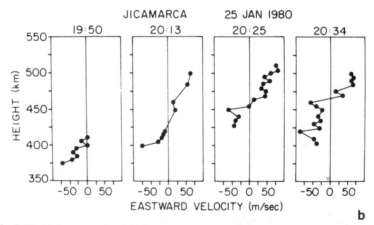

Fig. 3.12. (a) Summary plot giving the results of a number of barium ion cloud experiments in the equatorial zone. [After Fejer (1981). Reproduced with permission of Pergamon Press.] (b) Zonal drifts of plasma irregularities as a function of height and time using an interferometric technique. Note that the shear point where the velocity is zero moves up as the scattering layer rises. [After Kudeki *et al.* (1981). Reproduced with permission of the American Geophysical Union.]

it occurs where the plasma density decreases to values below the signal-to-noise threshold for plasma velocity measurements. The barium cloud technique can be used, however, and a number of experiments have been performed using sounding rockets. The results are summarized in Fig. 3.12a as a function of altitude and local time. The dashed line roughly indicates the base of the F layer. Below the F layer the plasma velocity is strongly westward. Thus the downward electric field is not only reduced but also reversed in sign. Simultaneous neutral wind measurements displayed velocities very different from the plasma velocities. These surprising results were verified by measurements of the velocity of plasma irregularities with the Jicamarca radar system used in an interferometer mode. In this technique coherent scatter echoes are used to trace the position of localized nonthermal scatterers in the radar beam. Four consecutive altitude profiles of the eastward irregularity velocity are plotted in Fig. 3.12b. The plasma velocities plotted here also change sign with height in agreement with the results from the barium experiment.

One explanation for the shear is a variation on the shorting effect discussed above. If the off-equatorial E-region dynamo creates a meridional electric field that is in the poleward direction, and if the E-region dynamo internal resistance is small enough, then the electric field will map to the equatorial plane as a vertically upward field below the F layer. This will in turn cause a westward plasma drift. However, the observed westward plasma drift velocity is higher than expected for such a process. For example, the weak westward plasma drift at the base of the F layer in the calculation shown in Fig. 3.11 (indicated by the change in sign of the vertical electric field) was due to the E-region dynamo. The reversed field is much lower than the measured values, however.

Haerendel *et al.* (1983) have suggested an alternative explanation in which the large vertical electric field in the F layer is due to partial closure of the vertical equatorial electrojet current in the F region. In this model the equatorial electrojet drives an upward current into the F-layer valley and creates an electric field E_z such that

$$E_z = J_z/\sigma_P$$

Since σ_P is small in the F-layer valley, it does not require much J_z to create a significant E_z. Further experiments are necessary to sort out this interesting problem in equatorial electrodynamics and to distinguish between the various competing theories.

3.3.   E-Region Dynamo Theory and the Daytime Equatorial Electrojet

To this point, we have discussed only the vertical equatorial electric field component and its morphology. The zonal component is smaller but is very important since it causes the plasma to move vertically. This motion greatly affects the plasma density since it causes the plasma to interact with quite different neutral densities as it changes altitude. This strongly affects the recombination rate and in turn the plasma content.

The zonal component of the electric field and the daytime vertical component are due primarily to the winds in the E region. This E-region dynamo is driven by tidal oscillations of the atmosphere. An excellent book on this topic has been written by Chapman and Lindzen (1970) and we only touch on some aspects of tidal theory here. The largest atmospheric tides are the diurnal and semidiurnal tides driven by solar heating. The semidiurnal lunar gravitational tide is next in strength in the upper atmosphere. It is interesting to note that this latter tidal mode is the strongest in the case of ocean tides.

One may legitimately question a terminology in which we discuss diurnal "tides" in the lower thermosphere (E region) but refer to a diurnal "wind" in the upper thermosphere (F region). The difference is that the tidal modes propagate into the lower thermosphere from below, whereas the upper thermospheric winds are driven by absorption of energy in the thermosphere itself. We could refer to the thermospheric response as due to an *in situ* diurnal tide but will stick with the traditional usage here.

Tidal theory is quite complex. The equations of the neutral atmosphere must be solved on a rotating spherical shell subject to the earth's gravitational field. Considerable insight is obtained by studying the free oscillations of the atmosphere—that is, the normal modes of the system. This is accomplished by reducing the set of equations to one second-order partial differential equation, which is often written in terms of the divergence of the wind field. The resulting equation is separable in terms of functions of latitude (θ), longitude (ϕ), altitude (z), and time (t). The longitude/time dependence is of the form $\exp[i(s\phi + ft)]$ where s must be an integer. For $s = 0$, the temporal behavior does not propagate with respect to the earth. For $s = 1$, the disturbance has one oscillation in longitude and propagates westward following the sun; this is the diurnal tide. The θ dependence can be expressed in terms of so-called Hough functions, which may be related to the spherical harmonic functions.

The height dependence of these normal modes is a crucial factor in tidal theory, since when forcing (e.g., solar heating) is included in the equations, the only modes excited are those that have a vertical structure which matches the vertical structure of the forcing. Although the mathematical models were well developed by 1900, controversy raged over the actual nature of the atmospheric tides until definitive measurements of the temperature structure of the earth's upper atmosphere became available from sounding rockets. The features of these temperature profiles, which are of most relevance to tidal theory, are due to the absorption of sunlight by ozone in the stratosphere and by water vapor in the troposphere.

When both the θ and z dependences of forcing functions and the normal modes are taken into account, many of the dominant features of the tidal oscillations can be explained. For our purposes we summarize these as follows:

1. Tidal oscillations propagate upward and the associated wind speed amplitude grows as they do so. (This amplification is due to the decreasing density of the atmosphere and is a consequence of energy conservation. This important feature of vertical wave propagation is discussed in detail in Chapter 5, where gravity waves are studied.)

2. Diurnal tides can propagate vertically only below 30° latitude. At higher latitudes they remain trapped in the stratosphere.

3. With the decreasing importance of the diurnal tide, the semidiurnal tide becomes dominant at latitudes higher than 30°.

Armed with this modest understanding of tidal theory, we may now investigate some aspects of the E-region dynamo.

In the E-region of the ionosphere, the conductivity tensor is not diagonal and the Hall terms must also be considered. Furthermore, the electric field cannot be taken as entirely self-generated by local wind fields as we assumed for the nighttime F-layer dynamo. This can be understood as follows. The entire dayside ionosphere is a good electrical conductor in which currents are driven by lower thermospheric tides. The tidal E-region wind field $\mathbf{U}(\mathbf{r}, t)$ is global in nature and it will drive a global current system given by $\mathbf{J}_w = \boldsymbol{\sigma}(\mathbf{r}, t) \cdot [\mathbf{U}(\mathbf{r}, t) \times \mathbf{B}]$. Now, since both $\boldsymbol{\sigma}(\mathbf{r}, t)$ and $\mathbf{U}(\mathbf{r}, t)$ depend on $\mathbf{r}$, the current $\mathbf{J}_w$ need not, in general, be divergence free. Thus, an electric field $\mathbf{E}(\mathbf{r}, t)$ must build up such that the divergence of the total current is zero and

$$\nabla \cdot [\boldsymbol{\sigma} \cdot (\mathbf{E}(\mathbf{r}, t) + \mathbf{U}(\mathbf{r}, t) \times \mathbf{B})] = 0 \qquad (3.12)$$

The resulting $\mathbf{E}(\mathbf{r}, t)$ is as rich and complex as the driving wind field and the conductivity pattern that produce it. The latter has a primarily diurnal variation over the earth's surface, while the dominant tidal winds change from diurnal to semidiurnal depending on latitude. As a first approximation we might then expect a diurnal electric field pattern at low latitudes and a mixture of diurnal and semidiurnal electric fields at higher latitudes. This crude analysis actually describes the situation fairly well.

The diurnal variations of the electric field components measured at Jicamarca have already been pointed out in Section 3.1. The global vertical magnetic perturbation pattern measured by ground-based magnetometers is shown in Fig. 3.13 and gives further evidence for this simple picture of the electric field. Such data have been used to construct the pattern of electrical currents in the entire ionosphere that is referred to as the Sq current system. (S stands for solar and q for quiet in this notation.) As noted in Chapter 2, ground magnetometers respond primarily to horizontal Hall currents since the magnetic field due to field-aligned cur-

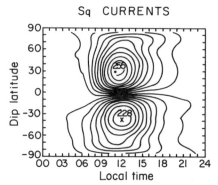

Fig. 3.13. Average contours of vertical magnetic field due to the Sq system measured during the International Geophysical Year. [After Matsushita (1969). Reproduced by permission of the American Geophysical Union.]

rents is canceled out by the Pedersen currents that link the field lines horizontally through the ionosphere. Since the electric field must be nearly the same at both ends of a magnetic field line, any differences between the two hemispheres create currents that flow between the hemispheres to maintain the field lines as equipotentials. However, to first order at equinox the tidal modes are symmetrical and the field-aligned currents are therefore minimal. In this case the Sq pattern due to Hall currents represents the true total currents in the ionosphere quite well.

Although the Sq current deduced from data such as those shown in Fig. 3.13 flows primarily on the dayside (where the conductivity is high), this does not mean that the electric field is confined to the dayside. In fact, the charge density which builds up at the terminators to force $\nabla \cdot \mathbf{J} = 0$ creates the diurnal zonal electric field pattern observed at Jicamarca. In this chapter we consider the effect of the zonal electric field at the equator. The effect of the E-region dynamo at mid-latitudes is discussed in Chapter 5.

Although small, the zonal electric field component has important consequences. It will drive a small Pedersen current along the dayside equator. More important, a vertically downward Hall current will also flow (see Fig. 3.15). Now, from $\nabla \times \mathbf{E} = 0$, we can deduce that

$$\partial E_x / \partial z = \partial E_z / \partial x$$

This means that the variation of the zonal component with altitude can be estimated by the ratio

$$\delta E_x = \delta E_z (\delta z / \delta x)$$

The horizontal scale size (δx) of conductivity variations is 100 times that of the vertical (δz) variations of conductivity, and both experiments and theory indicate

E_z is at most 10–20 times E_x. Taken together, this means that the zonal field can change only slightly in the E region itself, and we shall therefore assume that conductivity gradients rather than variations in E_x dominate the divergence of the vertical Hall current. To investigate this divergence, the altitude variations of typical noontime equatorial electrojet conductivities and densities are plotted in Fig. 3.14 (Forbes and Lindzen, 1976). Notice that the Hall conductivity dominates below 120 km and, indeed, is highly altitude dependent, even though the plasma density itself is nearly uniform with height.

As a first approximation to the physics of the E-region dynamo, we again consider a slab conductivity geometry, such as that illustrated in Fig. 3.15, subject to a constant zonal electric field. The Hall current cannot flow across the boundary and charge layers must build up. These generate an upward-directed electric field. In a steady state in this slab model, no vertical current may flow and the vertical Pedersen current must exactly cancel the Hall current. This implies that

$$\sigma_H E_x = \sigma_P E_z$$

and hence that

$$E_z = (\sigma_H/\sigma_P)E_x \tag{3.13}$$

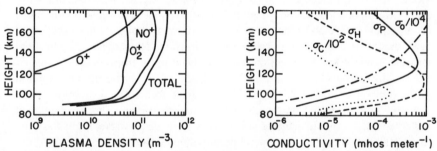

Fig. 3.14. Vertical profiles of daytime composition and plasma density (left) and conductivities (right) for average solar conditions. [After Forbes and Lindzen (1976). Reproduced by permission of Pergamon Press.]

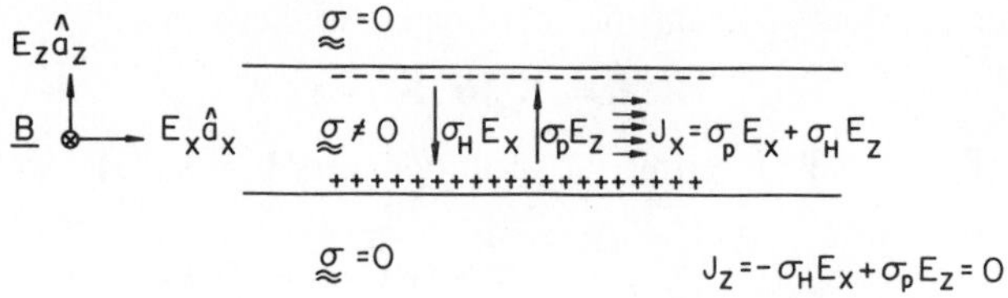

Fig. 3.15. The equatorial electrojet in a slab geometry.

Since $\sigma_H > \sigma_P$, the vertical electric field component considerably exceeds the zonal electric field component. In addition, $E_z(z)$ has the same z dependence as the function $\sigma_P(z)/\sigma_H(z)$. The zonal current is now given by

$$J_x = \sigma_H E_z + \sigma_P E_x \qquad (3.14)$$
$$J_x = [(\sigma_H^2)/(\sigma_P^2) + 1]\sigma_P E_x = \sigma_c E_x$$

where σ_c is the so-called Cowling conductivity. Notice that the local neutral wind does not enter this calculation at all; the electrojet is set up by the global tidal winds that create the diurnal zonal electric field component measured at the equator. In a more complete theory complications due to the zonal neutral wind may be included. Note that the meridional wind component does not enter at the magnetic equator, since for that component the cross product with the magnetic field vanishes.

In effect, (3.14) shows that the zonal conductivity is enhanced by the large factor $1 + \sigma_H^2/\sigma_P^2$, the Cowling conductivity factor, which leads to the intense current jet at the magnetic equator. This can be seen in Fig. 3.13, in which the magnetic field contours become very close together at the magnetic equator. This channel of electrical current is termed the equatorial electrojet. The Cowling conductivity is also plotted in Fig. 3.14 (divided by 100) and displays a peak at 102 km with a half-width of 8 km.

It is interesting to note that no one has yet directly measured the vertical electric field in the electrojet. In fact, even the zonal electric field data plotted in Fig. 3.2 were obtained in the F region. They may be used to determine E_x since the arguments made above based on $\nabla \times \mathbf{E} = 0$ imply that the zonal component of $\mathbf{E}$ should be continuous in the vertical direction. Numerous magnetic field measurements of the electrojet current have been made from rockets. Some of the data are reproduced here in Fig. 3.16. The plot displays the altitude variation of the jet over Peru with each profile normalized to a 100-nT variation of the magnetic field measured on the ground at Huancayo, Peru. That is, the actual perturbed magnetic field detected at Huancayo during each rocket flight was used to scale all data to a common ground perturbation of 100 nT.

An electrojet theory suitable for comparison with such detailed data must include realistic altitude and latitude variations of σ. Away from the equator, σ is often expressed in geographic rather than geomagnetic coordinates, and the resulting conductivity tensor is not as simple as (3.6). If $\mathbf{R}$ is the rotation matrix relating geographic and geomagnetic coordinates, we have

$$\mathbf{R} \cdot \mathbf{J} = \mathbf{R} \cdot (\boldsymbol{\sigma} \cdot \mathbf{E}')$$

where $\mathbf{E}' = \mathbf{E} + \mathbf{U} \times \mathbf{B}$. Inserting the identity matrix $\mathbf{I} = \mathbf{R}^{-1} \cdot \mathbf{R}$, we have

$$\mathbf{R} \cdot \mathbf{J} = (\mathbf{R} \cdot \boldsymbol{\sigma} \cdot \mathbf{R}^{-1}) \cdot \mathbf{R} \cdot \mathbf{E}'$$

and we see that a new tensor given by

$$\boldsymbol{\sigma}^R = \mathbf{R} \cdot \boldsymbol{\sigma} \cdot \mathbf{R}^{-1}$$

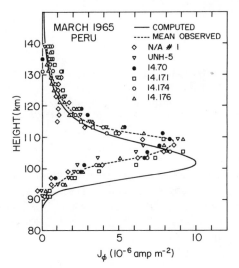

Fig. 3.16. Observed and computed eastward current density profiles near noon at the dip equator off the coast of Peru in March 1965, normalized to a magnetic field perturbation of 100 nT at Huancayo. Measured profiles are from Shuman (1970) (flight N/A #1), Maynard (1967) (flight UNH-5), and Davis *et al.* (1967) (flights 14.170, 14.171, 14.174, and 14.176). The theoretical profile is from Richmond's (1973a) theory. [After Richmond (1973b). Reproduced by permission of Pergamon Press.]

determines the relationship between $\mathbf{J}$ and $\mathbf{E}'$ in the new coordinate system. For the simple case when geomagnetic and geographic north coincide and the dip angle is I, $\mathbf{R}$ corresponds to a rotation by the angle I about the x axis. Following Forbes (1981), we denote the eastward direction as the λ axis, the northward direction as the θ axis, and the vertical as the y axis. Then

$$
\boldsymbol{\sigma}^{\mathrm{R}} = \begin{pmatrix} \sigma_{\lambda\lambda} & \sigma_{\lambda\theta} & \sigma_{\lambda z} \\ \sigma_{\theta\lambda} & \sigma_{\theta\theta} & \sigma_{\theta z} \\ \sigma_{z\lambda} & \sigma_{z\theta} & \sigma_{zz} \end{pmatrix} \tag{3.15}
$$

with

$$
\sigma_{\lambda\lambda} = \sigma_{\mathrm{P}}
$$

$$
\sigma_{\lambda\theta} = -\sigma_{\theta\lambda} = -\sigma_{\mathrm{H}} \sin I
$$

$$
\sigma_{\lambda z} = -\sigma_{z\lambda} = \sigma_{\mathrm{H}} \cos I
$$

$$
\sigma_{\theta\theta} = \sigma_{\mathrm{P}} \sin^2 I + \sigma_0 \cos^2 I \tag{3.16}
$$

$$
\sigma_{\theta z} = \sigma_{z\theta} = (\sigma_0 - \sigma_{\mathrm{P}}) \sin I \cos I
$$

$$
\sigma_{zz} = \sigma_{\mathrm{P}} \cos^2 I + \sigma_0 \sin^2 I
$$

Note that in this form if we set $I = 0$ we recover a matrix identical to $\boldsymbol{\sigma}$ in (3.6). Taking the divergence of $\mathbf{J}$ and substituting $\mathbf{E} = -\nabla\phi$, we have the dynamo equation

$$\nabla\cdot[\boldsymbol{\sigma}^R\cdot(-\nabla\phi + \mathbf{U}\times\mathbf{B})] = 0$$

where the quantities are all measured in the earth-fixed frame and expressed in the rotated coordinate system. This may be written

$$\nabla\cdot[\boldsymbol{\sigma}^R\cdot\nabla\phi] = \nabla\cdot[\boldsymbol{\sigma}^R\cdot(\mathbf{U}\times\mathbf{B})] \tag{3.17}$$

Even taking $\mathbf{U}$ and $\boldsymbol{\sigma}^R$ as known functions, this equation is a complicated partial differential equation with nonconstant coefficients. A common simplifying assumption is that no vertical current flows anywhere in the system; that is, all currents flow in a thin ionospheric shell. Setting J_z equal to zero (in geographic coordinates now) yields

$$\sigma_{z\lambda}E'_\lambda + \sigma_{z\theta}E'_\theta + \sigma_{zz}E'_z = 0$$

or

$$E'_z = -(\sigma_{z\lambda}E'_y + \sigma_{z\theta}E'_\theta)/\sigma_{zz}$$

For $I = 0$ this reduces to (3.13) at the equator. Using this expression to eliminate E'_z, $\mathbf{J}^R$ can now be written as a two-dimensional vector and the dynamo equation as a function of θ and λ only (Forbes, 1981),

$$\begin{pmatrix} J_\lambda \\ J_\theta \end{pmatrix} = \begin{pmatrix} \xi_{\lambda\lambda} & \xi_{\lambda\theta} \\ \xi_{\theta\lambda} & \xi_{\theta\theta} \end{pmatrix} \cdot \begin{pmatrix} E'_\lambda \\ E'_\theta \end{pmatrix}$$

where

$$\xi_{\lambda\lambda} = \sigma_P + (\sigma_H \cos I)^2/(\sigma_P \cos^2 I + \sigma_0 \sin^2 I)$$

$$\xi_{\lambda\theta} = -\xi_{\theta\lambda} = (-\sigma_0\sigma_H \sin I)/(\sigma_P \cos^2 I + \sigma_0 \sin^2 I)$$

$$\xi_{\theta\theta} = (\sigma_0\sigma_P)/(\sigma_P \cos^2 I + \sigma_0 \sin^2 I)$$

are the so-called layer conductivities. Insertion into $\nabla\cdot\mathbf{J} = 0$ in geographic coordinates yields

$$(\partial/\partial\lambda)[(\xi_{\lambda\lambda}/R_e \sin \theta)(\partial\phi/\partial\lambda) + (\xi_{\lambda\theta}/R_e)(\partial\phi/\partial\theta)]$$

$$+ (\partial/\partial\theta)\{\sin \theta[(\xi_{\theta\lambda}/R_e \sin \theta)(\partial\phi/\partial\lambda) + (\xi_{\theta\theta}/R_e)(\partial\phi/\partial\theta)]\} \tag{3.18}$$

$$= (\partial/\partial\lambda)[-\xi_{\lambda\lambda}uB_z + \xi_{\lambda\theta}vB_z] + (\partial/\partial\theta)\{\sin \theta[-\xi_{\theta\lambda}uB_z + \xi_{\theta\theta}vB_z]\}$$

Assuming E'_λ and E'_θ to be independent of height, (3.18) can be integrated over height to give the thin-shell dynamo equation:

$$(\partial/\partial\lambda)[(\Sigma_{\lambda\lambda}/R_c \sin \theta)(\partial\phi/\partial\lambda) + (\Sigma_{\lambda\theta}/R_c)(\partial\phi/\partial\theta)]$$

$$+ (\partial/\partial\theta)[\sin \theta(\Sigma_{\theta\lambda}/R_c \sin \theta)(\partial\phi/\partial\lambda) + (\Sigma_{\theta\theta}/R_c)(\partial\phi/\partial\theta)] \qquad (3.19)$$

$$= \frac{\partial}{\partial\lambda} \int [-\xi_{\lambda\lambda}u + \xi_{\lambda\theta}v]B_z \, dh$$

$$+ \frac{\partial}{\partial\theta} [\sin \theta \int [-\xi_{\lambda\theta}u + \xi_{\theta\theta}v]B_z \, dh]$$

where the Σ denote height-integrated layer conductivities, u and v are the zonal and meridional wind components, and R_c is the radius of the earth. It is important to note that assuming $\mathbf{E}' = \mathbf{E} + \mathbf{U} \times \mathbf{B}$ to be independent of height implies that u, v, and $-\nabla\phi$ are all independent of height, which is not very realistic. Nonetheless, comparisons between the electrojet currents obtained using thin-shell theory and experimentally measured currents have been made both by treating the zonal electric field as a free parameter (e.g., Sugiura and Cain, 1966; Untiedt, 1967) and by solving the dynamo equation by setting the zonal derivatives equal to zero. However, the peak current density seems to be underestimated in these models and details concerning the vertical and latitudinal extent of the electrojet are not reproduced. More seriously, without vertical currents at the equator, which are suppressed by the thin-shell model, it is very hard to satisfy the divergence requirement

$$\partial J_x/\partial x + \partial J_y/\partial y = 0$$

since the electrojet varies rapidly in latitude but only slowly in longitude. Forbes and Lindzen (1976) pointed out that allowing a vertical current could actually increase the electrojet intensity predicted by the models. For example, returning to the current equation at the equator and again ignoring the neutral wind,

$$J_x = \sigma_P E_x + \sigma_H E_z$$

$$J_z = -\sigma_H E_x + \sigma_P E_z$$

Eliminating E_z yields

$$J_x = \sigma_c E_x + (\sigma_H/\sigma_P)J_z \qquad (3.20)$$

Thus, if J_z was not zero (and positive) at the equator, J_x would exceed the Cowling current and the electrojet current would be stronger than $\sigma_c E_x$.

More realistic models which include such possibilities have been developed by Richmond (1973a,b) (e.g., Fig. 3.16) and Forbes and Lindzen (1976). Richmond's improvement over the thin-shell model involved integrating the divergence equation along a magnetic field line as discussed above for the F-layer dynamo. The zonal derivatives were all set to zero. The vertical electric field at the equator then becomes $(\Sigma_H/\Sigma_P)E_x$ rather than $(\sigma_H/\sigma_P)E_x$; that is, the integrated conductivities along the field line are used rather than the local values.

Above 100 km the integrated conductivity ratio exceeds the local value so $J_z \neq 0$ and both the eastward current and the vertical electric field are greater than the thin-shell model. The result seems to overestimate the current (see Fig. 3.16). Forbes and Lindzen (1976) relaxed the requirement of zonal invariance in their solution of the three-dimensional $\nabla \cdot \mathbf{J} = 0$ equation, but they did not include integration along $\mathbf{B}$. The next step is to combine these two approaches. The excellent review by Forbes (1981) details these and other electrojet studies including day-to-day variability, neutral wind, and lunar tidal effects that are beyond the scope of the present text.

Of more importance to the present text is the generation of large vertical electric fields in the electrojet. As discussed in detail in the next chapter, these fields create electron drifts that are unstable to the generation of plasma waves. Peak vertical electric fields in the range 10–20 mV/m have been predicted theoretically and would create electron drift speeds of 350–700 m/s. This exceeds the acoustic speed C_s, which is about 360 m/s, and leads to the generation of intense plasma waves via the two-stream instability. Indeed, as we shall see, very large Doppler shifts occur when radar signals are scattered from the electrojet waves. These Doppler shifts are limited to a maximum value equal to the ion acoustic speed, for reasons that are not yet fully understood. Nonetheless, these observations show that the vertical electric field reaches values at least as large as that given by $E_z = C_s B \simeq 9$ mV/m, and it is very likely that the field reaches values up to twice as large. This is the largest electric field found in the ionosphere below auroral zone latitudes.

3.4   Further Complexities of Equatorial Electrodynamics

The simple E- and F-region dynamos described above explain some of the observations presented earlier in Fig. 3.1. Since the E region controls the physics during the day, the small daytime vertical electric fields in the F region presumably mirror the meridional E-region polarization fields in the latitude ranges just north and south of the equator. The zonal field, both daytime and nighttime, is global in nature and is driven by the large-scale tides and winds in the sunlit hemisphere. In the postsunset period, the F-layer vertical field is enhanced due to the local F-layer dynamo discussed above. In the next paragraphs we point out some of the features not explained by these simple models.

3.4.1   The Prereversal Enhancement

The postsunset enhancement or prereversal enhancement of the zonal field occurs during all epochs and seasons studied except for the solar minimum solstices. The effect of this brief duration large eastward electric field can be quite significant since the F-layer plasma often is driven to very high altitudes, where

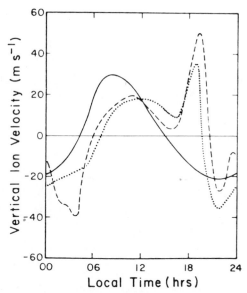

Fig. 3.17. Calculated vertical ion drift velocities for several driving wind components. The solid line includes only the tidal-driven E-region dynamo, while the dashed line includes the F-region dynamo as well. Typical measured vertical plasma drifts are indicated by the dotted line. [After Heelis *et al.* (1974). Reproduced with permission of Pergamon Press.]

recombination is slight and collisions are rare. Heelis *et al.* (1974) successfully predicted the postsunset effect in their model, which included *horizontal* conductivity gradients near sunset in the F-layer dynamo mechanism (in addition to the vertical gradients we have studied). In effect, near such a sharp east–west gradient, an enhanced zonal electric field is established to keep $\nabla \cdot \mathbf{J} = 0$. Experimental data support this explanation since the enhancement begins when the sun sets on either of the E layers in contact along $\mathbf{B}$ with the equatorial F region.

The evidence from their numerical simulation is presented in Fig. 3.17. The solid line shows the calculated vertical ion velocity in the F region (equivalent to a zonal electric field) when only a diurnal E-region tidal mode is considered. The basic day–night features are reproduced but no prereversal enhancement occurs. The dashed line shows a calculation including the F-region dynamo in the physics. The latter agrees quite well with the dotted line, which is a typical vertical drift profile for the plasma over Jicamarca.

In another simulation, Farley *et al.* (1986) completely suppressed the E-region dynamo although the E-region conductivity and its loading effect on the F-region electrodynamics were retained. They then injected a uniform 200-m/s eastward wind everywhere. The result was quite interesting and is shown in Fig. 3.18. The upper plot shows the zonal plasma drift speed. As expected, the

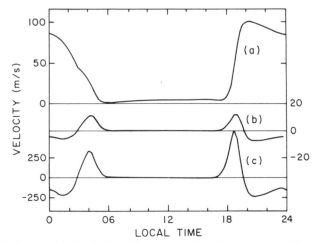

Fig. 3.18. Equatorial drift velocities driven by a uniform 200-m/s eastward F-region dynamo wind only: (a) the eastward F-region ion velocity at 350 km, (b) the upward F-region ion velocity, and (c) the westward E-region electron velocity. The E-region drift velocity corresponds to an altitude of about 105 km and to a ratio of Hall to Pedersen conductivity of 25. The sunset prereversal enhancements are clearly shown in (b) and (c). The model is not valid near sunrise. [After Farley *et al.* (1986). Reproduced with permission of the American Geophysical Union.]

plasma drift rises to a value less than but comparable to the wind speed at night but is practically zero during the daytime due to E-region shorting. The middle panel shows the vertical plasma drift. Enhancements are seen at both terminators but the simulation is not relevant at dawn, where the wind direction is reversed. The important point is that the enhancement is clearly an F-region dynamo effect. A simple sketch which may explain this effect is given in Fig. 3.19. The equatorial plane is shown as the vertical plane and its projection onto the southern hemisphere along **B** is also shown. The wind blows across the terminator, generating a vertical electric field E_z which is downward on both sides. E_z is much smaller on the dayside than on the nightside but it is not zero. Thus, E_z maps along **B** to an equatorward electric field component off the equator. This field drives a westward Hall current $J_{\theta\phi}$ on both sides of the terminator. However, even though E_z is smaller on the dayside by up to 90%, the Hall conductivity is more than 10 times the nighttime value. The result is that a negative charge density builds up near the terminator, creating the localized zonal E_ϕ perturbation shown in the figure. The current $J_{\phi\phi}$ cancels $J_{\theta\phi}$ in a steady state. The absence of a sunrise enhancement may be due to the lower value of E_z which occurs on the nightside of the dawn terminator relative to the nightside of the dusk terminator as well as the lower conductivities which occur in the predawn F region.

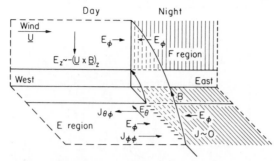

Fig. 3.19. Simplified model of the F-region prereversal enhancement driven by a uniform F-region wind **U**. Near the sunset terminator the F-region dynamo E_z is no longer shorted out and approaches $-\mathbf{U} \times \mathbf{B}$. This field maps to an equatorward E_θ in the E layer and drives a westward Hall current $J_{\theta\phi}$. But if no current flows in the nightside E region, a negative polarization charge must develop at the terminator, with E_ϕ as shown and $J_{\phi\phi}$ canceling $J_{\theta\phi}$. This E_ϕ maps back to the F region and causes first an upward (day) and then a downward (night) $\mathbf{E} \times \mathbf{B}$ plasma drift. [After Farley *et al.* (1986). Reproduced with permission of the American Geophysical Union.]

As we shall see, this vertical drift enhancement has many interesting consequences. It can produce conditions that are favorable to plasma instabilities, and an extremely structured ionosphere often results (see Chapter 4). Significant effects on neutral atmospheric dynamics also occur.

3.4.2   *High-Latitude Effects*

The equatorial plasma is not always as well behaved as indicated thus far. A number of interesting processes can affect equatorial electrodynamics, in addition to the neutral wind-generated electric fields already discussed which are driven by the solar heat input carried by photons. Energy also comes from the sun via particles in the solar wind and via the electromagnetic (Poynting) flux in the solar wind. These energy sources affect primarily the high-latitude polar cap and auroral zones but can also be detected at the equator. The sensitivity of radars to electric fields has, in fact, increased our knowledge of how the solar wind affects the magnetosphere/ionosphere system (Gonzales *et al.*, 1979, 1983; Somayajulu *et al.*, 1985, 1987; Earle and Kelley, 1987).

A classical event (Nishida, 1968) illustrating a clear relationship between the interplanetary magnetic field (IMF), which is of solar origin, and the magnetic field created by the equatorial electrojet is reproduced in Fig. 3.20. The two magnetic field measurements are clearly correlated although the magnetometers were separated by 10^5 km! Thus, although the magnetosphere and plasmasphere both act to shield out interplanetary and solar wind effects from low latitudes, the protection is not perfect. Some insight into the complex relationship between interplanetary and equatorial ionospheric phenomena has come from simultane-

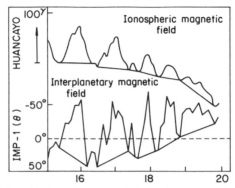

Fig. 3.20. Correlation of horizontal geomagnetic fluctuations observed at Huancayo on December 3, 1963, with changes in the direction of the interplanetary magnetic field component perpendicular to the sun–earth line observed from the satellite IMP-1 versus UT. [After Nishida (1968). Reproduced with permission of the American Geophysical Union.]

ous measurements of electric fields at Jicamarca and in the auroral zone. Since, as discussed elsewhere in this text, the much shorter chain of cause and effect between the IMF and auroral fields is *itself* not yet entirely understood, we can be satisfied here with some plausible explanations for the equatorial effects.

Forty-eight hours of Jicamarca electric field data are presented in the lower panel of Fig. 3.21. The eastward electric field component is plotted along with a lighter line which shows the average quiet-day value for the same solar cycle and season. The top trace is a superposition of magnetic field data from a number of auroral zone magnetometer stations which are used to generate the AU and AL auroral zone magnetic indices (see Chapter 6). A series of six bursts of auroral substorm activity is clear in this data set. The northward component of the interplanetary magnetic field is also plotted. The latter, unfortunately, has a number of data gaps, a problem endemic to studies of the type described here! One of these data gaps occurred during the first isolated substorm, and we ignore this period for the moment. The IMF turned southward at 1700 UT on January 18 and remained southward until about 0100 UT on January 19. During this time two substorms occurred in the auroral oval. At Jicamarca, which was in the postnoon sector, the period 1900–2300 UT was characterized by a westward perturbation from the quiet-time field. This period corresponds to the time between the two substorms. The period 0700–1200 UT on January 19 also fell between two substorms but was characterized by an eastward perturbation of the Jicamarca electric field measurement (postmidnight sector). Notice that there was almost exactly 12 h between these events, which means that the Jicamarca data were taken on the opposite side of the system when viewed in a magnetospheric or solar wind reference frame. The burst of eastward perturbation field

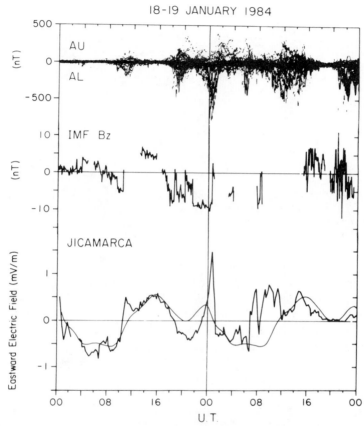

Fig. 3.21. Auroral zone and interplanetary magnetic field data along with zonal electric field measurements over Jicamarca, Peru, for a 48-h period. Local midnight occurs at 5 UT at Jicamarca. [After Fejer (1986). Reproduced with permission of Terra Scientific Publishing Co.]

at 0900–1000 UT on January 18 is now seen to be consistent with the postmidnight "substorm recovery" characteristics seen 24 h later (i.e., 0700–1200 UT on January 19). These eastward perturbations in the postmidnight sector are very commonly reported in the various studies of disturbance fields using the Jicamarca data base. In part, this is due to the fact that they are easily recognizable in that sector since the quiet-time field is usually steady and westward. (Notice, for example, the July 3, 1968 event identified in Fig. 3.3 is of this type.) As we shall see, however, there is also evidence that a local time dependence exists in the response and leads to enhanced perturbations in this sector.

Another event of this class is shown in Fig. 3.22. The shaded period is characterized as follows:

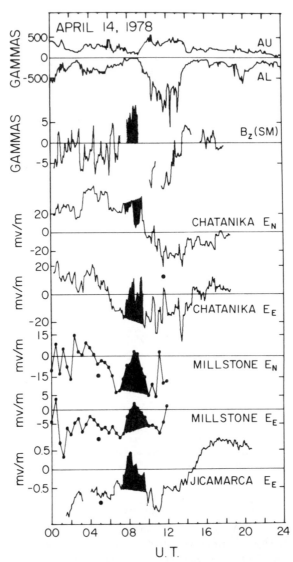

Fig. 3.22. Auroral (AU, AL) and interplanetary (B_z) magnetic field data along with auroral (Chatanika and Millstone) and equatorial (Jicamarca) electric field measurements. The shaded region corresponds to a rapid change in magnetospheric convection apparently triggered by a northward turning of the interplanetary field. [After Kelley *et al.* (1979). Reproduced with permission of the American Geophysical Union.]

1. An eastward electric field perturbation occurred in the postmidnight sector over Peru (Jicamarca data).

2. The IMF turned northward (panel 2) and magnetic activity decreased (panel 1).

3. The auroral zone electric field measured over Alaska (Chatanika data) and eastern Canada (Millstone Hill) decreased dramatically.

A qualitative explanation for such an event was proposed by Kelley et al. (1979). As we shall discuss in more detail in subsequent chapters, it is known that if a steady magnetospheric electric field exists, divergence of the ring current near $L = 4$ will eventually create a charge separation in such a way that the magnetospheric electric field will be shielded from low latitudes. This charge separation region is called the Alfvén layer. Now if the magnetospheric field rapidly increases or decreases, the charges will be temporarily out of balance with the new configuration and a brief low-latitude disturbance will result. An example of the effect of a rapid decrease in the external electric field is illustrated in Fig. 3.23. The transient state inside the plasmasphere when the external field vanishes is an eastward perturbation on the nightside and a westward perturbation during the day, just as observed in the data shown in Figs. 3.21 and 3.22. In other words, the low-latitude electric field may be due to Alfvén layer charges which still exist even though no external field is left in the outer regions to shield out! In circuit terminology, the finite inductance of the ring current maintains the voltage in the inner region.

An event similar to Nishida's original study shown in Fig. 3.20 was studied on February 17, 1976, using electric field data as well as magnetic measurements. The IMF, the auroral zone electric field, and the equatorial electric field

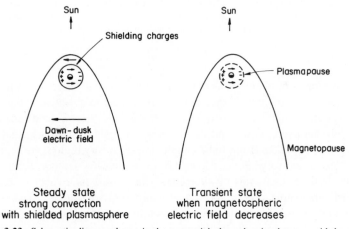

Fig. 3.23. Schematic diagram drawn in the equatorial plane showing how a rapid decrease in magnetospheric convection could create electric field changes inside the plasmasphere.

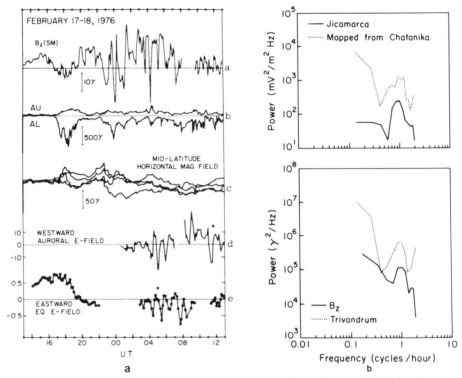

Fig. 3.24. (a) Interplanetary (B_z), auroral (AU, AL), and mid-latitude magnetic field data along with auroral and equatorial electric field data during a series of rapid interplanetary magnetic field changes. [After Gonzales *et al.* (1979). Reproduced with permission of the American Geophysical Union.] (b) Power spectra of the data plotted in (a). The Chatanika electric field data were reduced by the factor $L^{3/2}$ before analysis to show the value that would arise in the equatorial plane at $L = 5.5R_e$. [After Earle and Kelley (1987). Reproduced with permission of the American Geophysical Union.]

all display similar oscillations, as illustrated in Fig. 3.24a. These time variations are too quick to allow the charges in the Alfvén layer to change and thus seem to be consistent with the concept of very effective direct electric field penetration into the plasmasphere. Power spectra of the electric field (from Alaska and Peru) and magnetic field data (from India and the interplanetary region) are plotted in Fig. 3.24b and are very similar. Control of the earth's low-latitude electric field from hundreds of thousands of kilometers away in the interplanetary medium is indeed remarkable. The auroral zone data (from Chatanika, Alaska) were mapped to the equatorial plane before the Fourier analysis was done so that they could be directly compared to the Jicamarca data (which are also in the equatorial plane). The power is down by only a factor of four at Jicamarca, which means the field penetrating to the equator is reduced by only 50% from the magnetospheric (auroral zone) value in the equatorial plane. This is typical of the electric field values in the few-hour frequency range studied by Earle and Kelley (1987), who also have reported experimental evidence that the shielding process acts like a high-pass filter, allowing signals with periods shorter than about 8 h to penetrate to the equatorial ionosphere. This is in good agreement with the time constant predicted by ring current shielding theory (Vasyliunas, 1972). Furthermore, they show that for the zonal electric field component at Jicamarca, the signal in the frequency range corresponding to a period of a few hours exceeds the "geophysical noise" due to gravity waves at a moderate level of magnetic activity (K_p = 3). In other words, at frequencies above the ring current shielding frequency, high-latitude effects may always be present at all latitudes but cannot be detected unless they are above the fluctuation level due to neutral wind-driven electric fields.

Examples of the low-latitude effect of increasing magnetospheric electric fields are rarer, for reasons which are not altogether clear. One good example on March 23–24, 1971 is shown in Fig. 3.25. The IMF turned southward abruptly

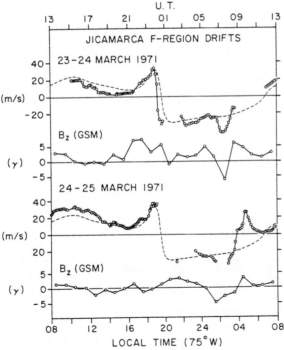

Fig. 3.25. F-region vertical drifts measured over Jicamarca (corresponding to an eastward electric field component) and IMF data. Note the large nighttime drift perturbations are well correlated with southward and northward IMF changes. Quiet-time patterns are shown by dashed lines. [After Fejer (1986). Reproduced with permission of Terra Scientific Publishing Co.]

at 0700 UT on March 27 after 5 h of northward field. A strong *westward* perturbation occurred over Jicamarca, which was in the postmidnight sector. Notice that 24 h later the more common eastward perturbation was seen in conjunction with a northward turning of the IMF.

Although several quantitative studies have been made of these processes, one of the most complete seems to be the early effort by Nopper and Carovillano (1978). They allowed for a strong day–night conductivity difference and modeled the auroral effect as the response of a resistive load to a current source controlled by the so-called region 1/region 2 field-aligned current systems (see Chapter 6). The former corresponds to the highest-latitude currents, which are closely controlled by and linked to the solar wind generator. The region 2 currents correspond to the inner magnetospheric portion of the auroral current system and link up with the ring current system. Their results are shown in Fig. 3.26 for three different cases. Their case (c) corresponds most closely to the situation wherein the external generator is abruptly reduced, since it corresponds to a set of region 2 currents equal in value to the region 1 currents. Such a choice yields a low latitude controlled by the region 2 currents and, we argue, would give the electric

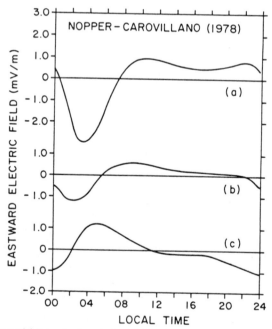

Fig. 3.26. Equatorial ionospheric perturbation electric field patterns calculated for three given field-aligned current distributions. Ionospheric dynamo effects were not included. [After Fejer (1986). Reproduced with permission of Terra Scientific Publishing Co.]

field perturbation sense if there were no polar generator present (a rapid decrease of the external source). Indeed, the perturbation is eastward in the postmidnight sector and westward postnoon. Conversely, the strong region 1 case (panel a) yields a westward perturbation in the postmidnight case (as also observed).

Another mechanism which creates low-latitude electrodynamic changes is the disturbance dynamo. If auroral zone heating and momentum sources are sufficiently strong, the low-latitude neutral atmospheric winds can also be affected. In turn, these winds will create different electric field patterns through the usual E- and F-region dynamo process. Unlike the virtually instantaneous changes shown above, the disturbance dynamo takes more time to develop and lags the auroral inputs by many hours. Gravity waves can also be generated by high-latitude processes and propagate to low latitudes. It seems likely that the wind fields associated with gravity waves will also create dynamo electric fields which vary in time with the wave period. Again, a time delay of many hours is expected if the gravity wave source is in the auroral zone (see Fig. 3.29 and associated text).

Very clear examples of both instantaneous and delayed equatorial perturbations are illustrated in Fig. 3.27. In the upper two panels the auroral AU and AL magnetic fields are plotted for a 48-h period on August 8–10, 1972. Below, the corresponding zonal electric field data measured at Jicamarca are given. In each of the latter plots the mean field pattern is shown as the solid line. The abrupt change in the Jicamarca field between 2300 and 0300 local time on August 8 has been studied by Gonzales *et al.* (1979) and has a waveform identical to that of the similar electric field measured simultaneously at Chatanika, Alaska. This is a clear case of rapid penetration of high-latitude electric fields to low latitudes. On the next day the Jicamarca zonal field component displayed a long-lived deviation from the average, commencing at 2200 local time and lasting for 8 h. Fejer *et al.* (1983) argue that such slowly developing electric field changes are due to a worldwide change in the thermosphere winds. These alter the equatorial F-region dynamo and thereby change the electric field at low latitudes from the normal pattern.

Taken together, we see that these and other high-latitude effects contribute to the "weather" in the low-latitude electric field and neutral wind patterns. Their study therefore yields information on both high- and low-latitude phenomena. Much can be learned about global electrodynamics and thermospheric physics through events of these types. The true test of global electrodynamic model calculations may be their ability to predict the low-latitude ionospheric effects of such diverse external influences. In Chapter 5 we discuss further the penetration of high-latitude electric fields to mid-latitudes. The origins of the high-latitude electric fields themselves are considered in some detail in Chapter 6.

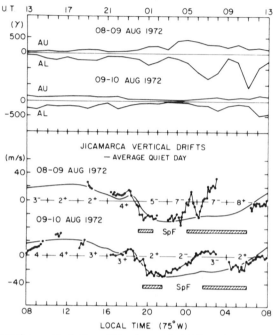

Fig. 3.27. Auroral magnetic fields and F-region vertical drifts at Jicamarca on August 8–10, 1972. The solid curves in the lower panel show the average quiet-time diurnal variation. Deviations from this pattern beginning at 2300 LT on August 8 are due to direct penetration effects, whereas the slower deviations starting at 2200 LT on August 9 are due to the disturbance dynamo. [After Fejer *et al.* (1983). Reproduced with permission of the American Geophysical Union.]

3.5   Feedback between the Electrodynamics and the Thermospheric Winds

We return now briefly to the problem of the large calculated thermospheric winds mentioned in Section 3.2. Part of the key to this dilemma lies in the crucial role of the plasma in controlling the thermospheric wind, even though the ionized component is only a very minor constituent (one part in 10^3 at 300 km). When a moving neutral particle strikes an ionized particle at rest in the thermosphere, some of the momentum imparted is converted into motion of the particle about the magnetic field and yields a net deflection in the $q(\mathbf{U} \times \mathbf{B})$ direction, where q includes the sign of the particle charge. Thus, unlike a collision with a neutral particle, the momentum transfer does not appear as linear momentum parallel to $\mathbf{U}$ in the ionized gas. A similarly dense minor neutral species would, after one or two collisions, merely accelerate to a velocity equal to the background wind speed, at which time no further momentum transfer would occur. The ionized particles, however, are "locked" onto the magnetic field lines instead of being accelerated parallel to $\mathbf{U}$ and can act as a steady drag on the neutral wind. We now show that the linear momentum lost per unit time by the neutral fluid due to this drag is equal to the electromagnetic force $\mathbf{F}_E = \mathbf{J} \times \mathbf{B}$ due to the wind-generated electric current which flows in the fluid. As shown in Chapter 2, the current due to a neutral wind $\mathbf{U}$ is given by

$$\mathbf{J} = \boldsymbol{\sigma} \cdot (\mathbf{U} \times \mathbf{B})$$

with

$$\boldsymbol{\sigma} = \begin{pmatrix} \sigma_P & 0 & 0 \\ 0 & \sigma_0 & 0 \\ 0 & 0 & \sigma_P \end{pmatrix}$$

being the appropriate tensor for an F-region process. Since $\boldsymbol{\sigma}$ is diagonal and $\mathbf{U} \times \mathbf{B}$ is perpendicular to $\mathbf{B}$, the triple cross product in $\mathbf{J} \times \mathbf{B}$ yields

$$\mathbf{F}_E = -\sigma_P B^2 \mathbf{U}_\perp \tag{3.21}$$

where $\mathbf{U}_\perp$ is the projection of the wind vector onto the plane perpendicular to $\mathbf{B}$. The plasma thus clearly acts as a drag on the neutrals, since the force acts in the opposite direction to $\mathbf{U}_\perp$. We have shown previously that in the F region we can approximate σ_P by the expression

$$\sigma_P = (ne^2 \nu_{in}/M\Omega_i^2) = n\nu_{in}M/B^2 \tag{3.22}$$

where n is the plasma density, M the ion mass, Ω_i the ion gyrofrequency, and ν_{in} the ion–neutral collision frequency. Substituting (3.22) into (3.21) yields

$$\mathbf{F}_E = -nM\nu_{in}\mathbf{U}_\perp \tag{3.23}$$

which even more clearly shows the draglike nature of the $\mathbf{J} \times \mathbf{B}$ force. At first glance this expression seems different from the ion drag term in (2.16). However, that expression involved the neutral–ion collision frequency ν_{ni} and the neutral density n_n. For equal-mass ions and neutrals it follows that the collision frequencies for momentum transfer are related by

$$n_n \nu_{ni} = n_i \nu_{in} = n\nu_{in}$$

and the two expressions are seen to agree with each other.

Referring back to Fig. 3.7, we may now interpret the other curves. The solid curve is ν_{ni}, the daytime neutral–ion collision frequency, and the vertical dotted line is the Coriolis parameter, $f = \Omega \sin \theta$, evaluated for 45° latitude. (Ω here is the rotation frequency of the earth, 7.35×10^{-4} rad/s.) The latter plays no role at the equator but the drag term does. For example, consider the steady-state solution to

$$d\mathbf{U}/dt = -(1/\rho)\nabla p - \nu_{ni}\mathbf{U} = \mathbf{F} - \nu_{ni}\mathbf{U}$$

which is given by

$$U = -F/\nu_{ni}$$

where typical values for F are also plotted in Fig. 3.7. Near the F-layer density peak at 250 km the equation yields a magnitude for U of only about 70 m/s. At 300 km the velocity rises to about 130 m/s. These values are much lower than those found by Lindzen (1966) without ion drag and show the great control the ionosphere has over the neutral atmosphere.

Lindzen realized this and repeated his calculations including ion drag (Lindzen, 1967). Using a typical daytime plasma profile (but keeping it the same day and night), Lindzen found that the zonal wind at 300 km was reduced from 550 m/s to only 120 m/s, which is quite reasonable. Much more sophisticated global thermospheric models now exist, as does more information on the ionospheric plasma morphology. In the next paragraph we describe some of the more recent calculations and their interpretation with respect to the observed variability in equatorial wind observations.

Given the control exhibited by the ionospheric plasma on the thermosphere, it is not surprising that the winds are so variable. One factor is the altitude of the nighttime ionospheric plasma layer, which is determined by the magnitude of the zonal component of the electric field. Anderson and Roble (1974) have calculated zonal neutral wind contours in conjunction with an empirical model of the electric field pattern for the two cases illustrated in Fig. 3.28. In the upper plot a diurnally varying electric field pattern is used which is typical of sunspot minimum conditions, while the lower plot includes the prereversal enhancement of the electric field which is largest during sunspot maximum. In the latter case the ionospheric plasma is driven to very high altitudes in the 2000–2400 LT sector. With the plasma out of the way the ion drag is greatly reduced and the thermospheric wind speeds nearly double. Since changes in the zonal electric field occur due to auroral disturbances, it is easy to see that weather in the equatorial thermosphere is also related to weather in the solar wind.

Another variable is the effect of enhanced high-latitude atmospheric heating during auroral storms. Models have shown that large-amplitude gravity waves can be created which propagate throughout the thermosphere, forming so-called traveling ionospheric disturbances. In extreme cases the worldwide wind patterns can be affected by heat input during such storms. Richmond and Matsushita (1975) have modeled the effect of an isolated magnetospheric substorm lasting for 2 h near 70° latitude. At the end of this time they calculated the patterns of wind and temperature variation shown in Fig. 3.29. A large disturbance is seen to propagate equatorward at a speed of about 750 m/s. The simulation shows that the effect reaches the equator with very little attenuation but requires many hours to arrive. The effect is larger in the meridional wind component than in the zonal component.

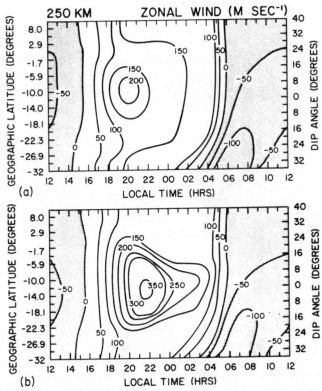

Fig. 3.28. Contours of the zonal neutral wind at 250 km, calculated from the thermospheric model as a function of latitude and local time: (a) without the postsunset enhancement of the $\mathbf{E} \times \mathbf{B}$ drift velocity and (b) including the postsunset enhancement. [After Anderson and Roble (1974). Reproduced with permission of the American Geophysical Union.]

A sophisticated model has been run by Sipler *et al.* (1983), who used it as a diagnostic tool in a numerical experiment to determine which processes dominate the wind variability. The measurements and model calculations of the wind are plotted in Fig. 3.30. In this study, the thermospheric global circulation model (TGCM) was run for August 21, 1978 and compared to the measured winds indicated by the heavy dashed line in Fig. 3.30. Then the lower atmospheric tides were added along with the high-latitude influence (light dashed line). Finally, ionospheric electric fields were added (dotted line). The conclusion was that the day-to-day variability was tied most closely to the tidal and electric field effects but that even in relatively quiet conditions the high-latitude plasma circulation played a role in the equatorial winds.

Finally, we return to the curious fact that the earth's upper atmosphere super-

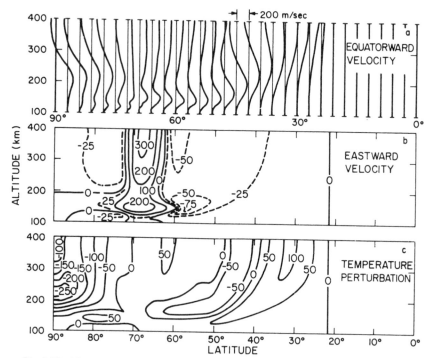

Fig. 3.29. "Snapshot" of the global disturbance wind at time = 2 h after a strong auroral event. (a) Equatorward neutral wind velocity profiles every 3° in latitude. (b) Eastward neutral wind velocity. The contour spacing is 100 m/s for the solid lines and 25 m/s for the dashed lines. (c) Neutral temperature perturbation. The contour spacing is 50 K. [After Richmond and Matsushita (1975). Reproduced with permission of the American Geophysical Union.]

rotates, that is, that the mean zonal wind in the rotating frame is eastward in the equatorial zone. A number of explanations have been put forth concerning this effect (e.g., the review by Rishbeth, 1972). Since a high eastward velocity is most common in the sunset-to-midnight period, theories which control the thermospheric winds via the ionospheric drag effect look very promising, since it is in this period that the winds are least opposed by the ionosphere.

We have already noted that the ionosphere often "gets out of the way" by rising to very high altitudes just after sunset. This reduces ion drag and allows much stronger thermospheric winds below the F layer. Viscosity could still couple winds in this channel to higher altitudes where the F-layer plasma acts as a drag. However, there is another effect which reduces the ion drag throughout the thermosphere. This is the generation of a polarization electric field in the thermosphere via the F-region dynamo process, which we discussed in detail in Section 3.2. The explanation is as follows. When the vertical electric field builds

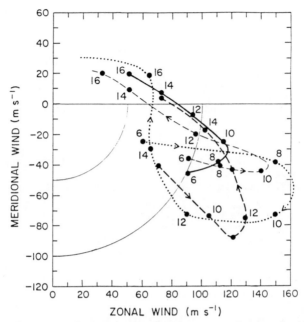

Fig. 3.30. Position of the tip of the measured wind vector as a function of universal time for August 21, 1978. The heavy dashed line gives the measured values and the solid line represents the TGCM calculations. Two other TGCM calculations are also presented: the fine dashed curve is a numerical experiment that includes tides from the lower atmosphere and enhanced (60-kV cross-tail potential) magnetospheric convection at high latitudes; the dotted line further includes enhanced $E \times B$ drifts representative of solar maximum conditions. [After Sipler *et al.* (1983). Reproduced with permission of the American Geophysical Union.]

up in the thermosphere, the electromagnetic braking force on the neutrals becomes modified to the form

$$\mathbf{J} \times \mathbf{B} = \boldsymbol{\sigma} \cdot (\mathbf{E} + \mathbf{U} \times \mathbf{B}) \times \mathbf{B} \qquad (3.24)$$

Now if $\mathbf{E} + \mathbf{U} \times \mathbf{B} = 0$, which occurs in a perfect F-region dynamo, the electromagnetic force vanishes and the thermospheric wind blows freely without ion drag no matter what the density or altitude of the plasma. In effect, the minor constituent plasma now is moving at the same velocity as the neutral wind and there is no momentum transfer. Both the observations and theory outlined above point to the near equality of these two velocities at night and hence to the drastic reduction of the nighttime ion drag. During the day this electric field cannot build up since it is "shorted out" by the low-altitude E-region conductivity, and hence F-region ion drag remains high during the day. Thus, the zonally averaged ion drag term clearly favors a net eastward average zonal wind and may indeed explain the observed superrotation of the earth's upper atmosphere at low latitudes.

References

Anderson, D. N., and Mendillo, M. (1983). Ionospheric conditions affecting the evolution of equatorial plasma depletions. *Geophys. Res. Lett.* **10**, 541.

Anderson, D. N., and Roble, R. G. (1974). The effect of vertical **E × B** ionospheric drifts on F-region neutral winds in the low-latitude thermosphere. *JGR, J. Geophys. Res.* **79**, 5231.

Chapman, S., and Lindzen, R. S. (1970). "Atmospheric Tides: Thermal and Gravitational." Gordon & Breach, New York.

Dalgarno, A., and Smith, F. J. (1962). Thermal conductivity and viscosity of atomic oxygen. *Planet. Space Sci.* **9**, 1.

Davis, T. N., Burrows, K., and Stolarik, J. D. (1967). A latitude survey of the equatorial electrojet with rocket-borne magnetometers. *J. Geophys. Res.* **72**, 1845.

Earle, G., and Kelley, M. C. (1987). Spectral studies of the sources of ionospheric electric fields. *JGR, J. Geophys. Res.* **92**, 213.

Farley, D. T., Bonelli, E., Fejer, B. G., and Larsen, M. F. (1986). The prereversal enhancement of the zonal electric field in the equatorial ionosphere. *J. Geophys. Res.* **91**, 13723.

Fejer, B. G. (1981). The equatorial ionospheric electric field. A review. *J. Atmos. Terr. Phys.* **43**, 377.

Fejer, B. G. (1986). Equatorial ionospheric electric fields associated with magnetospheric disturbances. *In* "Solar Wind–Magnetosphere Coupling" (Y. Kamide and J. A. Slavin, eds.), p. 519. Terra Sci. Publ. Co., Tokyo.

Fejer, B. G., Farley, D. T., Woodman, R. F., and Calderon, C. (1979). Dependence of equatorial F-region vertical drifts on season and solar cycle. *JGR, J. Geophys. Res.* **84**, 5792.

Fejer, B. G., Farley, D. T., Gonzales, C. A., Woodman, R. F., and Calderon, C. (1981). F-region east–west drifts at Jicamarca. *JGR, J. Geophys. Res.* **86**, 215.

Fejer, B. G., Larsen, M. F., and Farley, D. T. (1983). Equatorial disturbance dynamo electric fields. *Geophys. Res. Lett.* **10**, 537.

Forbes, J. M. (1981). The equatorial electrojet. *Rev. Geophys. Space Phys.* **19**, 469.

Forbes, J. M., and Lindzen, R. S. (1976). Atmospheric solar tides and their electrodynamic effects. II. The equatorial electrojet. *J. Atmos. Terr. Phys.* **38**, 911.

Gonzales, C. A., Kelley, M. C., Fejer, B. G., Vickrey, J. F., and Woodman, R. F. (1979). Equatorial electric fields during magnetically disturbed conditions. 2. Implications of simultaneous auroral and equatorial measurements. *JGR, J. Geophys. Res.* **84**, 5803.

Gonzales, C. A., Kelley, M. C., Behnke, R. A., Vickrey, J. F., Wand, R. H, and Holt, J. (1983). On the latitudinal variations of the ionospheric electric field during magnetospheric disturbances. *JGR, J. Geophys. Res.* **88**, 9135.

Haerendel, G., Bauer, O. H., Cakir, S., Foppl, H., Rieger, E., and Valenzuela, A. (1983). Coloured bubbles—an experiment for triggering equatorial spread F. *Eur. Space Agency* [*Spec. Publ.*] *ESA SP* **ESA SP-195**, 295–298.

Harris, I., and Priester, W. (1965). On the diurnal variation of the upper atmosphere. *J. Atmos. Sci.* **22**, 3.

Heelis, R. A., Kendall, P. C., Moffett, R. J., Windle, D. W., and Rishbeth, H. (1974). Electrical coupling of the E and F-regions and its effect on F-region drifts and winds. *Planet. Space Sci.* **22**, 743.

Kelley, M. C., Fejer, B. G., and Gonzales, C. A. (1979). An explanation for anomalous ionospheric electric fields associated with a northward turning of the interplanetary magnetic field. *Geophys. Res. Lett.* **6,** 301.

King-Hele, D. G. (1970). Average rotational speed of the upper atmosphere from changes in satellite orbits. *Space Res.* **10,** 537.

Kudeki, E., Fejer, B. G., Farley, D. T., and Ierkic, H. M. (1981). Interferometer studies of equatorial F-region irregularities and drifts. *Geophys. Res. Lett.* **8,** 377.

Lindzen, R. S. (1966). Crude estimate for the zonal velocity associated with the diurnal temperature oscillation in the thermosphere. *J. Geophys. Res.* **71,** 865.

Lindzen, R. S. (1967). Reconsideration of diurnal velocity oscillation in the thermosphere. *J. Geophys. Res.* **72,** 1591.

Matsushita, S. (1969). Dynamo currents, winds, and electric fields. *Radio Sci.* **4,** 771.

Maynard, N. C. (1967). Measurements of ionospheric currents off the coast of Peru. *J. Geophys. Res.* **72,** 1863.

Meriwether, J. W., Moody, J. W., Biondi, M. A., and Roble, R. G. (1986). Optical interferometric measurements of nighttime equatorial thermospheric winds at Arequipa, Peru. *JGR, J. Geophys. Res.* **91,** 5557.

Nishida, A. (1968). Geomagnetic D_P2 fluctuations and associated magnetospheric phenomena. *J. Geophys. Res.* **73,** 1795.

Nopper, R. W., and Carovillano, R. L. (1978). Polar equatorial coupling during magnetically active periods. *Geophys. Res. Lett.* **5,** 699.

Richmond, A. D. (1973a). Equatorial electrojet. I. Development of a model including winds and instabilities. *J. Atmos. Terr. Phys.* **35,** 1083.

Richmond, A. D. (1973b). Equatorial electrojet. II. Use of the model to study the equatorial ionosphere. *J. Atmos. Terr. Phys.* **35,** 1105.

Richmond, A. D., and Matsushita, S. (1975). Thermospheric response to a magnetic substorm. *JGR, J. Geophys. Res.* **80,** 2839.

Rishbeth, H. (1972). Superrotation of the upper atmosphere. *Rev. Geophys. Space Phys.* **10,** 799.

Rishbeth, H., and Garriott, O. K. (1969). "Introduction to Ionospheric Physics," Int. Geophys. Ser., Vol. 14. Academic Press, New York.

Shuman, B. M. (1970). Rocket measurement of the equatorial electrojet. *J. Geophys. Res.* **75,** 3889.

Sipler, D. P., and Biondi, M. A. (1978). Equatorial F-region neutral winds from nightglow OI 630.0 nm Doppler shifts. *Geophys. Res. Lett.* **5,** 373.

Sipler, D. P., Biondi, M. A., and Roble, R. G. (1983). F-region neutral winds and temperatures at equatorial latitudes: Measured and predicted behaviors during geomagnetically quiet conditions. *Planet. Space Sci.* **31,** 53.

Somayajulu, V. V., Reddy, C. A., and Viswanathan, K. S. (1985). Simultaneous electric field changes in the equatorial electrojet in phase with polar cusp latitude changes during a magnetic storm. *Geophys. Res. Lett.* **12,** 473.

Somayajulu, V. V., Reddy, C. A., and Viswanathan, K. S. (1987). Penetration of magnetospheric convective electric fields to the equatorial ionosphere during the substorm of March 22, 1979. *Geophys. Res. Lett.* **14,** 876.

Sugiura, M., and Cain, J. C. (1966). A model equatorial electrojet. *J. Geophys. Res.* **71,** 1869.

Untiedt, J. (1967). A model of the equatorial electrojet involving meridional currents. *J. Geophys. Res.* **72,** 5799.

Vasyliunas, V. M. (1972). The interrelationship of magnetospheric processes. *In* "Earth's Magnetospheric Processes" (B. M. McCormac, ed.), p. 29. Reidel, Boston, Massachusetts.

Wharton, L. E., Spencer, N. W., and Mayr, H. G. (1984). The earth's thermospheric superrotation from Dynamics Explorer 2. *Geophys. Res. Lett.* **11,** 531.

Woodman, R. F. (1970). Vertical drift velocities and east–west electric fields at the magnetic equator. *J. Geophys. Res.* **75,** 6249.

Chapter 4 | Equatorial Plasma Instabilities

In this chapter we study some of the plasma wave phenomena which occur at low latitudes in the earth's ionosphere. Most of the information we have concerning these processes comes from radio wave scattering or reflection experiments conducted from the surface of the earth. *In situ* measurements made with rocket- and satellite-borne sensors have also contributed significantly to our present understanding. As in the previous chapter, instabilities occurring in the F region are addressed first, followed by consideration of lower-altitude phenomena. Where convenient, M and m have been used to denote the ion and electron masses, respectively.

4.1 F-Region Plasma Instabilities: Observations

Plasma instability phenomena occurring in the equatorial F-region ionosphere are grouped under the generic name equatorial spread F (ESF). This stems from the earliest observations using ionosondes, which showed that on occasion the reflected echo did not display a well-behaved pattern but was "spread" in range or frequency. The phenomenon occurs primarily at night, although isolated daytime events occur (Woodman *et al.*, 1985). We now know that one name cannot do justice as a descriptor of all the processes which contribute to nonthermal equatorial F-region plasma scatter, but we will nonetheless stick to tradition. The modern era in ESF studies began on a "low" theoretical note in 1970 with publication of the first compilation of measurements made by the Jicamarca Radar Observatory in Peru (Farley *et al.*, 1970). The authors concluded that no theory published to date could explain the data. Fortunately, considerable progress has occurred since then, and the theory has kept pace with continuing additions and improvements in the data.

In the early 1970s, gray-scale radar maps came into vogue as a method of following the position and intensity of ESF plasma density irregularities. One of the more spectacular examples of the maps obtained at the Jicamarca Observatory is reproduced in Fig. 4.1 and a more typical map is reproduced in Fig. 4.2. The Jicamarca radar transmits nearly vertically, and at its 50-MHz frequency it is sensitive to backscatter from waves which satisfy the Bragg matching condition. This condition requires that $\mathbf{k}_r = \mathbf{k}_s + \mathbf{k}_m$ where $\mathbf{k}_r$ is the radar wave vector, $\mathbf{k}_s$ the scattered wave vector, and $\mathbf{k}_m$ the wave vector in the medium. (See Appendix A for more details.) Since $\mathbf{k}_s = -\mathbf{k}_r$ for backscatter, it follows that $\mathbf{k}_m = 2\mathbf{k}_r$. Thus the Jicamarca 50-MHz radar ($\lambda = 6$ m) detects only waves with vertical wave vectors corresponding to a 3-m wavelength. In the F region of the ionosphere (above 200 km), both figures display intense echoes over a 100-km altitude extent for much of the time. The gray scale shows the echo strength to be up to 50 dB higher than that caused by thermal fluctuations at the 3-m wavelength. This "thermal backscatter level" is the source of the incoherent scatter echoes often used to determine ionospheric parameters (see Appendix A). The thick irregularity layer moves up or down with time rather periodically in Fig. 4.1 and occasionally is interrupted by a period of intense backscatter signal which extends to very high altitudes. In Fig. 4.2 this happens only once, at about 2040 local time, while in the other example six such excursions are apparent.

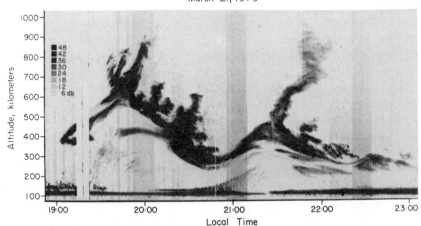

Fig. 4.1. Range–time–intensity map displaying the backscatter power at 3-m wavelengths measured at Jicamarca, Peru. The gray scale is decibels above the thermal noise level. [After Kelley *et al.* (1981). Reproduced with permission of the American Geophysical Union.]

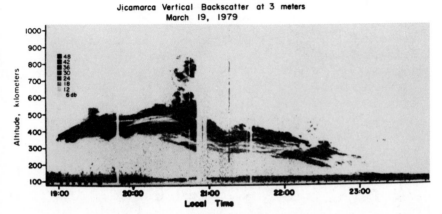

Fig. 4.2. Range–time–intensity map displaying the backscatter power at 3-m wavelengths measured at Jicamarca, Peru. The gray scale is decibels above the thermal noise level. [After Kelley *et al.* (1981). Reproduced with permission of the American Geophysical Union.]

These towering echoing features have been termed "plumes." As we shall see, this name is quite apt for structures which are probably the result of a convective instability driven by gravitational forces. The horizontal time scale is such that the mean horizontal plasma drift speed (125 m/s toward the east) multiplied by time yields the same horizontal distance scale as given in the vertical altitude scale. Thus the picture yields an accurate geometry of the ionospheric scatterers as viewed looking south antiparallel to the magnetic field if the irregularities are frozen into a medium which moves with constant velocity. Since the probing radar wavelength is very small compared to the scales involved in the phenomena, they can be considered only as tracers and care must be taken in the interpretation of the observed features in these maps. Nonetheless, the inspired interpretation made by Woodman and LaHoz (1976) that the plumes represent upwelling of plasma "bubbles" has stood the test of time and is consistent with the *in situ* data.

Several rockets and satellites have now penetrated both the irregularity layers and the plume structures. For such studies it has proved very useful to correlate the *in situ* probing with Jicamarca or with the scanning radar located on the island of Kwajalein in the South Pacific. An example of a radar map made at 0.96-m wavelength with the Altair radar on Kwajalein is presented in Fig. 4.3a, along with the plasma density profile made simultaneously on board the PLUMEX I rocket (Rino *et al.*, 1981). The profile is highly irregular throughout the rocket trajectory, but the most crucial observation is that the intensely scattering top, or "head," of the radar plume is colocated with a region of depleted plasma, a bubble. Another example correlating AE-E satellite plasma

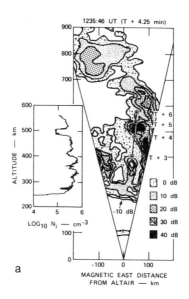

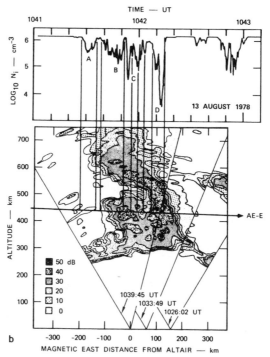

Fig. 4.3 (a) Simultaneous vertical rocket plasma density profile and backscatter map made with the Altair radar on the island of Kwajalein. Dots show the rocket trajectory. [After Rino *et al.* (1981). Reproduced with permission of the American Geophysical Union.] (b) Simultaneous horizontal satellite plasma density profile and backscatter map made with the Altair radar. [After Tsunoda *et al.* (1982). Reproduced with permission of the American Geophysical Union.]

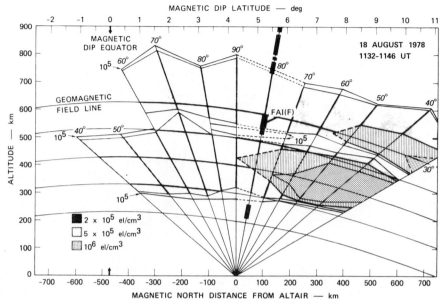

Fig. 4.4. Incoherent backscatter map of the equatorial zone during an equatorial spread F event. The depleted plasma region follows the contour of the magnetic field lines. [After Tsunoda (1980b). Reproduced with permission of Pergamon Press.]

density measurements with the Altair data is presented in Fig. 4.3b and shows that a horizontal satellite cut through a radar plume also indicates that it is colocated with a very structured region of plasma in which the mean density is much less than the value outside the volume of backscatter (Tsunoda *et al.*, 1982). As a final piece of radar data, Fig. 4.4 shows that the depleted regions are elongated along the direction of the magnetic field for hundreds of kilometers (Tsunoda, 1980b). The AE-E satellite *in situ* plasma drift detector has shown that the plasma inside a bubble typically moves upward and westward relative to the background medium. Since the plasma drifts eastward in the nighttime, the resulting plume structures are often "tilted" when displayed on a gray-scale map.

A wave number power spectrum of the plasma density measured along the PLUMEX I rocket trajectory is given in Fig. 4.5 (Kelley *et al.*, 1982). Such data help to bridge the gap in wavelength between the large-scale plumes and the small radar wavelength. These data show that spread F-related turbulent structures occur over seven orders of magnitude in spatial scale, from more than 10^5 m to less than 0.1 m.

Forward scatter or "scintillation" measurements have also provided informa-

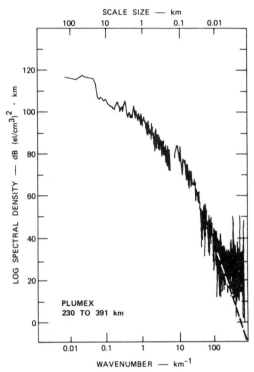

Fig. 4.5. Power spectrum of the plasma density detected along the rocket trajectory shown in Fig. 4.3. Note that wave number here actually means inverse wavelength which equals $k/2\pi$. [After Kelley *et al.* (1982). Reproduced with permission of the American Geophysical Union.]

tion on structure in the intermediate wavelength range (100–1000 m), since at the radio frequencies typically used for these experiments the scale of the irregularities which do the scattering is in that range (Basu *et al.*, 1978). In the example shown in Fig 4.6, the VHF scintillation intensity (lower trace) continues for a considerable time after the most intense 3-m backscatter ends. This shows that the large-scale features remain in the medium long after the smallest structures disappear. Decaying neutral fluid turbulence also behaves in this manner; that is, the shortest-scale features decay first when the driving forces are removed.

Although the high-altitude excursions of structured plasma are certainly spectacular, there is ample evidence that the process begins at low altitudes. Examples of measurements made below the F peak are presented in Figs. 4.7a and b. The former shows rocket data and therefore is a nearly vertical cut through the structures. The detrended plasma density profile in the upper panel is very structured and displays shocklike features with very steep edges. The spectrum of this

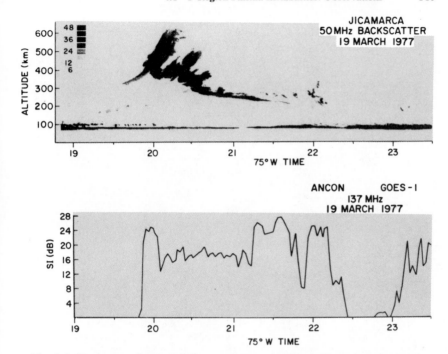

Fig. 4.6. Simultaneous Jicamarca backscatter power and VHF scintillation intensity due to irregularities in the same scattering volume. [After Basu *et al.* (1978). Reproduced with permission of the American Geophysical Union.]

data set is shown below it. A power law of the form k^{-n} describes the spectrum quite well for wave numbers larger than 6 km^{-1}. The index n has been determined in a number of experiments of this type and ranges from 2.1 to 2.7 for the one-dimensional measurement made by a spacecraft instrument when it traverses a fluctuating medium (see Appendix A). The satellite data shown in Fig. 4.7b are quite different. The waveform is nearly sinusoidal and, in fact, has been termed a bottomside sinusoidal structure (BSS) by Valladares *et al.* (1983). The spectrum is peaked near a 1-km wavelength in this event with a large spectral index n for large k values. The spectral peak for these *horizontal* cuts through the structured bottomside region ranges from 800 to 3000 m in the various examples reported. These data were not taken simultaneously, of course, but there are theoretical simulations which suggest that such an anisotropic behavior might occur (See Section 4.3).

Although not exhaustive, the data presented in these few figures are sufficient to motivate a discussion of plasma instability theory as it applies to ESF.

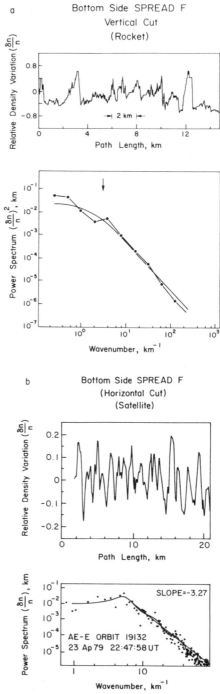

Fig. 4.7 (a) Detrended plasma density profile (top) and its power spectrum (bottom) obtained during bottomside spread F. The arrow shows the wave number corresponding to 2 km. [After Kelley *et al.* (1979). Reproduced with permission of the American Geophysical Union.] (b) Example of a bottomside sinusoidal structure detected by instruments on board the AE-E spacecraft. [After Valladares (1983).]

4.2   Development and Initiation of Equatorial Spread F

4.2.1   Linear Theory of the Rayleigh–Taylor Instability

Dungey (1956) first proposed the gravitational Rayleigh–Taylor (GRT) instability as the process driving ESF. This mechanism was temporarily rejected along with all the other candidate theories by Farley *et al.* (1970) since, as we shall see, it seemed capable only of generating structure on the bottomside of the F-region plasma density profile. The manner in which the GRT instability can cause irregularities to grow in the equatorial ionosphere is illustrated in Fig. 4.8a, using a two-dimensional model. Here the steep upward-directed gradient which develops on the bottomside of the nighttime F layer (see Chapter 3) is approximated by a step function. The density is equal to n_1 above the interface and vanishes below. The gravitational force is downward, antiparallel to the density gradient, and the magnetic field is horizontal, into the paper. An initial small sinusoidal perturbation is also illustrated, and we assume for the present that the plasma is nearly collisionless, that is, that κ is large. From (2.36c) we can now determine the electrical current which flows by considering the ion and electron velocities due to the pressure gradients and gravity. First, we note that the pressure-driven current does not create any perturbation electric fields since the current is everywhere perpendicular to the density gradient. The current thus flows parallel to the modulated density pattern with no divergence.

Turning to the gravitational term in (2.36c), the species velocity is proportional to its mass, and hence the ion term dominates. A net current flows in the x direction with magnitude

$$J_x = nMg/B$$

Since the current is in the $\mathbf{g} \times \mathbf{B}$ direction, which is strictly horizontal, J_x will be large when n is large and small when n is small. There is thus a divergence and charge will pile up on the edges of the small initial perturbation. As a result, perturbation electric fields ($\delta \mathbf{E}$) build up in the directions shown. These fields in turn cause an upward $\delta \mathbf{E} \times \mathbf{B}$ drift of the ions and electrons in the region of plasma depletion and a downward drift in the region where the density is enhanced. Lower (higher) density plasma is therefore advected upward (downward) in the depleted region, creating a larger perturbation, and the system is unstable. An analogous hydrodynamic phenomenon is illustrated in the series of sketches in Fig. 4.8b. These have been derived from photographs of the hydrodynamic Rayleigh–Taylor instability when a light fluid supports a heavier fluid against gravity. Initial small oscillations in the surface grow "in place," pushing the lighter fluid upward. In the ionospheric case the "light fluid" is the magnetic field, which prevents the plasma from freely falling. The system is unstable when $\mathbf{g}$ and ∇n are oppositely directed. We include the $\mathbf{E}_0$ term below.

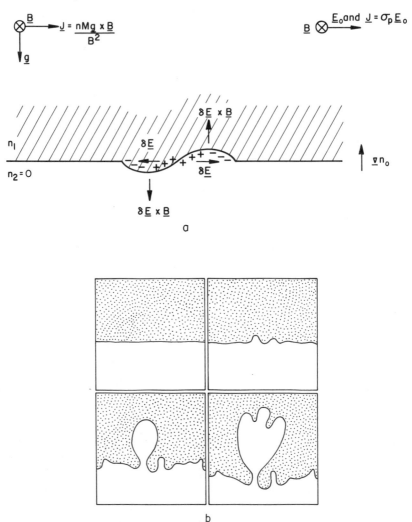

Fig. 4.8. (a) Schematic diagram of the plasma analog of the Rayleigh–Taylor instability in the equatorial geometry. (b) Sequential sketches made from photos of the hydrodynamic Rayleigh–Taylor instability. A heavy fluid is initially supported by a transparent lighter fluid.

We now calculate the growth rate of the RT instability assuming a small initial perturbation in plasma density and electric field. Inertial terms are dropped, which corresponds to the condition $\nu_{in} \gg \partial/\partial t$. This approximation breaks down at high altitudes but is certainly valid on the bottomside and up to 450 km or so. We also follow tradition and ignore viscosity, which may not be valid. After some manipulation of (2.34), the steady velocity of each species under the influence of electric and gravitational fields is given by

$$\mathbf{V}_j = [1/(1 + \kappa_j^2)]\mathbf{W}_j + [\kappa_j/(1 + \kappa_j^2)][\mathbf{W}_j \times \hat{B}] \quad (4.1)$$

where

$$\mathbf{W}_j = b_j\mathbf{E} + \mathbf{g}/\nu_{jn} - D_j\nabla n/n \quad (4.2)$$

For the electrons κ_e is so large that the second term dominates in (4.1) and

$$\mathbf{V}_e = \mathbf{E} \times \mathbf{B}/B^2 - (m/eB^2)(\mathbf{g} \times \mathbf{B}) + (k_B T_e/eB^2 n)(\nabla n \times \mathbf{B}) \quad (4.3)$$

For the ions $\kappa_i \gg 1$ in the F region but it is not so great that the first term in (4.1) can be ignored, as was the case for the electrons. For the F region that term can be written as $\mathbf{W}_i/\kappa_i^2$ and we have

$$\mathbf{V}_i = (b_i/\kappa_i^2)\mathbf{E} + \mathbf{g}/\nu_{in}\kappa_i^2 - (D_i/\kappa_i^2)(\nabla n/n) + \mathbf{E} \times \mathbf{B}/B^2$$
$$+ (M/eB^2)(\mathbf{g} \times \mathbf{B}) - (k_B T_i/eB^2 n)(\nabla n \times \mathbf{B}) \quad (4.4)$$

The electron mass is so small that the $\mathbf{g} \times \mathbf{B}$ term in (4.3) can be dropped. Likewise, the second and third terms in the right-hand side of (4.4) are small in the ion case due to the fact that $\kappa_i \gg 1$. Now, since the $\mathbf{E} \times \mathbf{B}$ term is identical for ions and electrons, no current flows due to those terms and $\mathbf{J}$ is given by

$$\mathbf{J} = ne(\mathbf{V}_i - \mathbf{V}_e)$$
$$= \sigma_P\mathbf{E} + (ne/\Omega_i)\mathbf{g} \times \hat{B} - (k_B/B^2)(T_i + T_e)(\nabla n \times \mathbf{B}) \quad (4.5)$$

Here we have used the fact that for large κ_i, $\sigma_P = neb_i/\kappa_i^2$. The gravity term also is rewritten in terms of Ω_i. Notice that the gravitational current flows even in a collisionless plasma, while the electric field term exists only if $\sigma_P \neq 0$, that is, in a partially collisional plasma.

We now study the linear stability of a vertically stratified equatorial F layer under only the influence of gravity, that is, the pure Rayleigh–Taylor case. We set $\mathbf{E}_0 = 0$ but retain a first-order electric field perturbation $\delta\mathbf{E}$ if it arises. The continuity and current divergence equations from Chapter 2 will be used in the analysis. Ignoring production and loss, which is reasonable in the postsunset time period when the F layer is very high, the continuity equation is

$$\partial n/\partial t + \mathbf{V} \cdot \nabla n + n(\nabla \cdot \mathbf{V}) = 0 \quad (4.6)$$

where for $M \gg m$ the plasma velocity $\mathbf{V}$ may be approximated by $\mathbf{V}_i$. First consider the "compressibility" term $(\nabla \cdot \mathbf{V})$. From (4.4) with $\mathbf{E} = 0$ and κ_i large,

$$\nabla \cdot \{(M/eB^2)(\mathbf{g} \times \mathbf{B}) - (k_B T_i/enB^2)(\nabla n \times \mathbf{B})\} = 0 \quad (4.7)$$

since g and B do not vary in the $\mathbf{g} \times \mathbf{B}$ direction, and since we also have $\nabla \cdot (\nabla n \times \mathbf{B}) = 0$ and $(\nabla n \times \mathbf{B}) \cdot \nabla n = 0$. Equation (4.7) states that the plasma flow is incompressible. This is usually a good approximation and is often taken to be valid *a priori* for any ionospheric F-region calculation. However, care must be taken in applying this result since it is not a fundamental principle and must be checked in each case. It is certainly not true in the E region, where compressibility plays an important role in the formation of images of F-region phenomena (see Chapter 8). Setting $(\nabla \cdot \mathbf{V}) = 0$, the equations we shall linearize are

$$\partial n/\partial t + \mathbf{V}\cdot\nabla n = 0 \tag{4.8a}$$

$$\nabla\cdot\mathbf{J} = 0 \tag{4.8b}$$

To study the electrostatic instability of these equations in the presence of a vertical zero-order density gradient, we can write the electric potential and the plasma density as

$$\phi = \phi' e^{i(\omega t - kx)} \tag{4.9a}$$

$$n = n_0(z) + \delta n e^{i(\omega t - kx)} \tag{4.9b}$$

where the initial perturbation propagates in the x direction. Note that we have assumed charge neutrality so that $n_e = n_i = n$. Using (4.5) with $\mathbf{E}$ replaced by $\delta\mathbf{E}$, (4.8b) becomes

$$\nabla\cdot[(ne/\Omega_i)\mathbf{g}\times\hat{B} + (ne^2\nu_{in}/M\Omega_i^2)\delta\mathbf{E}] = 0 \tag{4.10}$$

where we have used the fact that $\nabla\cdot(\nabla n\times\mathbf{B}) = 0$ to set the divergence of the pressure-driven current equal to zero. Here $\delta\mathbf{E}$ is the perturbation electric field associated with the potential ϕ and we have substituted (2.40b) for σ_P in the F region. Since $\delta\mathbf{E} = -\nabla\phi$, the vector inside the square bracket in the equation above only has an x component and taking the x derivative yields

$$(eg/\Omega_i)(\partial n/\partial x) - (e^2\nu_{in}/M\Omega_i^2)(\partial n/\partial x)(\partial\phi/\partial x) - (ne^2\nu_{in}/M\Omega_i^2)(\partial^2\phi/\partial x^2) = 0$$

The second term is of second order, and hence the linear form of this equation is

$$(eg/\Omega_i)(\partial n/\partial x) - (ne^2\nu_{in}/M\Omega_i^2)(\partial^2\phi/\partial x^2) = 0 \tag{4.11a}$$

Making the substitutions

$$P = M\nu_{in}/B^2, \qquad Q = mg/B$$

(4.11a) becomes

$$Q(\partial n/\partial x) - Pn(\partial^2\phi/\partial x^2) = 0 \tag{4.11b}$$

Turning now to the continuity equation (4.8a) we have

$$(\partial n/\partial t) + V_x(\partial n/\partial x) + V_z(\partial n/\partial z) = 0$$

The two velocity components may be obtained from (4.4) if we remember that the electric field $\delta\mathbf{E}$ is of first order, κ_i is large, and the pressure-driven velocity does not contribute to $\mathbf{V} \cdot \nabla n$. Since $\partial n/\partial x$ is of first order, only the zero-order V_x contributes to the second term. Hence, $V_x = Mg/eB = Q/e$. In the third term, $\partial n/\partial z$ is of zero order due to the inclusion of a vertical density gradient in the plasma. We must then include the first-order vertical velocity given by $V_z = \delta E_x/B$. The linearized continuity equation is therefore

$$\partial n/\partial t + (Q/e)(\partial n/\partial x) - (1/B)(\partial \phi/\partial x)(\partial n/\partial z) = 0 \qquad (4.12)$$

Using the plane wave solutions, (4.11b) and (4.12) may be written

$$-ikQ \; \delta n + n_0 k^2 P \phi' = 0 \qquad (4.13a)$$

$$(i\omega - ikQ/e) \; \delta n + (ik/B)(\partial n_0/\partial z)\phi' = 0 \qquad (4.13b)$$

where n_0 and dn/dz are evaluated at some reference altitude of interest, for instance, $n_0 = n_0(z_0)$. These are two equations in two unknowns, δn and ϕ', and may be solved by setting the determinant of coefficients equal to zero. This yields the dispersion relation

$$\omega = (kQ/e) - i(g/\nu_{\text{in}})(1/n_0)(\partial n_0/\partial z) \qquad (4.14)$$

The real part of ω, ω_r, shows that the plane waves propagate eastward with phase velocity V_ϕ given by

$$V_\phi = \omega_r/k = Q/e = Mg/eB \qquad (4.15a)$$

For an atomic oxygen plasma at the equator $V_\phi \approx 6$ cm/s, which is quite small. The imaginary part of ω is

$$\omega_i = -(g/\nu_{\text{in}})[(1/n_0)(\partial n_0/\partial z)] \qquad (4.15b)$$

When $\partial n_0/\partial z$ is positive (corresponding here to the density gradient antiparallel to $\mathbf{g}$), ω_i is negative and

$$e^{i\omega t} = e^{i\omega_r t} e^{\gamma t}$$

where γ is positive and hence yields a growing solution. The number γ is the growth rate of the instability and is given by

$$\gamma = g/L\nu_{\text{in}} \qquad (4.16)$$

where L is the inverse gradient scale length ($L = [(1/n_0) \; dn_0/dz]^{-1}$).

Although quite simple, this result offers explanations for a number of properties of ESF. First, in the initial development of spread F there is a strong tendency for VHF radar to obtain echoes confined to the height range where the density gradient is upward. In fact, the early Jicamarca study showed that the onset of nonthermal backscatter usually began at a density level about 1% of the plasma density at the F peak. Several rockets have been flown during bottomside ESF at times when no radar echoes were obtained above the F peak, and indeed intense irregularities were found below the peak but a smooth profile was found above. These cases are thus in agreement with the linear theory, in that the latter predicts instability only when **g** is antiparallel to ∇n. Another feature predicted by the theory is a height dependence for γ due to the collision frequency term in the denominator of the growth rate. The higher the layer, the lower ν_{in} and the larger the instability growth rate. As mentioned earlier, Farley *et al.* (1970) noted a strong tendency for ESF to be generated when the layer was at a high altitude, where the collision term is small. Notice that the plume structure in Fig. 4.2 and the initial plume in Fig. 4.1 were both generated when the echoing layer was at its peak altitude. This also suggests that the most spectacular effects occur when the layer is high.

4.2.2   The Generalized Rayleigh–Taylor Process

Gravity is not the only destabilizing influence in the equatorial ionosphere. If we return to (4.5) we can include the effect of the ambient electric field $\mathbf{E}_0$ as well as the neutral wind. The latter can easily be included by remembering (see Chapter 3) that when electric fields and winds both exist, the current density is $\boldsymbol{\sigma} \cdot \mathbf{E}_0'$, where $\mathbf{E}_0' = \mathbf{E}_0 + \mathbf{U} \times \mathbf{B}$. First we note that for a zonally eastward electric field, the zero-order Pederson current is in the same direction ($\mathbf{g} \times \mathbf{B}$) as the gravity-driven current. The derivation outlined above, which considered only gravity, can be generalized to include the effect of an electric field by replacing g/ν_{in} with $g/\nu_{in} + E_{x0}'/B$ in the growth rate, where E_{x0}' is the zonal component of the electric field in the neutral frame of reference. A zonally eastward electric field drives a Pedersen current parallel to the boundary between plasma and vacuum. Any undulation of the boundary will intercept charge just as in the gravitational case and cause the perturbation to grow. A zonally westward field will be stabilizing on the bottomside since in this case $(E_{x0}') \, \partial n_0/\partial z$ is negative. The general condition for instability is that the $\mathbf{E}_0' \times \mathbf{B}$ direction be parallel to the plasma density gradient. As discussed in the previous chapter, the zonal electric field component at the equator often increases to a large eastward value just after sunset, driving the F layer to very high altitudes. This uplift contributes in two ways to the destabilization of the plasma. Not only is the electric field in the right direction for instability but also the g/ν in term becomes large due to the high altitude of the layer. The growth rates of the gravitational and electric field-driven processes are plotted as a function of height in Fig. 4.9. For a 0.5-mV/m electric field the two sources of instability are equal at an altitude of 360 km. The gravitational term dominates above this height and increases exponentially with altitude.

Since a large-scale neutral wind is usually horizontal, $(\mathbf{U} \times \mathbf{B}) \times \mathbf{B}$ is also horizontal and hence can have no component parallel to ∇n if the ionosphere is vertically stratified. Hence E_{x0}' above is due entirely to the zonal electric field,

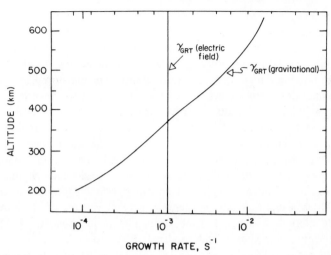

Fig. 4.9. Linear growth rates for the gravitational and electric field-driven Rayleigh–Taylor instabilities in the equatorial ionosphere for typical conditions. [After Kelley *et al.* (1979). Reproduced with permission of the American Geophysical Union.]

E_{x0}. However, another term can be added to the linear growth rate by considering the possibility of a horizontal component of ∇n. In fact, since the layer does change height during the course of any given night, there is reason to expect that the layer can be tilted with respect to the vertical. In such a case, the linear growth rate may be written in the following form:

$$\gamma_{RT} = \frac{E'_{x0}}{LB} \cos \sigma + \frac{g}{\nu_{in}L} \cos \sigma + \frac{E'_{z0} + uB}{LB} \sin \sigma \qquad (4.17)$$

where L is the gradient scale length on the bottomside, σ is the tilt angle, which is measured positive from east toward the zenith, the z axis is upward, the x axis is eastward, and we ignore any dip angle of the magnetic field. This is the growth rate of the generalized Rayleigh–Taylor instability as it is applied to the equatorial ionosphere. Kelley *et al.* (1981), for example, used it to interpret Fig. 4.1 and concluded that the preference for plumes to be generated during "negative" slopes of the Jicamarca radar profile was due to the contribution of an eastward neutral wind u to the instability growth rate. That is, if the ionosphere is tilted such that the neutral wind blows toward the east into a region of increasing plasma density, the electric field in the neutral frame ($\mathbf{E}'_0$) is upward and $\mathbf{E}'_0 \times \mathbf{B}$ is antiparallel to the density gradient, a stable configuration. If the wind blows antiparallel to the plasma gradient, which is believed to occur during the negative slopes in Fig. 4.1, it is destabilizing. Tsunoda (1981) has shown quite clearly using the scanning radar at Kwajalein that wind effects are important. Since the electric field is generated by both local and remote wind fields (see the previous chapter) the generalized growth rate is quite a complicated function.

Linear instability theory may also be used to explain the day–night asymmetry of ESF since the instability is influenced by the effects of the E region on the charge buildup which leads to the instability. The equatorial ionosphere is not two-dimensional since electric currents can easily flow down the magnetic field lines if there is a conducting path through the E region. The physics is identical to that discussed in Chapter 3, which showed that the F-region dynamo was suppressed during the day, when the low-altitude conductivity dominates. During the daytime, any perturbation electric fields due to the instability are shorted out in the E region. This explains why spread F does not begin until well after sunset, even though **g** and ∇n are antiparallel on the bottomside during the day as well as at night. A full explanation of the morphology of ESF thus must include the diurnal, seasonal, and solar cycle effects on the electric field, on the neutral density, temperature, and wind patterns, and on the conductivities of the E and F regions. Developments in the ability to create realistic models of the equatorial ionosphere are occurring rapidly and many of the effects mentioned above can be included in the near future (see Chapter 3).

4.2.3 The Seeding of Equatorial Spread F

In addition to the factors mentioned above, there is very likely an additional random factor in the occurrence of ESF which may forever hinder attempts to develop predictive capabilities. This stems from the ability of internal gravity waves to organize the equatorial plasma into high- and low-density regions with the same horizontal wavelength as the gravity wave. Once such a perturbation occurs, the generalized RT instability can take over and cause the oscillation to grow. Thus, the wave can act as a seed for the instability and greatly decrease the time needed to develop a large-amplitude disturbance. An example of the organization of plasma into structures with scales of several hundred kilometers is given in Fig. 4.10. The data are from a transequatorial HF radio propagation experiment. If the ionosphere were uniform, refraction would yield a single "great circle" path with some minimum time delay. When the plasma density has east–west structure, however, additional, larger time delays occur which can be used to characterize the structure. The data show that paths other than the great circle route occur in regular intervals which are found to be typical of internal gravity wave wavelengths. One explanation for this effect, proposed by Beer (1973), was that the temperature variation in the gravity wave would change the recombination rate of the ionospheric plasma and thereby create a density modulation. The effect would be particularly strong if the plasma drift speed matched the phase velocity of the wave, since in such a case the perturbation would always act in the same sense on a given parcel of plasma. This is a varia-

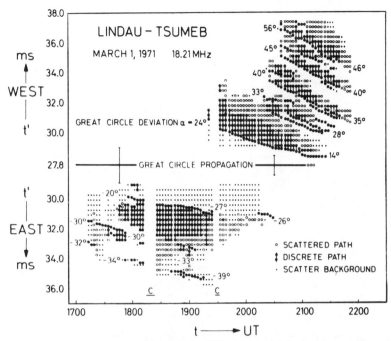

Fig. 4.10. Forward scatter measurements in the African sector showing regular wavelike regions of enhanced plasma density. [After Röttger (1973). Reproduced with permission of Pergamon Press.]

tion on the so-called spatial resonance theory for ESF suggested first by Whitehead (1971). Wavelike modulations of ordinary clouds are created by this effect as the local temperature is raised above or lowered below the dew point by a gravity wave.

Klostermeyer (1978) also appealed to the spatial resonance effect, but pointed out that the internal wind field δU in a gravity wave must also drive electrical currents δJ. Due to the finite wavelength of the gravity waves, the associated winds are not uniform in space. The divergence of the wind-driven electrical current is therefore not zero and an electric field δE must build up with a wavelength equal to that of the gravity wave. This process is illustrated in Fig. 4.11. If the E region is a perfect insulator, no field-aligned currents can flow and the $\nabla \cdot J = 0$ equation may be replaced by the more restrictive equation $\delta J = 0$, where δJ has components due to the gravity wave wind and the electric field. This requirement has been used to generate the diagram. The resulting electric field pattern has alternating eastward and westward components, which, due to the $\delta E \times B$ drift, will cause portions of the ionosphere to rise and portions to fall. Now, if the plasma has a zero-order vertical density gradient, the $V \cdot \nabla n$

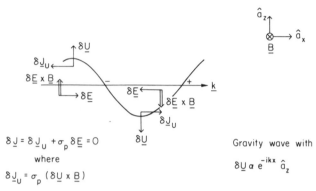

Fig. 4.11. Schematic diagram showing how the perturbation winds in a gravity wave generate electric fields.

term in the continuity equation will lead to a sinusoidal density pattern with the same horizontal wavelength as the gravity wave. These east–west oscillations may then act as an initial perturbation which is amplified by the Rayleigh–Taylor process. An additional attractive feature of the gravity wave theory is that after sunset the ionosphere begins to descend. This is also the direction of the phase velocity of a gravity wave carrying energy upward from below (remember that an upward gravity wave energy flux has a downward phase velocity). Spatial resonance could thus occur with a zonally eastward-propagating wave, since it is well known that the plasma also drifts in the eastward direction (see Chapter 3). Furthermore, the magnitudes of the typical eastward and downward plasma drifts are roughly comparable to the typical phase velocities of large-scale gravity waves in the upper atmosphere. A more quantitative analysis of these factors proceeds as follows.

If we assume a nonconducting E region and a gravity wave at F-region heights, then, as argued above, the divergence of the wind-driven current will set up polarization charges and an electric field such that

$$\delta \mathbf{J} = \sigma_P(\delta \mathbf{E} + \delta \mathbf{U} \times \mathbf{B}) = 0$$

and hence

$$\delta \mathbf{E} = -(\delta \mathbf{U} \times \mathbf{B})$$

where σ_P is the Pedersen conductivity, $\delta \mathbf{E}$ the perturbation electric field in the earth-fixed frame, $\delta \mathbf{U}$ the perturbation wind velocity due to the gravity wave, and $\mathbf{B}$ the magnetic field. The net effect of the dynamo is to cause the plasma and the neutral gas to move together in the gravity wave wind field since $\delta \mathbf{V} = \delta \mathbf{E} \times \mathbf{B}/B^2 = \delta \mathbf{U}$. To determine quantitatively the effect of such a perturbation, we first consider the case of perfect spatial resonance in which the mean unperturbed plasma is at rest in a fixed phase front pattern associated with the gravity wave; that is, the plasma drift matches the wave phase velocity exactly. Since the plasma behaves as an incompressible fluid, the density variation with time at a fixed point in space is given by the advective derivative

$$\partial n/\partial t = -\delta \mathbf{V} \cdot \nabla n \tag{4.18}$$

where n is the plasma density and $\delta \mathbf{V}$ the perturbation plasma velocity. Here we again use a right-hand coordinate system with x positive in the eastward direction and y positive toward north. To proceed, we consider gravity waves propagating exactly horizontally (eastward), which is reasonably valid for the long-period waves of interest here, and thus we set $\delta \mathbf{U} = w_0 e^{i(\omega t - kx)} \hat{a}_z$. For perfect spatial resonance the perturbation plasma velocity in the plasma reference frame is time independent (the wave frequency is Doppler shifted to zero) and (4.18) becomes

$$\partial n/\partial t = -w_0 e^{-ikx}(\partial n/\partial z) \tag{4.19}$$

For an initial density profile of the form $n_0 e^{z/L}$ the solution to (4.19) is

$$n(x, z, t) = n_0 e^{z/L} e^{\gamma(x)t} \tag{4.20a}$$

where

$$\gamma(x) = -(w_0/L)e^{-ikx} \tag{4.20b}$$

These equations describe a spatial pattern of rising and falling density contours in the plasma reference frame. For $w_0 = 2$ m/s and $L = 20$ km, the peak value of γ is 10^{-4} s^{-1}. This "perfect" spatial resonance theory gives an infinite response as $t \to \infty$, while a perturbation of 5% occurs in 550 s. Note, however, that although an infinite response eventually occurs, the perturbation plasma ion velocity due to the presence of the wave *never exceeds* the amplitude of the wave-induced neutral velocity. This constraint severely limits the altitude modulation of the F layer due to a pure gravity wave-driven process. In Klostermeyer's (1978) nonlinear approach to this problem, an anomalous diffusion due to plasma microinstabilities was modeled, and the saturation amplitude of the density perturbation for perfect spatial resonance was found to be at most about one order of magnitude (e.g., see his Figure 4). This corresponds to an uplift of about 3 plasma scale heights or about 50 km. It is now clear that much larger perturbations occur (e.g., see Fig. 4.1) and a pure gravity wave theory explaining ESF is not tenable.

This simple model can be extended to address the question of how "resonant" a wave must be to create a significant effect. It is crucial that the vertical velocities match, but Kelley *et al.* (1981) have shown that the horizontal phase velocity of the gravity wave has to be within only about 100 m/s of the horizontal plasma drift velocity to produce a 5% seeding effect in one-half of a wave period. This is not a very severe constraint and makes it very plausible that gravity waves can

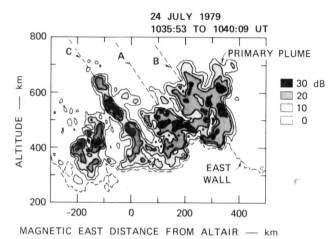

Fig. 4.12. Backscatter map showing a primary plume and three other plumelike features on the west wall of a plasma upwelling. [After Tsunoda (1983). Reproduced with permission of the American Geophysical Union.]

be responsible for seeding spread F. On the other hand, if we require that a large effect be due to spatial resonance, very accurate velocity matching must occur. For example, if an order-of-magnitude change is required in the electron density, the two horizontal velocities must be matched to within 3 m/s.

Returning again to Fig. 4.1, Kelley *et al.* (1981) concluded that a gravity wave could have seeded the event but that the large uplifts (~ 80 m/s) observed must have resulted from an amplification of the induced perturbation by the Rayleigh–Taylor process. They also argued that the multiple plumes located on the descending phase of the oscillation were due to a neutral wind blowing eastward across a structure tilted such that it had a westward density gradient. Tsunoda (1983) has shown this effect convincingly, using the scanning Altair radar. In the data shown in Fig. 4.12, for example, several plumes grow from a structure tilted in just this fashion. (Note that for $\mathbf{U}$ east, $\mathbf{E}' \times \mathbf{B}$ has a component toward the west, which in turn then has a component parallel to ∇n in the tilted ionosphere. This is the instability criterion for a wind-driven process.)

Prakash and Pandy (1980) have pointed out that the perturbation electric fields which develop in response to the wave-driven electric current will be shorted out if magnetic field lines can link the regions of positive and negative charge. For pure (magnetic) zonal propagation this will not occur, but if a finite meridional component of the gravity wave vector exists, the seeding process may be limited to locations very near the dip equator. Those gravity waves generated as a result of auroral activity [large-scale traveling ionospheric disturbances (TIDs)] propagate from the poles to the equator, while medium-scale TIDs may propagate in any direction. Thus, gravity waves generated in the auroral zone may be less effective for seeding equatorial spread F than waves generated by tropospheric sources such as thunderstorms and frontal systems.

Finally, we note that a small slippage between the neutral zero-order wind and plasma drift is important for the spatial resonance mechanism to be effective. If there was no difference between the two velocities, then spatial resonance would occur at an atmospheric "critical level" where the horizontal wave phase velocity and the neutral wind speed are equal. The vertical component of the gravity wave's phase velocity would then be zero, and the vertical component of the plasma drift could not be matched (Kelley *et al.*, 1981).

4.2.4 Role of Velocity Shear in Equatorial Spread F

The question of gravity wave seeding and the general problem of the very long wavelengths which occur in equatorial spread F must be viewed in a context which includes the velocity shear which we now know exists in and below the F layer (e.g., see Chapter 3). Extending earlier work dealing with the $\mathbf{E} \times \mathbf{B}$ instability (Perkins and Doles, 1975), Satyanarayana *et al.* (1984) showed that velocity shear acts to stabilize the Rayleigh–Taylor instability. This is illustrated in Fig. 4.13, where the normalized growth rate $\hat{\gamma} = \gamma \nu_{in} L/g$ is plotted versus the normalized wave number $\hat{k} = 2\pi L/\lambda$ for various values of the shear parameter

$$\hat{S} = \frac{V_0/L}{(g/L)^{1/2}}$$

In their analysis V_0 is the maximum in the plasma flow velocity and the scale length L characterizes both the velocity shear and the density gradient scale length. $\hat{S} = 0$ corresponds to no shear and the corresponding curve yields the dependence of the Rayleigh–Taylor growth rate on $\hat{k}$. This result differs from the simple analysis given above, for which $\hat{\gamma} = 1$ for all values of k until diffusion

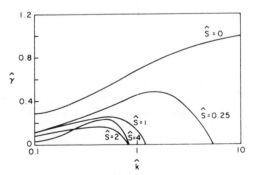

Fig. 4.13. Normalized growth rate plotted versus normalized wave number for various shear strengths. [After Satyanarayana *et al.* (1984). Reproduced with permission of the American Geophysical Union.]

sets in. This more accurate linear theory shows that $\hat{\gamma}$ equals unity only for $\hat{k} \geq$ 10 or $\lambda \geq 2\pi L/10 \approx 0.6L$. Since $L = 15$ km, the pure RT growth rate is small for $\lambda \geq 10$ km, where much of the largest perturbation occurs. For $\hat{S} > 0$ the growth rate decreases at all $\hat{k}$, *but* a peak in $\hat{\gamma}$ begins to evolve for small values of $\hat{k}$. This means that velocity shear is stabilizing but that it does push the most unstable waves to long wavelengths. It is interesting to note that for $\hat{S} > 2$ the growth rate begins to rise again, indicating that a Kelvin–Helmholtz instability has set in. The parameter $\hat{S}$ may be written

$$(\hat{S})^{-1} = \left[\frac{(g/\rho)\, \partial p/\partial z}{(\partial V/\partial z)^2} \right]^{1/2} = (R'_i)^{1/2}$$

where we have defined a plasma Richardson number R'_i which is analogous to the atmospheric Richardson number. It is well known in atmospheric science that a shear is unstable for $R_i < 1/4$. This corresponds to $\hat{S} > 2$. However, very intense shears would be required to meet this criterion. For example, if $L = 15$ km and $\hat{S} = 2$, $V_0 = 2(gL)^{1/2} = 760$ m/s, which is much larger than the typical plasma velocity. The shear for $\hat{S} = 2$ is thus

$$V_0/L = 5 \times 10^{-2} \text{ s}^{-1}$$

which is considerably larger than the observed large-scale vertical shears reported in the literature, although it may be that the observations are all made after some limit has been set on the process, such as might occur at marginal stability.

Kelley *et al.* (1986) turned this argument around and asked the following question. Given the observed shears, what would $\hat{k}$ be for maximum growth? They found the expected values of $\hat{k}$ to be in the range 0.5–2, which corresponds to λ values in the range πL to $4\pi L$. Again for $L = 15$ km, λ ranges from 45 to 180 km. These wavelengths are much shorter than the outer scale limit for the horizontal spread F structure reported by Röttger (1973) and indicated by the studies of Kelley *et al.* (1981, 1986) and Tsunoda (1983). However, these values of $\hat{k}$ are in good agreement with multiple plume spacings reported in the same three papers (see also Figs. 4.1 and 4.12). It thus may well be the case (Kelley *et al.*, 1986) that gravity waves determine the outer horizontal scale for equatorial spread F by preferentially seeding a region of k space and creating perturbations with relatively low growth rates. Velocity shear and neutral winds then combine to structure the westward-directed horizontal density gradients, creating the observed multiple plumes.

4.2.5 Summary

To summarize, it seems clear that the generalized Rayleigh–Taylor instability plays a crucial role in the production of equatorial spread F and that each of the destabilizing terms can be important at times. In addition, the initial conditions which lead to an event on a given night depend in a complicated way on the state of the ionosphere and the neutral atmosphere and on the patterns of gravity waves in the upper atmosphere. Finally, the preferred scale for the generation of multiple plumes may depend on the velocity shear which occurs in the bottomside of the F layer.

4.3 Nonlinear Theories of ESF

4.3.1 Two-Dimensional Computer Simulations

Considerable effort has gone into the development of computer simulations of the Rayleigh–Taylor instability. To describe this work we temporarily change back to a coordinate system in which $\mathbf{B}$ is in the z direction, since the simulations have used this orientation. The set of equations which are solved in the two dimensions perpendicular to $\mathbf{B} = B\hat{a}_z$ is a subset of the full governing equations. They consist of two continuity equations for the electrons and ions,

$$(\partial n_j/\partial t) + \nabla \cdot (n_j \mathbf{V}_j) = -R(n_j - n_{j0}) \tag{4.21a}$$

the electron velocity equation with $\kappa_e \gg 1$,

$$\mathbf{V}_e = \frac{\mathbf{E} \times \hat{a}_z}{B} \tag{4.21b}$$

the ion velocity equation for intermediate κ_i,

$$\mathbf{V}_i = (1/B)[(M/e)\mathbf{g} + \mathbf{E}] \times \hat{a}_z + (\nu_{in}M/eB^2)[(M/e)\mathbf{g} + \mathbf{E}] \tag{4.21c}$$

the charge continuity equation,

$$\nabla \cdot \mathbf{J} = 0 = \nabla \cdot (n_i e \mathbf{V}_i - n_e e \mathbf{V}_e) \tag{4.21d}$$

and

$$\mathbf{E} = -\nabla \phi \tag{4.21e}$$

R is the recombination rate. In obtaining these equations we have assumed there is no neutral wind, we have neglected the inertial terms and the pressure-driven terms in the equations for conservation of momentum, and we have assumed that the $\mathbf{g} \times \mathbf{B}$ electron velocity term is small due to the small electron–ion mass ratio.

The electrostatic potential ϕ is divided into a zero-order term ϕ_0 and a perturbation term ϕ'. If we require the zero-order ion velocity to be zero, then we must have $\nabla \phi_0 = Mg/e$. The electron continuity equation evaluated in a frame of reference moving with the $\mathbf{E}_0 \times \mathbf{B} = (M/e)\mathbf{g} \times \mathbf{B}$ velocity is

$$\partial n/\partial t - (1/B)(\nabla \phi' \times \hat{a}_z) \cdot \nabla n = -R(n - n_0) \tag{4.22}$$

where we have dropped the subscript on the n due to the quasi-neutrality condition $n_e \approx n_i \approx n$. The charge continuity equation (4.21d) is evaluated using (4.21b, c) and the condition $\nabla\phi_0 = Mg/e$, giving

$$\nabla \cdot (\nu_{in} n \nabla\phi') - (\mathbf{g} \times \mathbf{B}) \cdot \nabla n = 0 \tag{4.23}$$

If we set $R = 0$ and linearize the density terms in (4.22) and (4.23) and appropriately convert to the other coordinate system, we recover the dispersion relation (4.14) with $Q = 0$, which is a result of the change of reference frames made here. Equations (4.22) and (4.23) are the two equations which have been solved numerically for n and ϕ' subject to a variety of geophysical conditions and initial perturbations (see review by Ossakow, 1981).

In a typical simulation, such as the ones illustrated in Figs. 4.14a–c, the vertical density profile is specified, as are the altitude of the layer and an initial perturbation in density. After a time the order of 1000 s the initial modulation grows to a sufficient amplitude that the low-density region (bubble) pushes through to the topside of the density profile, where the linear theory predicts a stable system (Ossakow, 1981). One of the crucial problems posed by Farley *et al.* (1970) is thus resolved. Although the system is linearly unstable only on the bottomside, the nonlinear evolution yields structure well above the F peak.

Computer simulations of plumes have been made to determine the ionospheric parameters to which plume development is most sensitive. Not too surprisingly, the higher the layer and the steeper the vertical density gradient, the faster the bubbles grow. This is also clear from the linear GRT growth rate. In addition, the simulations illustrated here show that longer horizontal wavelengths result in larger density depletions and larger uplift velocities. For example, at a fixed height the topside density drops by only 10% for a 3-km horizontal wavelength, while it drops by two orders of magnitude for a 150-km initial perturbation. Comparison of the electric potential distribution around the upwelling plasma shows why this is the case. The charge buildup which leads to the upward flow creates fringing electric fields which penetrate deeply into the bottomside region. Low-density plasma thus convects upward to altitudes above the F peak. In fact, since the flow field is incompressible in the two-dimensional approximation used in the simulation, the density at a given point changes only when a low-density region is advected to that height by the $\delta\mathbf{E} \times \mathbf{B}_0$ drift. Each density contour in the set of figures could thus be specified by its initial height as well as by the density. This idea explains the frequent observations of metallic and molecular ions at F-region altitudes in the equatorial zone. In other words, the chemical composition is also to some extent a label of the altitude. Returning to Fig. 4.1, the penetration of electric fields predicted by the simulations can be seen from the upwelling of features as low as 120 km in conjunction with the large-amplitude topside modulations.

Another simulation is shown in Fig. 4.14c. Here, a constant eastward neutral

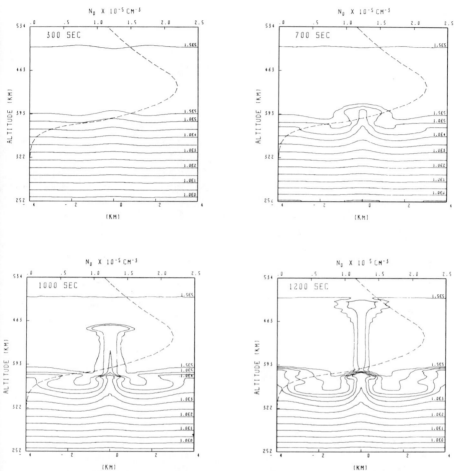

Fig. 4.14. Contour plots showing computer simulations of the Rayleigh–Taylor instability for (a) a 2-km scale perturbation and (b) a 100-km scale perturbation, both initially of 5% magnitude. Contours are labeled in units of reciprocal cubic centimeters. [After Zalesak and Ossakow (1980). Reproduced with permission of the American Geophysical Union.] (c) Computer simulation including a background eastward neutral wind and finite Pedersen conductivity in the off-equatorial E region. [After Zalesak *et al.* (1982). Reproduced with permission of the American Geophysical Union.] (*Figure continues.*)

b

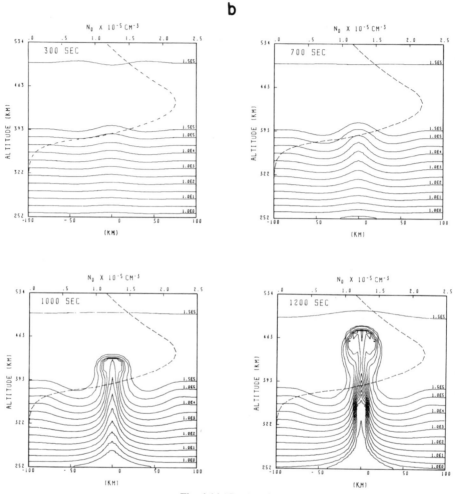

Fig. 4.14 (*Continued*)

wind is included, as well as the effect of finite conducting "end plates" in the northern and southern hemispheres. As discussed in Chapter 3, the finite E-region conductivity causes a competition for control of the equatorial F-region background electric field and a shear in the plasma flow can result. This shear yields a tilted bubble structure, as often observed in the Jicamarca and Altair data (e.g., Fig. 4.12).

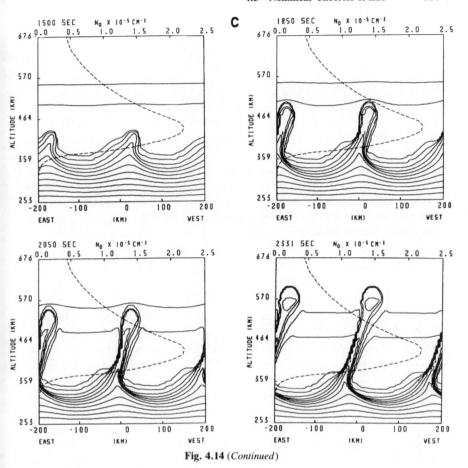

Fig. 4.14 (*Continued*)

4.3.2   *Comparisons between* in situ *Spectral Measurements and Theory*

We turn now to the studies of the power spectra of equatorial spread F and examine the information they contain concerning the physics of these phenomena. First, the relationship between density perturbation and electric field structure may be derived from (4.13a) which may be written

$$-ikQ\delta n = -ik\phi'(Pk/i) \tag{4.24a}$$

or

$$\delta E_x = \left(\frac{Q}{P}\right)\frac{\delta n}{n}$$

which may be written

$$\delta E_x = -\left(\frac{gB}{\nu_{in}}\right)\frac{\delta n}{n} \tag{4.24b}$$

where δE_x represents the electric field amplitude.

In (4.24b) the electric field is proportional to the density perturbation and hence the power spectrum of the two quantities, δE_x and $\delta n/n_0$, should be the same. As we shall see below, the relationship between δE_x and $\delta n/n_0$ is different for different plasma instabilities and may be used as a test for sorting out what processes dominate the physics as a function of altitude and wave number. In this context, it is worth pointing out that the electrical signature of a "fossil" or established plume should be quite different from that of a developing structure. Once the depleted region comes into equilibrium with the background medium and flows with the same velocity, the internal electric field will just be that due to ambipolar diffusion. If the F region has a higher conductivity than the E region, the electric field due to diffusion (see Chapter 8) is given by

$$\delta E = (k_B T/e)(\nabla n/n) \tag{4.24c}$$

rather than by (4.24b). This expression predicts that the electric field and density fluctuation power spectrum differ by a factor of k^2 due to the presence of the gradient operator. As we shall see, (4.24c) also arises when the irregularities are described by a Boltzmann relationship such as holds when drift waves are generated.

The computer simulations have been used to predict the power spectrum of the density structures for comparison with experiments. Keskinen et al. (1980a, b) found isotropic fluctuations at intermediate scales with wave number spectra which displayed a power law of the form k^{-n} with $n \approx 2.5$. This result is in good agreement with bottomside vertical power spectral measurements on rockets, but not with the 1-km sinusoidal structure (Fig. 4.7b) found in the satellite data. Also, in the regions of very strong turbulence such as the data shown in Fig. 4.5, the measurements do not agree with the simulations. That is, the smaller negative index (near $n = 1.5$) and the break in the spectrum to a very steep slope (≥ 4) are not reproduced in the simulations.

The shallow part of the observed spectrum is perplexing since at the long wavelengths the physics should be described quite well by the equations used in the simulations. A clue may lie in the observation that the shallower form correlates well with the intensity of the observed turbulence (Livingston et al., 1981). LaBelle and Kelley (1986) have studied a number of possible origins of this spectral difference. One possibility involves preferential injection of energy near the kilometer scale length. LaBelle and Kelley (1986) pointed out two mechanisms of this type:

1. Generation of a local maximum in the linear growth rate $\gamma(k)$ near 1 km due to the formation of images in the E region. Vickrey *et al.* (1984) previously showed evidence for this process in an equatorial F-region experiment and LaBelle (1985) has extended the theory. The whole process of image formation is discussed in detail in Chapter 8 of this text.

2. Preferential generation of kilometer-scale waves on the walls of the wedges driven upward by the primary instability. The scale of these waves would be shorter than the primary instability since very steep horizontal gradients exist on the bubble walls (e.g., a few kilometers scale length, L_w). Also, the growth rates on the walls would be large since neutral winds are the origin of the instability present on a vertical "wall" of ionization. In this case $\gamma \approx u/L_w \approx$ 0.05 s^{-1}, which is much larger than typical bottomside growth rates. These structures with vertical wave vectors would be analogous to BSS structures on the bottomside where ∇n is vertical.

LaBelle and Kelley (1986) also discussed a nonlinear theory which has been invoked to explain shallow spectra. This idea was first presented by Sudan (1983) for electrojet turbulence and later applied to the spread F case by Sudan and Keskinen (1984). The Sudan electrojet theory is discussed in some detail in the second part of this chapter, which deals with instabilities in the electrojet. The essence of the Sudan hypothesis is that as long as the "eddy growth rate" $\Gamma(k)$ of plasma structuring dominates the linear growth rate $\gamma(k)$, then the structure at some wave number k_1 will be due to energy passed to it from values of $k < k_1$ rather than to the linear process. This argument is a plasma analog of the neutral turbulence theory of Kolmogorov (1941). This theory states that if a neutral fluid is stirred at some wave number k_0, structure will form in this region of k space but energy will cascade to larger and larger values of k, creating a whole spectrum of turbulence. The cascade will cease at some wave number k_m where energy dissipation occurs due to molecular viscosity. Between k_0 and k_m there is an "inertial subrange" where energy is passed from eddy to eddy with no net gain or loss. In an isotropic homogeneous medium the omnidirectional spectral density of the flow energy, $E^2(k)$, has a power law spectrum of a form proportional to $k^{-5/3}$, the famous 5/3 power law for fluid turbulence.

Sudan (1983) found that for two-dimensional plasma turbulence such as spread F or electrojet turbulence, a Kolmogorov-type spectrum should occur for $\Gamma > \gamma$. Absolute fluctuation levels are predicted by the calculation. In applying this theory to the spread F case, LaBelle and Kelley (1986) found that reasonable agreement with the data was possible but *only* if an anomalous dissipation process exists. For example, the Sudan prediction is presented as curve b in Fig. 4.15 under the hypothesis that the plasma diffusion coefficient is 100 times the classical value. This curve should be compared to the actual plasma density fluctuation spectrum (curve a) measured during one of the Project CONDOR

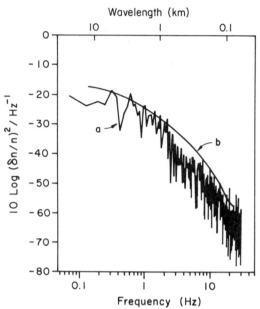

Fig. 4.15. (a) Spectrum of density irregularities measured at 390 km on the upleg of an equatorial rocket flight. (b) Spectrum of density irregularities predicted by Sudan and Keskinen (1984). The wave number-to-frequency transformation has been approximated using the rocket velocity, and the theoretically predicted spectrum has been integrated over wave number to obtain the same units as the measured spectrum. [After LaBelle and Kelley (1986). Reproduced with permission of the American Geophysical Union.]

rocket experiments. The agreement in spectral shape and absolute spectral intensity is very good above wavelengths of about 500 m. In particular, note that the spectral slope in this range is near $-(5/3)$. The requirement for an anomalous diffusion is quite interesting and is discussed in more detail in the next section since such diffusion may be driven by short-wavelength drift waves.

In summary, it may be that a process such as that suggested by Sudan dominates for strong plasma turbulence, turbulence strong enough that $\Gamma > \gamma$. In this range a Kolmogorov-type $k^{-5/3}$ spectrum pertains. At lower turbulence levels the spectrum relaxes to a one-dimensional $k^{-2.5}$ form similar to those found in the simulations and in the experiments outside strongly turbulent regions. How turbulent the plasma must be may be examined as follows.

The nonlinear eddy growth rate $\Gamma(k)$ is determined by the amplitude of the density fluctuation spectrum itself, since it is determined by the turbulent transfer of energy. Sudan (1983) gives the approximate relationship as

$$\Gamma(k) = (g/\nu)k^2 I_k^{1/2}(k)$$

where I_k is the (two-dimensional) spectrum of the density fluctuations. Since $\gamma = g/\nu L$ we have

$$(\Gamma(k)/\gamma) = Lk^2 I_k^{1/2}(k)$$

Consider the density fluctuation strength at wave number k_1, $(\Delta n/n)_{k_1}$, integrated over a range of wave numbers Δk where $k_1 - (k_1/2) \leq k \leq k_1 + (k_1/2)$. For this range $(\Delta n/n)_{k_1} \simeq k_1 I_k^{1/2}(k_1)$. To have $\Gamma(k_1)/\gamma > 1$ then requires

$$(\Delta n/n)_{k_1} > \lambda/2\pi L$$

For a 1-km wave in a region driven by a density gradient with $L \simeq 15$ km, $(\Delta n/n)_{k_1}$ need only be greater than about 1% to be strongly turbulent.

An alternative explanation for the observed spectra has been proposed by Zargham and Seyler (1987) and Kelley et al. (1987). Unlike the results of Keskinen et al. (1980a, b), their simulation shows a very strong anisotropy in the development of the irregularities, with nearly sinusoidal structures developing horizontally and shocklike structures in the direction of ∇n. One-dimensional cuts were made through the simulation for near-vertical (Fig. 4.16a) and near-horizontal trajectories (Fig. 4.16b). The predicted waveforms and their spectra are in excellent agreement with the rocket and satellite data, respectively, which were shown earlier in Figs. 4.7a and b. In this simulation the preferred horizontal scale evolves out of a weak maximum in the linear growth rate. Valladares et al. (1983) and LaBelle and Kelley (1986) have given a number of reasons why the linear growth rate should display such a maximum. It thus seems clear that a preferred scale should develop perpendicular to ∇n. The shocklike structures parallel to ∇n are a nonlinear result.

Kelley et al. (1987) also point out that the shallow spectra observed in the rocket data near the F peak may just be due to a changing angle of attack between the rocket and the anisotropic irregularities. A cut through the simulation at an intermediate angle is shown in Fig. 4.16c. The spectrum is shallow at small values and steeper than k^{-3} for large k, much as shown in Fig. 4.5. The anisotropy displayed by the Zargham and Seyler simulation is similar to the early barium cloud simulation work by McDonald et al. (1980), who solved a similar set of equations with a different numerical scheme.

4.4   Short-Wavelength Waves in Equatorial Spread F

The discussion thus far has concentrated on the seeding and generation of long- and intermediate-wavelength flute mode irregularities in the F region. On the other hand, the observational data base is determined primarily by scatter from short-wavelength waves—for example, the 3-m waves to which the Jicamarca radar is sensitive. Recent measurements with the Altair and Tradex radars on

a

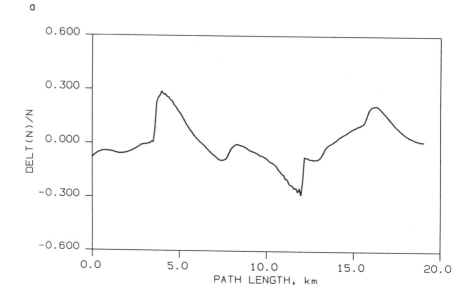

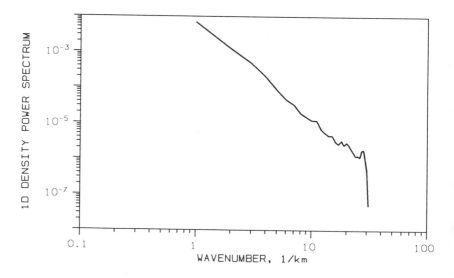

THETA   =   80.00

Fig. 4.16. (a) One-dimensional cuts through a spread F simulation corresponding to (a) a rocket trajectory, (b) a satellite trajectory, and (c) an intermediate angle of attack. The spectral units are reciprocal kilometers. [After Kelley *et al.* (1987). Reproduced with permission of the American Geophysical Union.]

b

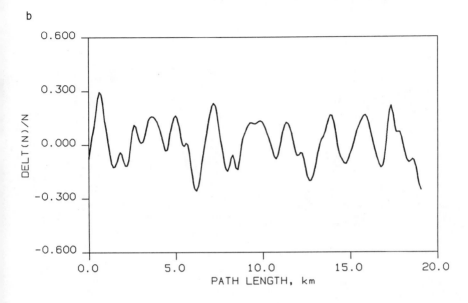

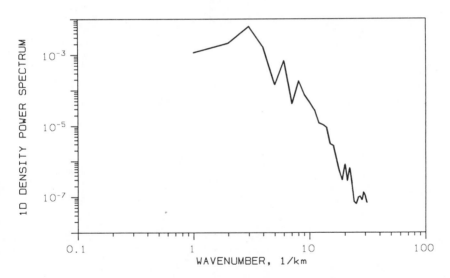

THETA   =   10.00

Fig. 4.16 (*Figure continues*)

C

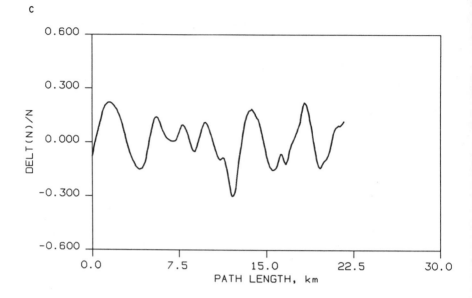

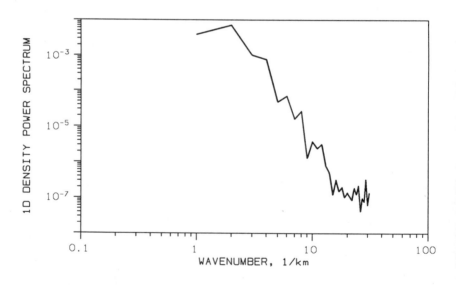

THETA   =   60.00

Fig. 4.16 (*Continued*)

Kwajalein have shown that the ionospheric medium is irregular at wavelengths as short as 0.11 m. It is essential, therefore, to address the question of wave generation at wavelengths well below the ion gyroradius (~4 m).

That the medium is linearly stable at such scales in the direct RT process can be seen by comparing the growth $g/\nu_{in}L$ to the classical perpendicular diffusive damping rate, given by $k^2 D_{c\perp}$. Before doing this, the correct diffusion coefficient $D_{c\perp}$ must be chosen. We discuss this matter in some detail in Chapter 8. In brief, the diffusion coefficient for the jth species perpendicular to the magnetic field, $D_{j\perp}$, can be written $D_{j\perp} = r_{gj}^2 \nu_j$, which corresponds to a random walk of one gyroradius (r_{gj}) per collision. As discussed in detail in Chapter 8, with a non-conducting E region, cross-field diffusion in the F region is determined by the electron diffusion coefficient, which is smaller than the ion coefficient. This is due to the ambipolar electric field which is generated when the more rapidly diffusing ions "get away" from the electrons perpendicular to **B**. The resulting charge separation electric field retards the ion diffusion velocity in such a way that the diffusion coefficient $D_{c\perp} = D_{e\perp}$ for the plasma as a whole. If the E-region conductivity is large, the charge separation (ambipolar) electric field is shorted out by current flow along the magnetic field lines which close through the E layer. In this case there is an enhanced perpendicular plasma diffusion coefficient, whose value approaches $D_{i\perp}$ as the E-region conductivity becomes much larger than the F-region conductivity. Hence, $D_{c\perp}$ may range from a minimum of $D_{e\perp}$ to as large as $D_{i\perp}$ depending on the conductivity of the E region.

The wavelength at which the RT process is marginally stable, that is, where linear growth equals diffusive damping, can be determined from the following equation:

$$\frac{g}{\nu_{in}L} - k^2 D_{c\perp} = 0$$

with the result that

$$\lambda_c = 2\pi (D_{c\perp} \nu_{in} L/g)^{1/2}$$

where λ_c is the critical wavelength at marginal stability. The numerical value of $D_{e\perp}$ is about 1 m^2/s, which for $g/\nu_{in}L$ between 10^{-4} and 10^{-2} yields λ_c in the range 60–600 m. The RT process is thus linearly stable in the range where most of the radar observations have taken place, that is, for backscatter from waves with $\lambda \leq 3$ m. If no other wave generation process exists, the 3-m waves must receive energy from the longer scales via a cascade process of some sort. In the terminology of neutral fluid turbulence, the 3-m waves lie in the viscous sub-range. The radar Doppler spectra at 3 m, however, often display a very broad range of phase velocities, some approaching the sound speed, which suggests that a turbulent wave process is operating. Furthermore, the study by LaBelle *et al.* (1986) suggests that an anomalous diffusion coefficient much larger than even

$D_{i\perp}$ exists under topside spread F conditions. The latter requires that some wave process other than the pure RT mode must be operating. The microphysics is therefore more complicated than the analogy to fluid turbulence suggests.

Because of these factors, considerable effort has gone into studying the possible generation of short-wavelength waves by the larger primary waves. That is, the plasma conditions set up by the long-wavelength RT process may be unstable to production of other waves. Both theory and experiment show that steep gradients can develop in the medium, and most studies have involved gradient-driven instabilities which are lumped under the generic term "drift waves."

As we noted in Section 4.2, for a purely two-dimensional system perpendicular to **B** the divergence of the pressure gradient-driven drift current vanishes. In three dimensions this need not be the case and the possibility of further unstable waves results. (In this section we continue to use the more common convention of orienting our coordinate system with the z axis parallel to **B**.) Inclusion of the z direction (formerly the y direction in Section 4.2.1) is equivalent to allowing a finite wavelength λ_z and thus nonzero k_z. A complete analysis requires use of kinetic theory. However, considerable insight can be attained from a low-frequency ($\omega \ll \Omega_i$) fluid approach (Chen, 1974). Consider the configuration in Fig. 4.17a, in which a density gradient exists in the x direction and an electrostatic wave propagates in the y–z plane as a plane wave. In this context the direction of the ∇n in the plane perpendicular to **B** is arbitrary. Indeed, in fully developed topside spread F, sharp density gradients probably occur in any direction. Following (4.9), we write the electrostatic potential and the density as $\phi = \phi' \exp(i\omega t - ik_y y - ik_z z)$ with $\delta\mathbf{E} = -\nabla\phi$ and $n = n_0(x) + n_1$, where $n_1 = \delta n \exp(i\omega t - ik_y y - ik_z z)$. Quasi-neutrality implies $n_{e0} = n_{i0} = n_0$ and we indicate the total electron and ion densities by n_e and n_i, respectively. In this model we neglect the effect of gravity. The cross-field electron velocity (2.3c) for low frequencies ($\delta\mathbf{V}_{e\perp}/\partial t = 0$) and the small collision frequencies in the F region ($\kappa_e \gg 1$) is

$$\mathbf{V}_{e\perp} = -(k_B T_e/eBn_0)(\partial n_0/\partial x)\hat{a}_y + (ik_y\phi/B)\hat{a}_x$$
$$- (ik_y n_{e1} k_B T_e/Ben_0)\hat{a}_x \qquad (4.25a)$$

where the first (third) term is the zero-order (first-order) electron drift due to the density gradient and the second is the first-order $\delta\mathbf{E} \times \mathbf{B}$ drift in the perturbation potential. In the z direction (parallel to **B**) there are only first-order terms since the driving gradient is perpendicular to **B** and, in the limit of no collisions, the equation of motion for electrons to first order is

$$n_0 m dV_{ez}/dt = -(\partial p_e/\partial z) - n_0 e\, \delta E_z = -k_B T_e\, \partial n_e/\partial z + n_0 e\, \partial\phi/\partial z$$
$$= ik_z k_B T_e n_{e1} - ik_z n_0 e\phi \qquad (4.25b)$$

Due to the low frequencies of interest here, our neglect of field-aligned drift, and the small electron mass, the left-hand side vanishes and

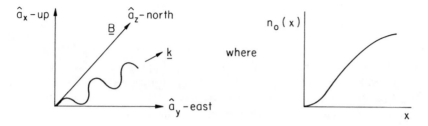

Drift Wave Geometry

a

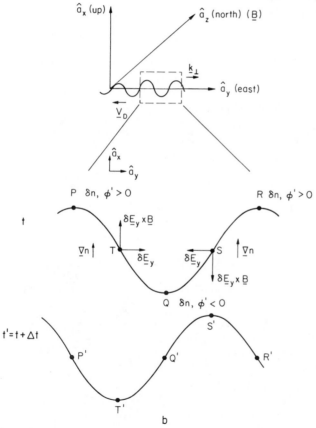

b

Fig. 4.17. (a) Geometry for investigation of drift wave generation on a local density gradient. (b) Perturbation parameters in a drift wave.

$$e\phi'/k_BT_e = \delta n_e/n_0 \qquad (4.26)$$

In effect, the electrons can move quickly enough along the magnetic field lines to stay in an equilibrium Boltzmann distribution in the potential ϕ'. In the electron reference frame a fixed pattern of high- and low-potential patterns is seen parallel to **B** with wavelength λ_z. Each region of high electron density is balanced by a region of depleted electron density ($\lambda_z/2$) away along **B**. The corresponding ion equations are, after substituting in the plane wave assumptions,

$$\mathbf{V}_{i\perp} = (k_BT_i/n_ieB)(\partial n_0/\partial x)\hat{a}_y + (ik_y\phi/B)\hat{a}_x + (ik_yn_{i1}k_BT_i/Ben_0)\hat{a}_x \quad (4.27)$$

and including ion inertia in the parallel equation and assuming $V_{iz} \propto \exp(i\omega t)$ gives

$$in_0\omega MV_{iz} = ik_zn_0e\phi + ik_zk_BT_in_{i1}$$

which yields

$$V_{iz} = (ek_z/\omega M)\phi + (k_BT_ik_z/\omega M)(n_{i1}/n_0)$$

Putting these ion velocity terms into the ion continuity equation,

$$(\partial n_i\partial t) + n_i(\nabla\cdot\mathbf{V}_i) + \mathbf{V}_i\cdot\nabla n_i = 0$$

and linearizing yields

$$(n_{i1}/n_0)[1 - (k_BT_i/m)(k_z^2/\omega^2)] = (e\phi'/M)(k_z^2/\omega^2) - (k_y\phi'/\omega Bn_0)(\partial n_0/\partial x)$$

We now define the diamagnetic drift frequency $\omega^* = \mathbf{k}_\perp\cdot\mathbf{V}_{e\perp}^0$, giving

$$\omega^* = -k_y(k_BT_e/eB)(1/n_0)(\partial n_0/\partial x) = -(k_yk_BT_e)/BeL \qquad (4.28)$$

which is just the perpendicular wave number k_y times the electron diamagnetic drift speed V_D. Using this notation and assuming that either the parallel phase velocity ω/k_z is much greater than the ion thermal velocity or $T_i = 0$ gives

$$(\delta n_i/n_0) = [(k_z^2k_BT_e)/M\omega^2) + \omega^*/\omega](e\phi'/k_BT_e)$$

Now by the quasi-neutrality condition $\delta n_{i1}/n_0 \simeq \delta n_{e1}/n_0$ and thus using (4.26)

$$(k_z^2k_BT_e)/(M\omega^2) + \omega^*/\omega = 1$$

This expression yields the dispersion relation

$$\omega^2 - \omega\omega^* - k_z^2k_BT_e/M = 0 \qquad (4.29)$$

which may be solved for ω,

$$\omega = [\omega^* \pm (\omega^{*2} + 4k_BT_ek_z^2/M)^{1/2}]/2$$

For large k_z and for propagation nearly parallel to **B,** this yields the usual acoustic wave

$$\omega/k_z = \pm(k_B T_e/M)^{1/2}$$

which is the correct dispersion relation for $T_i = 0$. In the small-k_z limit, ω either vanishes or yields the more interesting drift wave solution with

$$\omega = \omega^* = k_y V_D$$

The perpendicular phase velocity of this mode is V_D, the electron diamagnetic drift speed.

The drift wave propagation mechanism can be understood from the sketch in Fig. 4.17b, in which a drift wave with wave vector k_y propagates along a gradient in plasma density. The electron diamagnetic drift is in the y direction. Suppose at points P and R the density perturbation is positive. From (4.26) the potential ϕ is also positive. At point Q both are negative. The associated electric fields, δE_y, thus point from P to Q and from R to Q. These fields create plasma drifts up at point T and down at point S. But since

$$\partial n/\partial t = -\mathbf{V} \cdot \nabla n$$

the density will decrease with time at T and increase with time at S. Hence, at some later time, the disturbance looks like the lower curve in Fig. 4.17b. The wave has propagated in the direction of the electron diamagnetic drift.

This far we have investigated only the real part of the dispersion relation. Instability arises if the electric field pattern is shifted slightly in the $-y$ direction relative to $\delta n/n_0$. Then the $\mathbf{E} \times \mathbf{B}$ drifts will peak in the $\hat{a}_x$ direction at a position where there is *already* a slight positive perturbation in $\delta n/n_0$. The $\delta n/n_0$ amplitude thus grows slightly relative to the existing pattern and an instability results. One source for such a phase shift is a finite parallel resistivity that retards the electron flow and delays the potential buildup. Collisions were neglected in (4.25b). Thus, in a collisional plasma a finite parallel resistivity has a destabilizing influence for waves with $\lambda \gg r_{gi}$ and the "resistive drift wave mode" may compete with the RT mode (Hudson and Kennel, 1975).

Our interest here, however, lies with the shorter-wavelength waves ($kr_{gi} \simeq 1$). This regime requires a kinetic theory approach. Costa and Kelley (1978a, b) carried out a collisionless drift wave calculation and found that the growth rate (γ) peaked at $kr_{gi} \simeq 1.5$ ($\lambda = 20$ m) and found typical values of γ ranging from 0.1 to 40 s^{-1} using experimental values for the gradient scale lengths in the *disturbed* spread F environment (see Fig. 4.18). We emphasize "disturbed" since without the large-scale flute mode instability, the steep gradients necessary to generate drift waves so quickly would not exist. The drift wave phase velocities associated with the observed steep gradients plotted in Fig. 4.18, for example, range from 20 to 100 m/s. Since a number of such gradients may be in a radar field of view at any one time, the observed large values and the wide range of Doppler spectra in turbulent spread F may therefore be explained by these modes. The high phase velocity of drift waves is strong evidence for their im-

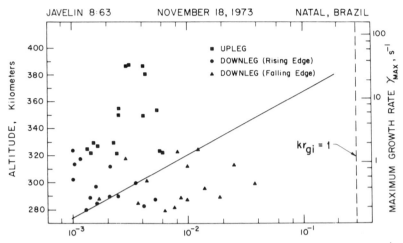

INVERSE GRADIENT SCALE LENGTH (dℓn n/dx) OF PLASMA DENSITY, meters^{-1}

Fig. 4.18. Measured values of local disturbed density gradients during spread F conditions (indicated by the filled symbols) along with the growth rate of the collisionless drift wave (indicated by the straight line). [After Costa and Kelley (1978a). Reproduced with permission of the American Geophysical Union.]

portance since the large-scale eddies themselves do not have large associated velocities. For example, consider the perturbation electric field in a 50% depleted region at 350 km where $\nu_{in} = 1$. From (4.24b), $|\delta \mathbf{E}| = (\delta n/n)(gB/\nu_{in}) = 5$ m/s, which is much smaller than the typical spectra at 3 m indicate. Electric field power spectra can be used to investigate the wave mode in the transition range (10 m $< \lambda <$ 100 m). Using (4.26),

$$\delta \mathbf{E}(k) = -\nabla \phi(k) = -i\mathbf{k}\left(\frac{k_B T_e}{e}\right)\frac{\delta n_e}{n_0} \tag{4.30}$$

Simultaneous electric field and density spectra in ESF are shown in Fig. 4.19. The spectrum at the left is $(\delta n/n_0)^2$, while that on the right is $\delta E^2(k)$. Notice that the latter shows no change in spectral form but remains roughly $k^{-2.5}$ throughout. The density spectrum does break, however, to a $k^{-4.5}$ form. This is strong evidence that drift waves occur in spread F since it agrees with (4.30).

Theories for short-wavelength drift waves have been extended to include collisions, and the even shorter-wavelength drift cyclotron and lower hybrid drift (LHD) modes have also been treated. Huba and Ossakow (1981a, b) showed that collisions stabilize the low-frequency pure drift mode discussed above for $\lambda \leq$ 6 m but that the lower hybrid drift instability could occur in low-density topside bubbles. A density depletion is favored since a high absolute plasma density stabilizes the LHD mode (Huba and Ossakow, 1981a). Since these waves grow

March 14, 1983

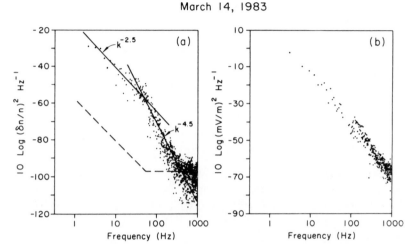

Fig. 4.19. Spectra of density and electric field irregularities in equatorial spread F as observed from a rocket launched March 14, 1983 from Punta Lobos, Peru: (a) spectrum of density irregularities; (b) spectrum of electric field irregularities. The dashed line indicates the background spectrum measured during a quiet period. [After LaBelle *et al.* (1986). Reproduced with permission of the American Geophysical Union.]

fastest for $r_{ge} < \lambda < r_{gi}$, they are an attractive explanation for backscatter at 0.11 m as reported by Tsunoda (1980a). Also, backscatter calculations by Kelley *et al.* (1982) show that extrapolating the observed $k^{-4.5}$ spectrum (probably due to low-frequency drift waves) to the 1-m scale yields agreement with radar data *outside* plume heads, but that *inside* the scatter intensity is much higher than the extrapolation predicts. This implies that another wave mode must be acting in the low-density plasma, such as the LHD wave.

Intense 3-m backscatter at Jicamarca usually ends by midnight in an ESF event although scintillation of satellite signals due to kilometer-scale structures may continue (e.g., Fig. 4.6). One explanation is that the steep gradients due to a strongly driven RT process are eventually smoothed out by generation of drift waves. The smooth structures which result would not create drift waves and hence they would then decay at the classical rate. For a nonconducting E region, the perpendicular diffusion rate is $D_{e\perp} = r_{ge}^2 \nu_e \simeq 1$ m²/s. The decay time constant for the waves is $\tau = (k^2 D_{e\perp})^{-1}$, which for $\lambda = 1$ km yields $\tau = 8$ h. A kilometer-scale feature can thus easily survive until the next sunrise, when production should begin filling the flux tube from below.

Analysis by LaBelle *et al.* (1986) has shown that the electric field fluctuation due to the observed drift waves in equatorial spread F can indeed create an anomalous diffusion 100 times as great as classical cross-field diffusion. They determined this from a calculation of the $\delta E \times B$ drifts due to the drift waves and showed that the result is in good agreement with the anomalous diffusion coefficient required by simple conservation arguments.

4.5 ESF Summary

Great strides have occurred in recent years concerning equatorial spread F. Reasonable hypotheses exist for the entire range of observed structures spanning at least six orders of magnitude in spatial scale ($0.1 - 10^5$ m). In order of descending scale, the possible contributing processes are

1. Gravity wave seeding and electrodynamic uplift ($\lambda > 200$ km).
2. Shear effects (200 km $> \lambda > 20$ km).
3. The generalized Rayleigh–Taylor instability (0.1 km $< \lambda < 20$ km).
4. Low-frequency drift waves (1 m $< \lambda < 100$ m).
5. Lower hybrid drift waves ($\lambda < 1$ m).

As an event progresses, the free energy released by the RT process is dissipated via drift waves of several types. These waves destroy the gradients which create them, leaving features with $\lambda \simeq 1$ km intact in the postmidnight period. Classical diffusion and plasma production via sunlight eventually smooth out the ionosphere. A discussion of questions such as "Why does spread F occur some nights and not on others?" and "Why are certain seasons preferred at certain locations?" is beyond the scope of the present text. The E-region conductivity, the plasma uplift by electric fields, and the neutral atmospheric density and dynamics which are discussed here all influence the probability of occurrence of this phenomenon and must enter any predictive theory.

4.6 E-Region Plasma Instabilities: The Observational Data Base

As discussed in Chapter 3, the equatorial electrojet is part of the worldwide system of electric fields and currents driven by the dynamo action of the neutral wind. The dynamo currents are confined to the E region, where the conductivity is greatest, and the current is essentially horizontal except perhaps at high altitudes and high latitudes. The basic reason for the existence of the equatorial electrojet is the large value of the Cowling conductivity close to the dip equator. We return now to the more natural coordinate system in which $\mathbf{B} = B\hat{a}_y$. In the simplest electrojet model the east–west dynamo electric field E_x sets up a vertical polarization field E_z which completely inhibits the vertical Hall current everywhere. The polarization electric field points upward during the day and has about the same magnitude but points downward at night. Thus, the drift velocity of the plasma at night is of the same order of magnitude as the daytime drifts, but the electrical current is much smaller due to the low electron density. For the case of zero vertical current, the vertical polarization electric field at the magnetic equator is given by

$$E_z = (1 + \sigma_H/\sigma_P)E_x \tag{4.31a}$$

which can also be written as

$$E_z \simeq (\nu_i/\Omega_i)[E_x/(1 + \nu_e\nu_i/\Omega_e\Omega_i)] = [\kappa_i^{-1}/(1 + \Psi_0)]E_x \tag{4.31b}$$

where Ω_e, Ω_i, ν_e, and ν_i are the usual electron and ion collision frequencies and gyrofrequencies and $\Psi_0 = \nu_e\nu_i/\Omega_e\Omega_i$. In the equatorial region the east–west electric field is about 0.5 mV/m, and the maximum vertical polarization field (at about 105 km) is estimated from (4.31) to be of the order of 10–15 mV/m. The electrons are magnetized and therefore the $\mathbf{E} \times \mathbf{B}$ drift under the influence of this vertical field yields an east–west electron drift velocity on the order of 400–600 m/s. The drift direction is westward during the day and eastward at night. This flow is sometimes supersonic; that is, the electrons move faster than the acoustic velocity in the medium.

In this section we follow closely the recent review by Fejer and Kelley (1980) and ask the reader to refer to that publication for the numerous references to the experimental work in this area. The occurrence of an anomalous scattering region in the ionospheric E region close to the dip equator was observed initially from ionosonde records. These echoes were called "equatorial sporadic E" (E_{sq}) echoes because of their apparent similarity to the sporadic E phenomenon (called E_s) occurring at other latitudes. However, the characteristics and generating mechanisms of these two phenomena are known now to be quite different. The intensity of the E_{sq} is well correlated with the electrojet strength, and VHF forward-scattering experiments showed that these echoes are field aligned and are caused by scattering from electron density irregularities immersed in the electrojet. The most important results concerning the physics of the electrojet scattering region have been obtained from VHF radar measurements performed at the Jicamarca Radar Observatory near Lima, Peru, since 1962. Rocket observations at Thumba, India; Punta Lobos, Peru; and Kwajalein Island have also provided valuable information on the electrojet irregularities from e-field and density profiles. Recently, multifrequency HF and VHF radar observations have been performed in Central and East Africa.

Equatorial E-region irregularities are present during both day and night. The irregularities are field aligned; that is, the wave number along the magnetic field ($k_\parallel$) is much less than $k_\perp$. Radar echoes are observed only when the wave vector is nearly perpendicular to the earth's magnetic field. The width of the north–south angular spectrum is typically 1° for 3-m irregularities. This property clearly indicates that a plasma process is occurring since neutral atmospheric turbulence is isotropic at the 3-m scale. Both electron density and electric field but not magnetic field fluctuations have been detected, indicating an electrostatic wave process. Radar spectral studies have shown the existence of two classes of irregularities, called type 1 and type 2, associated with the electrojet. The characteristics of the type 1, or two-stream (the reason for this name will be clear later) irregularities, were determined in the early measurements in Peru. The type 2 irregularities have been studied in detail only with the advent of improved spectral measurements at Jicamarca. The line of echoes centered near 100 km in Figs. 4.1 and 4.2 is due to the electrojet waves.

The two-stream or type 1 irregularities have a narrow spectrum with a Doppler shift (120 ± 20 Hz for a 50-MHz radar) that corresponds approximately to the ion acoustic velocity (about 360 m/s) in the electrojet region. These echoes appear nearly simultaneously over a large range of zenith angles when the electron drift velocity is larger than a certain value. Figure 4.20 shows a series of spectra taken during a period of strong scattering. The mean Doppler shift of the peak

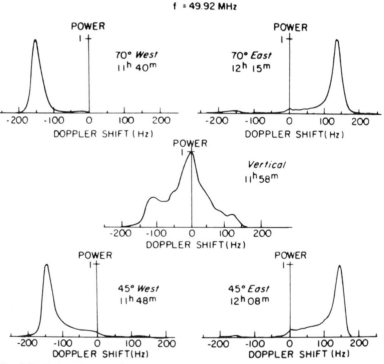

Fig. 4.20. Series of Doppler spectra from the equatorial electrojet irregularities at different elevation angles obtained at Jicamarca during a period of relatively strong scattering. The spectra are normalized to a fixed peak value. [After Cohen and Bowles (1967). Reproduced with permission of the American Geophysical Union.]

in the spectrum was constant when the zenith angle was between 45° and 70° both to the east and to the west. The data were normalized to a fixed peak value, since the echo power varies over a large range (up to about 40 dB) throughout the day. Thus, the area below the spectral curves is not proportional to the relative signal strength. In particular, the scattered power from the oblique type 1 echoes is appreciably larger than the power from the overhead echoes coming from vertically traveling "non-two-stream" irregularities. During the daytime the Doppler shift of the dominant peak is positive when the antenna is directed toward the east and negative when the antenna is looking to the west. The reverse is true at night. Therefore the phase velocity of the type 1 irregularities has a component in the direction of the electron flow, since during the day (night) the electron flow is westward (eastward). The fact that the Doppler shift is independent of zenith angle shows that the irregularities are not just advecting with the zero-order electron flow. The waves propagate at the ion acoustic velocity no matter what their angle to the current. Often, in fact, the vertical antenna detects echoes with both positive and negative spectral peaks corresponding to the acoustic speed, that is, propagating perpendicular to the flow both upward and downward.

The average phase velocity of the type 2 irregularities is smaller than the ion acoustic velocity and is approximately proportional to the cosine of the radar elevation angle (Balsley, 1969). The spectral width is much broader than that of the type 1 echoes and is often greater than the mean Doppler shift. Figure 4.21 shows the variation of the type 2 spectra and average phase velocity with zenith angle. The solid curves in the bottom right panel indicate the expected variation of the average velocity V_{obs} as a function of zenith angle θ on the basis of the relation $V_{obs} = V_D \sin \theta$ for three values of the drift velocity V_D. The excellent agreement between the experimental and theoretical variations shows that the type 2 phase velocity is proportional to the projection of the electrojet drift velocity on the radar line of sight.

Some additional characteristics of the type 1 and type 2 irregularities deduced from radar data are as follows:

1. Threshold: The type 2 irregularities are observed even for very small values of the eastward drift velocity during daytime. Ionosonde and radar measurements have shown that the irregularities are absent during the occasional daytime periods of westward current flow (counterelectrojet conditions). In this case the irregularities disappear when the driving electric field changes from eastward to westward. At nighttime the type 2 irregularities are almost always observed. One exception occurs after sunset, when the electric field reverses sign. Another case of this effect can be seen in Fig. 4.1. When the F-region scattering region reached "apogee" at around 2120 LT, the E-region irregularities ceased, only to reappear when the echoing region above started to fall again. It seems clear that

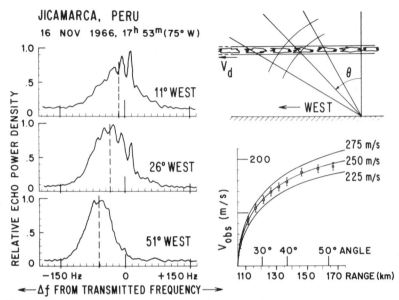

Fig. 4.21. Type 2 spectra measured at 50 MHz simultaneously at different antenna zenith angles. The dashed lines indicate the average Doppler shifts. The geometry of the experiment is shown in the top right panel. The results of the experiment, together with three theoretical curves for which a sine dependence of the average phase velocity with zenith angle was assumed, are shown in the bottom right panel. [After Balsley (1969). Reproduced with permission of the American Geophysical Union.]

the zonal electric field which causes the F layer to move vertically switched from east to west at this time, going through zero in the process. A dramatic example of the effect of electric field changes is shown in Fig. 4.22. Before 2044, the radar echoes came from two height ranges. For about 10 minutes they disappeared completely, only to return again with one echoing layer exactly midway between the original two echo heights. Thus, a very small threshold electric field seems to be required for type 2 echoes. There is a definite electric field threshold for the excitation of type 1 irregularities, however. These echoes are observed only when the electron drift velocity is somewhat larger than the ion acoustic velocity (about 360 m/s).

2. Scattering cross section: The scattering cross section of the type 2 irregularities is approximately proportional to the square of the drift velocity but is independent of zenith angle, while the type 1 scattering cross section increases rapidly with zenith angle, peaking near the horizon. This is where the radar beam is almost parallel to the electron flow in the electrojet.

3. Altitude dependence of the electrojet echoes: Fejer *et al.* (1975) studied

JICAMARCA
18 FEBRUARY 1971

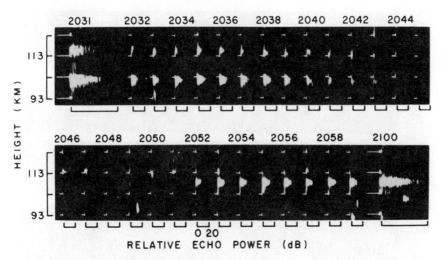

Fig. 4.22. Example of the change in altitude of the scattering regions before and after the electrojet reversal, which occurred at about 2048–2050. [After Fejer *et al.* (1975). Reproduced with permission of the American Geophysical Union.]

the vertical structure of the electrojet scattering region in detail and reported a considerable difference in the echoing region between day and night. Figure 4.23 shows some selected profiles of the electrojet scattered power on February 18–19, 1971, which were observed at Jicamarca using an altitude resolution of 3 km. During daytime, when the electron drift velocity is westward, echoes are generated between 93 and 113 km. The daytime power profile as a function of height shows a single peak that remains at a constant altitude (about 103 km). The nighttime scattering is much more structured.

4. Wavelength dependence: The vast majority of data concerning the electrojet is, of course, at 3-m wavelength. Interferometric techniques have recently been introduced at Jicamarca, however, which show that the scattering cells are often organized into kilometer-scale wave structures which propagate horizontally at about one-half the acoustic speed. An example is shown in Fig. 4.24. The most intense backscatter (darkest regions) is related to vertical 3-m waves traveling at the acoustic speed. It seems clear that the large-scale wave is organizing and creating the conditions which produce the vertically propagating acoustic waves. Notice that one wave period requires about 15 s to pass over Jicamarca in this example.

JICAMARCA
18-19 February 1971

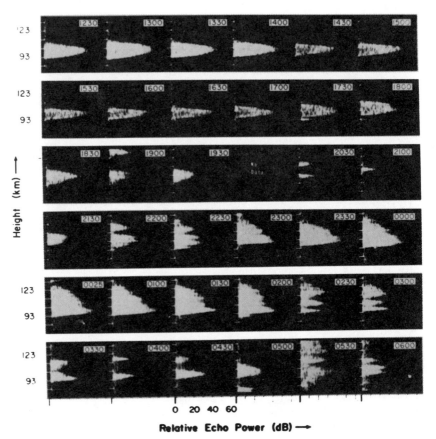

Fig. 4.23. Sample of the 50-MHz backscattering power profiles from the electrojet irregularities measured with the large vertically directed incoherent scatter antenna at Jicamarca. Spread F echoes contaminated the data between 0405 and 0550 and perhaps at 1900. [After Fejer *et al.* (1975). Reproduced with permission of the American Geophysical Union.]

We turn now to complementary data sets obtained from sounding rockets flown through the daytime and nighttime electrojets. Most of the early rocket experiments (prior to 1983) were flown with relatively high apogees and hence passed through the electrojet very quickly. Such high velocities preclude detection of the large-scale waves discussed above. Nonetheless, some important results were obtained, particularly due to the pioneering efforts of the Indian rocket group (e.g., Prakash *et al.*, 1972). Examples of simultaneous density profiles and density fluctuation measurements taken during daytime and nighttime conditions are shown in Fig. 4.25. During the day, the fluctuations in electron den-

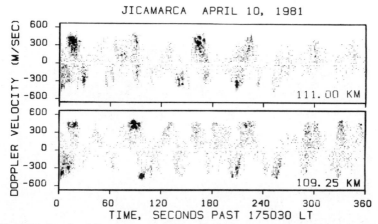

Fig. 4.24. Doppler shift spectrogram of the vertical backscatter signal measured at Jicamarca. Each spectrogram is normalized to its own peak power. The power values are divided into nine linearly spaced levels, with the darkest shades corresponding to the largest power values. Negative Doppler velocities indicate downgoing waves. [After Kudeki *et al.* (1982). Reproduced with permission of the American Geophysical Union.]

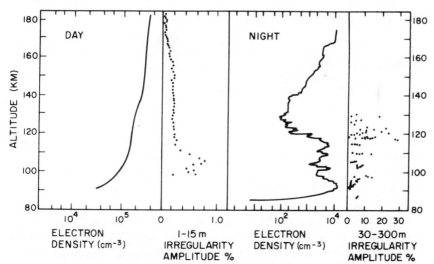

Fig. 4.25. Height variations of the electron density and irregularity amplitudes measured at Thumba, India, around noon and midnight. [Adapted from Prakash *et al.* (1972). Reproduced with permission of the American Geophysical Union.]

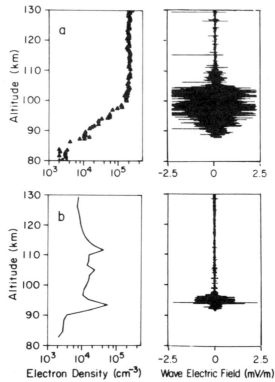

Fig. 4.26. (a) Electron density data from a swept Langmuir probe together with high-frequency (34–1000-Hz) electric field data for a daytime rocket flight. (b) Electron density data (correct to within a factor of 2) from a fixed-bias Langmuir probe on a nighttime flight along with high-frequency (60–1000-Hz) electric field data. [After Pfaff *et al.* (1982). Reproduced with permission of the American Geophysical Union.]

sity were located where the gradient is upward, were peaked at 103 km, and were relatively unstructured. The nighttime zero-order density profile is very structured on the negative gradient at about 120 km, with secondary short-wavelength fluctuations also peaking whenever the zero-order gradient shows steep downward-directed gradients. Electric field fluctuation data taken with similar rocket trajectories and similar conditions are shown in Fig. 4.26. Here, the upper panel is daytime data and shows strong electric field signals both on the positive gradient and just above. The nighttime data show fluctuations on the downward gradient. To summarize, the waves are clearly electrostatic (δE and $\delta n/n$ fluctuations exist), are organized by the direction of the gradient, and have broadband electric field fluctuations of the order of several millivolts per meter, which is comparable to the vertical polarization field.

We turn now to results of the CONDOR daytime rocket launch carried out on March 12, 1983 from Peru (Kudeki *et al.*, 1987; Pfaff *et al.*, 1988a, b). The rocket apogee was kept low to maximize the time and horizontal distance in the electrojet. The zero-order density profile is presented on the left side of Fig. 4.27 for the daytime electrojet flight, during which the magnetic deflection at Huancayo was 130 nT. A sonogram of the signals from a plasma wave receiver for electric field fluctuations is plotted on the right side. The data divide clearly into two segments. Above the peak in electron density the profile is smooth and the wave data extend to high frequencies. In the region of upward electron density gradient the plasma density is structured and the wave data come in strong bursts which do not extend to very high frequencies. The total east–west electric field has been measured as a function of time and altitude during this flight and is shown in the upper panel of Fig. 4.28. The data show very intense alternating electric fields organized in large-scale features very reminiscent of the radar data

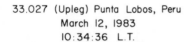

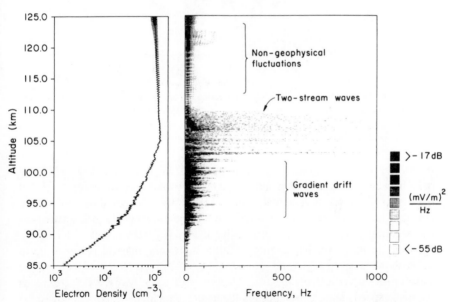

Fig. 4.27. Frequency–height sonogram of the horizontal component of the irregularities measured during the upleg of rocket 33.027 during strong electrojet conditions. This instrument had a low-frequency roll-off (3 dB) at 16 Hz. The electron density profile (left) shows the presence of large-scale irregularities. Both panels show nongeophysical "interference" above about 110 km. [After Pfaff *et al.* (1988a). Reproduced with permission of the American Geophysical Union.]

Horizontal Electric Fields (Upleg)
33.027 – Punta Lobos, Peru

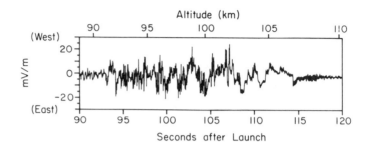

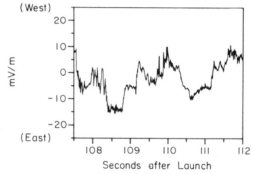

Fig. 4.28. Horizontal electric fields observed during the upleg traversal of the electrojet. Note the steepened waveforms and the "flat-topped" nature of the fields in the upper portion of the electrojet (102–107 km), as also seen in the enlargement in the lower panel. [After Pfaff *et al.* (1988a). Reproduced with permission of the American Geophysical Union.]

shown in Fig. 4.24. In fact, large-scale waves of this type were detected by the radar between 103 and 106 km and were a primary launch criterion for the rocket flight. It is of interest to note that as the rocket altitude increased from 92 to 106 km, the signal changed qualitatively from a turbulent sinusoid to a squared-off and seemingly saturated waveform. Similar features were detected during the downleg. In the lower half of Fig. 4.28, a short time interval is blown up to better show the steep edges and saturationlike waveforms. Note that within these waveforms the vertical *perturbation* electron drift velocity, $\delta E \times B/B^2$, exceeds the sound speed. The electron density fluctuation data in this height range display similar, but not identical, waveforms with peak $\delta n/n$ values in the 10–15% range. The electric field and density fluctuations in these data are interpreted as horizontally polarized electrostatic waves with wavelengths of about 2 km.

At about 106-km altitude, the vertical density gradient changes sign and the large-scale waves abruptly cease. An expanded view of this height range is shown in Fig. 4.29. Notice an almost evanescent behavior of the large-scale waves between 106.5 and 107.3 km. Above this height the total electric field in the second panel shows a nearly sinusoidal waveform with a 2–3-mV/m amplitude. The sonogram above this shows that the frequency of this peak changes with time in a similar fashion, decreasing from about 80 Hz to less than 20 Hz in a 2-km height range. The lower panels are the raw dc electric field signals, which show that the superimposed acoustic wave signal is nearly a pure sinusoid. It appears only when the detector is east–west aligned. The spectrum of this

Primary Two-Stream Waves (Upleg)
March 12, 1983 – Punta Lobos, Peru

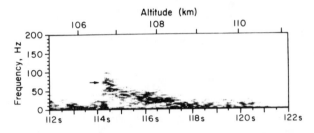

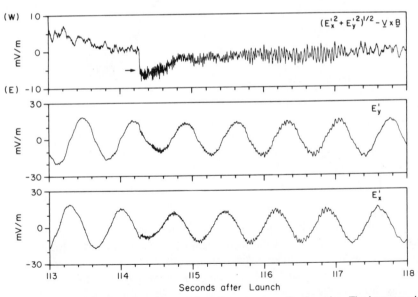

Fig. 4.29. Electric field observations of the two-stream waves for the upleg. The lower panels show the raw dc-coupled data, above which is plotted the square root of the sum of the squares of these waveforms. The upper panel shows a sonogram of these waves. (Note the change in scale of the time axis.) An arrow indicates the onset of the strong burst of primary two-stream waves. [After Pfaff *et al.* (1988b). Reproduced with permission of the American Geophysical Union.]

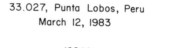

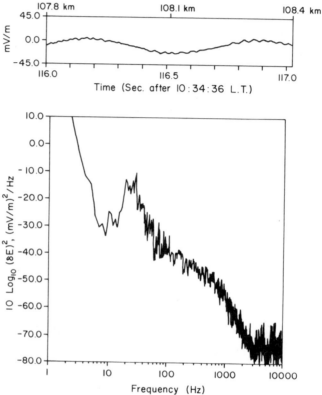

Fig. 4.30. Raw data and the spectrum of pure two-stream waves in the daytime electrojet. The peak signal occurs when the detector is parallel to the current and the spectrum maximizes at a few meter wavelength. [After Pfaff *et al.* (1988b). Reproduced with permission of the American Geophysical Union.]

signal is shown in the lower part of Fig. 4.30. The peak in the spectrum occurs near 2.5 m. This means that the wavelength of the peak in the instability spectrum is shifted by nearly three orders of magnitude from the spectral peak due to the large-scale waves found in the region of the zero-order vertical gradient at a slightly lower altitude.

As we shall see below, these data are remarkably well organized by the linear theory into a pure "two-stream" instability at and above the peak in plasma density, a pure "gradient drift" instability well below the peak, and a mixture of the two just below the peak.

4.7   Linear Theories of Electrojet Instabilities

Two mechanisms can efficiently amplify thermal density fluctuations in the equatorial electrojet. These mechanisms result from plasma instabilities known as the two-stream instability and the gradient drift (also known as $\mathbf{E} \times \mathbf{B}$, or cross-field, instability).

Many features of the type 1 irregularities are explained by a modified two-stream instability theory developed independently by Farley (1963), using kinetic theory, and by Buneman (1963), using the Navier–Stokes equation. They have shown that the plasma is unstable for waves propagating in a cone of angle ϕ about the plasma drift velocity, such that $V_D \cos \phi > C_s$. For smaller drift velocities the plasma can still be unstable provided there is a plasma density gradient of zero order oriented in the right direction relative to the electric field driving the electrons. This instability was first studied by Simon (1963) and Hoh (1963) for laboratory plasmas and is termed the gradient drift instability. Many features of the type 2 radar echoes are explained by this instability.

We first investigate the linear theory for the two-stream instability. Consider the zero-order condition in which the electrons stream east with a velocity $V_D \hat{a}_x$ driven by a vertically downward zero-order polarization electric field $-E_{z0}\hat{a}_z$ (see Chapter 3). This corresponds to nighttime conditions. The ions are assumed to be at rest to zero order due to the high collision frequency ($\nu_{in} \gg \Omega_i$). For the moment we ignore any zero-order plasma gradient. Perturbations in density, electrostatic potential, and velocity are of the form $\delta n e^{i(\omega t - kx)}$, $\phi' e^{i(\omega t - kx)}$, and $\delta V e^{i(\omega t - kx)}$, and we have taken $\mathbf{k}$ to be in the $\hat{a}_x$ direction ($\mathbf{k} = k\hat{a}_x$), which is perpendicular to $\mathbf{B} = B\hat{a}_y$. The linearized electron continuity equation, ignoring production and loss, which are assumed to be in equilibrium in the zero-order equations, is

$$i\omega \, \delta n + V_D(-ik \, \delta n) - ikn_0 \, \delta V_{ex} = 0 \qquad (4.32)$$

where the second term comes from the $(\mathbf{V} \cdot \nabla n)$ term and the third from $n(\nabla \cdot \mathbf{V})$. In this expression the common factor $e^{i(\omega t - kx)}$ has been factored out. Notice that since $\mathbf{k}$ is horizontal and there is no vertical gradient, the vertical perturbation drift does not change the electron density. Equation (4.32) can be rewritten as

$$\delta V_{ex} = (\omega/k - V_D)(\delta n/n) \qquad (4.33a)$$

where the subscript zero has been dropped from the mean density n_0. Notice that we have already assumed quasi-neutrality by using δn as a variable rather than δn_e. In the electron momentum equation the term $m \, d\mathbf{V}_e/dt$ is dropped because of the small electron mass, leaving

$$0 = (-e/m)(\mathbf{E} + \mathbf{V}_e \times \mathbf{B}) - (k_B T_e/m)(\nabla n/n) - \nu_e \mathbf{V}_e$$

where again we ignore gravity and where ν_e is the electron–neutral collision frequency. In deriving this equation an isothermal process has been assumed and hence no term involving ∇T arises. This simplifies the derivation since no independent heat equation is needed, but may hide some important physics, particularly at high latitudes. The linearized version of this in component form is, in the z direction,

$$0 = -\Omega_e \, \delta V_{ex} - \nu_e \, \delta V_{ez} \tag{4.33b}$$

In this derivation we have chosen to follow tradition and write all gyrofrequencies as positive numbers (i.e., $\Omega_e = +eB/m$). In the z component of the momentum equation, the zero-order electric field and electron drift terms cancel in the equilibrium state about which we are making our perturbations. For the terms in the x direction we have

$$0 = (-eik/m)\phi' + \Omega_e \, \delta V_{ez} + ik(k_B T_e/m)(\delta n/n) - \nu_e \, \delta V_{ex}$$

or, equivalently,

$$-\Omega_e \, \delta V_{ez} + \nu_e \, \delta V_{ex} = -ik[(e\phi'/m) - (k_B T_e/m)(\delta n/n)] \tag{4.33c}$$

where again the zero-order terms cancel. The corresponding ion equation derived from continuity is

$$i\omega(\delta n/n) - ik \, \delta V_{ix} = 0 \tag{4.33d}$$

The ion momentum equation yields

$$(i\omega + \nu_i) \, \delta V_{ix} = ik[(e\phi'/M) + (k_B T_i/M)(\delta n/n)] \tag{4.33e}$$

where we have used $\Omega_i \ll \nu_i = \nu_{in}$. Since the ions are assumed to be at rest to zero order, the zero-order polarization field is not included in the ion equation of motion. Equations (4.33a)–(4.33e) constitute five equations in the five unknowns δV_{ex}, δV_{ez}, δV_{ix}, ϕ', and $\delta n/n$. The resulting determinant of coefficients is

$$
\begin{vmatrix}
1 & 0 & 0 & 0 & (V_D - \omega/k) \\
-\Omega_e & -\nu_e & 0 & 0 & 0 \\
\nu_e & -\Omega_e & 0 & ike/m & -ik(k_B T_e/m) \\
0 & 0 & -ik & 0 & i\omega \\
0 & 0 & (i\omega + \nu_i) & -ike/M & -ik(k_B T_i/M)
\end{vmatrix} = 0 \tag{4.34}
$$

This can be evaluated in a straightforward manner to find the relationship between ω and k (e.g., Sudan *et al.*, 1973),

$$(\omega - kV_D) = (-\Psi_0/\nu_i)[\omega(i\omega + \nu_i) - ik^2 C_s^2] \tag{4.35}$$

where $\Psi_0 = \nu_e \nu_i/\Omega_e \Omega_i$ and $C_s^2 = k_B(T_e + T_i)/M$. If we now set $\omega = \omega_r - i\gamma$ and require $\gamma \ll \omega_r$ and ν_i, we find the real and imaginary parts of ω to be

$$\omega_r = kV_D/(1 + \Psi_0) \tag{4.36a}$$

$$\gamma = (\Psi_0/\nu_i)(\omega_r^2 - k^2 C_s^2)/(1 + \Psi_0) \tag{4.36b}$$

When γ is positive, the waves will grow, and thus the requirement for instability is

$$\omega_r^2 > k^2 C_s^2 \quad \text{or} \quad V_D > (1 + \Psi_0)C_s$$

This is the origin of the term "two-stream" since the electrons must drift through the ions at a speed exceeding the sound speed to generate the waves.

We can now use these mathematical results to gain insight into the physical instability mechanisms involved. First, from the electron continuity equation we have

$$\delta V_{ex} = (\omega/k - V_D) \, \delta n/n \tag{4.37}$$

But using the two electron momentum equations [(4.33b) and (4.33c)] we also must have

$$\delta V_{ex} = -[\nu_e/(\Omega_e^2 + \nu_e^2)][(ike\phi'/m) - (ikk_BT_e/m)(\delta n/n)] \tag{4.38}$$

If we study the properties of the wave near marginal stability where $\gamma = 0$ (or equivalently assume $\omega_r \gg \gamma$), we can replace $(\omega/k - V_D)$ by $(\omega_r/k - V_D)$, which in turn becomes $-(\Psi_0/[1 + \Psi_0])V_D$, using (4.36a). Setting (4.37) and (4.38) equal and using this result yields

$$[\nu_e/(\Omega_e^2 + \nu_e^2)](ike\phi'/m) = [\nu_e(ikk_BT_e/m)/(\Omega_e^2 + \nu_e^2)$$
$$+ \, \Psi_0 V_D/(1 + \Psi_0)](\delta n/n)$$

Now if we note that $\delta E_x = ik\phi'$ and $V_D = E_{z0}/B$, where E_{z0} is the zero-order vertical electric field, this may be written

$$[\nu_e e/m(\Omega_e^2 + \nu_e^2)] \, \delta E_x = [\nu_e(ikk_BT_e/m)/(\Omega_e^2 + \nu_e^2) + \Psi_0 E_{z0}/B(1 + \Psi_0)] \, \delta n/n$$

and finally,

$$\delta E_x/E_{z0} = \{\Psi_0(\Omega_e^2 + \nu_e^2)/[\nu_e\Omega_e(1 + \Psi_0)] + ikk_BT_e/eE_{z0}\} \, \delta n/n \tag{4.39}$$

This expression shows that the relationship between δE_x and $\delta n/n$ is wavelength dependent. For long wavelengths, the second term in the bracket is negligible and δE_x and δn are either in phase or 180° out of phase, depending on the sign of E_{z0}. Using the good approximation that $\Omega_e \gg \nu_e$ and $\Psi_0 = \nu_e\nu_i/\Omega_e\Omega_i$ we have for this case

$$\frac{\delta E_x}{E_{z0}} = \frac{\nu_i}{\Omega_i(1 + \Psi_0)} \frac{\delta n}{n} \tag{4.40}$$

This is very similar to the result found earlier for the Rayleigh–Taylor instability [i.e., (4.24b)]. The wavelength at which the magnitudes of the first and second terms in (4.39) are equal for $T_i = T_e$ (i.e., $C_s^2 = 2k_BT_e/M$) is given by

$$\lambda_c = \frac{\pi C_s^2(1 + \Psi_0)}{\nu_i V_D}$$

Setting $\Psi_0 = 0.22$ at an altitude of 105 km, $C_s = V_D = 360$ m/s and $\nu_i = 2.5 \times 10^3$ s^{-1} yields $\lambda_c \simeq 0.4$ m. This wavelength is thus quite small and for most wavelengths of interest we may use (4.40) to relate the perturbed electric field to the density.

Using these results, the instability process can now be understood from Fig. 4.31a. The figure is drawn for nighttime conditions with the vertical electric field downward. The sinusoidal wave can thus represent both $\delta n/n$ and δE_x for $\lambda \gg \lambda_c$. That is, the eastward perturbation δE_x is positive when $\delta n/n$ is positive. The two quantities are in phase with net positive or negative charges built up as shown, where these charges are associated with the perturbation electric field (i.e., $\rho_c = \varepsilon_0 (\nabla \cdot \mathbf{E}) = -\varepsilon_0 ik\,\delta E_x$).

Now, the wave will grow if more plasma moves into a region of high density than leaves that region. Consider the ion motion expressed by (4.33e). The real

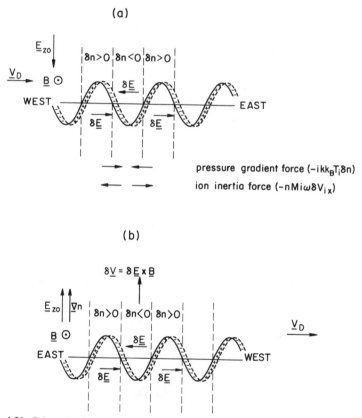

Fig. 4.31. Schematic diagrams showing the linear instability mechanism in (a) the two-stream process for nighttime conditions and (b) the gradient drift process for daytime conditions.

part of the equation corresponds to the ion velocity in phase with the wave. The $(+ik\phi')$ term is just the electric field, δE_x, which is also in phase and hence is included in the real part of the expression. The imaginary part gives the out-of-phase motion at the spatial positions where $\delta n/n$ and δE_x both vanish and hence the part of the ion motion which causes the wave to grow or decay. The two vectors, pressure gradient force and inertial force, are shown in Fig. 4.31a. For growth, the ion inertial force must be larger than the pressure term so that the plasma moves horizontally from low density to high density. This in turn requires that

$$\omega_r\,\delta V_{ix} > k(k_B T_i/M)(\delta n/n) \tag{4.41}$$

In other words, the ion inertial force must be greater than the pressure gradient force which is trying to smooth out the density enhancement by diffusion. Equation (4.41) can also be written, using (4.33d) to eliminate δV_{ix}, as

$$\omega_r\omega/k^2 > (k_B T_i/M)$$

Near marginal stability, $\gamma \ll \omega_r$, so $\omega \simeq \omega_r$, and using the dispersion relation (4.36a) for ω_r/k we have the requirement that

$$V_D/(1 + \Psi_0) > (k_B T_i/M)^{1/2} \tag{4.42}$$

This is almost but not quite the required threshold condition for growth. The exact result, $V_D > (1 + \Psi_0)C_s$, came out of the detailed determinant analysis since the electron pressure term adds to the ion pressure via the ambipolar diffusion effect. Ion inertia is thus the destabilizing factor in the two-stream instability, while ambipolar diffusion causes damping.

Turning to the gradient drift instability, we must include the change of density due to the vertical perturbation electron drift in the electron continuity equation. The term comes from $\mathbf{V} \cdot \nabla n$, which in linearized form adds a term ik/L to the first row, second column of the determinant, where $L = [(1/n)(dn/dz)]^{-1}$ is the zero-order vertical gradient scale length. This added term changes only the imaginary part of ω; the real part is identical. This means that (4.39) also relates the perturbed field δE_x to the density perturbation δn for the gradient drift mode. To study the stability condition, consider Fig. 4.31b, which shows *daytime* conditions and includes a zero-order density gradient which is upward. From Fig. 4.31b it is clear that the gradient is a destabilizing factor when it is upward since then the upward perturbation drift $\delta V_{ez} = \delta E_x/B$ occurs in a region where the density is already depleted ($\delta n < 0$). That is, a low-density region is convected upward into a region of higher background density, causing a growth in the relative value of $\delta n/n$. If the gradient is reversed in sign but the E_{z0} left the same, the perturbation electric fields will cause high-density plasma to drift into higher-density regions, which is stabilizing. Thus, for instability the zero-order vertical electric field must have a component parallel to the zero-order density gradient.

At the equator during normal daytime conditions (see Chapter 3), the unstable gradient direction is upward since E_{z0} is upward, while at night the unstable gradient is downward.

There is no threshold drift requirement for the gradient drift instability in the physics thus far. However, if recombination is considered in the continuity equations, a term of the form $-\alpha n^2$ must be included which, when linearized, becomes of the form $-2\alpha n \, \delta n$. A finite threshold exists for the gradient drift process even when the electron drift velocity is finite. Neglecting the two-stream term in the damping rate [e.g., the first term in (4.44b) below], Fejer *et al.* (1975) found that the electron drift threshold velocity for instability to occur is given by

$$V_D > L[2\alpha n_0(1 + \Psi_0)(\Omega_i/\nu_i) + (k^2 C_s^2 \nu_e/\nu_i \Omega_e)(1 + \Psi_0)] \qquad (4.43)$$

Finally, following Fejer *et al.* (1975), the complete linear theory must include a finite zero-order ion drift velocity $\mathbf{V}_{Di}$ and the possibility of an arbitrary $\mathbf{k}$ vector, yielding the following set of expressions for the linear theory of both the gradient drift and two-stream modes:

$$\omega_r = \mathbf{k}\cdot(\mathbf{V}_D + \Psi\mathbf{V}_{Di})/(1 + \Psi) \qquad (4.44a)$$

$$\gamma = (1 + \Psi)^{-1}\{(\Psi/\nu_i)[(\omega_r - \mathbf{k}\cdot\mathbf{V}_{Di})^2 - k^2 C_s^2]$$
$$+ (1/Lk^2)(\omega_r - \mathbf{k}\cdot\mathbf{V}_{Di})(\nu_i/\Omega_i)k_\parallel\} - 2\alpha n_0 \qquad (4.44b)$$

with

$$\Psi = \Psi_0[(k_\perp^2/k^2) + (\Omega_e^2/\nu_e^2)(k_\parallel^2/k^2)] \qquad (4.45)$$

and under the assumption that $\gamma \ll \omega_r$.

If the latter assumption is relaxed, Kudeki *et al.* (1982) have shown that waves with $\lambda \simeq L$ propagate more slowly; that is, a more exact result for the linear theory with $\mathbf{k}\cdot\mathbf{B} = \mathbf{V}_{Di} = 0$ is given by

$$\omega_r + i\gamma = \mathbf{k}\cdot\mathbf{V}_D(1 + ik_0/k)/[(1 + \Psi_0)(1 + k_0^2/k^2)] \qquad (4.46)$$

where $k_0 = (\nu_i/L\Omega_i)/(1 + \Psi_0)$. Finally, we note that recombination is much more important during daytime conditions than nighttime, since the zero-order density is large and $2\alpha n_0$ is sizable.

The linear theory outlined here has been applied to data from three rocket flights and the results are reproduced in Fig. 4.32 (Pfaff *et al.*, 1985). Panel (a) shows the current density and electron density profiles. Panel (b) shows the electron drift velocity, the phase velocity for two-stream waves (which must exceed C_s for instability to occur), and the approximate value for C_s. Using these parameters and the gradient of the electron density profile, the growth rate has been calculated as a function of altitude and wave number in panel (c) using a gray scale. Two-stream waves result when the threshold for the two-stream instability

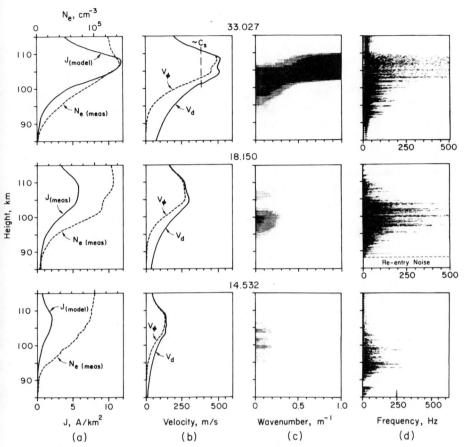

Fig. 4.32. (a) Measured or model current and smoothed electron density profiles for each of three experiments; (b) calculated electron drift and phase velocity profiles computed using the values in (a); (c) gray-scale representation of the growth rate calculated from the linear dispersion relation for horizontal waves using the model parameters discussed in the text. Darker shades represent higher values of γ, whereas white indicates that $\gamma < 0$. (d) Frequency–height sonograms showing rocket-measured irregularities. [After Pfaff *et al.* (1985). Reproduced with permission of Pergamon Press.]

is exceeded. This condition occurred only in the CONDOR rocket flight shown on top and is evident in the linear calculation [panel (c)] as well as the wave observations in panel (d). In the calculation, waves are predicted to grow at high wave numbers. Indeed, sonograms of the rocket wave fluctuation data in panel (d) show strong high-frequency (in the rocket frame) waves in the region where the density gradient vanishes or reverses sign, but only for the CONDOR data. Such waves are weak in the middle data set and are not seen at all in the lowest set of data. In the center data set the conditions were close to two-stream instability and the wave activity was stronger than in the case of the lowest data set. However, in both the center and lowest data set the waves were restricted primarily to the region of upward vertical density gradient. Those data are thus in excellent detailed agreement with linear theory.

The fluid theory discussed thus far breaks down at short wavelengths and a kinetic approach is necessary. Such calculations have been carried out by Farley (1963) and more recently by Schmidt and Gary (1973). Pfaff *et al.* (1988b) compared Schmidt and Gary's calculation of $\gamma(k)$ with the spectrum of electric field fluctuations measured in the CONDOR rocket flight during pure two-stream conditions (above the density gradient) as shown in Fig. 4.33. The two plots are remarkably similar. This strongly suggests that pure two-stream waves grow in a narrow range of wave numbers near the peak in their linear growth rate.

To summarize, the linear theory explains (a) the acoustic speed drift velocity threshold for type 1 irregularities, (b) the low drift velocity threshold for type 2 waves, (c) the association of waves with upward density gradients during the day and downward ones at night, (d) the wavelength and angular dependence of the phase velocity of type 2 waves, (e) the slow phase velocity of large-scale waves, and (f) the peak in the two-stream threshold spectrum.

The linear theory does not explain, among other things, (a) 3-m wave generation by the gradient drift process (since it is linearly stable at that wavelength), (b) the dominance of and square-wave nature kilometer-scale waves, (c) the observation of vertically propagating two-stream waves (perpendicular to the current), (d) the constant phase velocity of type 1 waves at any angle to the current, and (e) details of the observed wave number spectra.

4.8   Nonlinear Theories of Electrojet Instabilities

4.8.1   Two-Step Theories for Secondary Waves

Although strictly speaking not a nonlinear theory, the discussion by Sudan *et al.* (1973) provides a conceptual framework from which several properties of the fully developed turbulence can be understood. The basic idea is that "primary" waves reach sufficiently large amplitudes that the perturbation electric fields and density gradients are themselves large enough to drive "secondary" waves unstable. For example, referring to the observational data from the CONDOR flights in Fig. 4.28, the primary δE is horizontal and hence, if it is large enough, $\delta E/B$ will drive two-stream waves in the vertical direction. Indeed, the rocket electric field measurements show that the perturbation electric fields can be of the order of $C_s B$. This is illustrated in Fig. 4.34, in which the horizontal electric field fluctuation strength has been divided by B and plotted as a function of altitude for the upleg and downleg of the rocket flight. The dotted lines show the

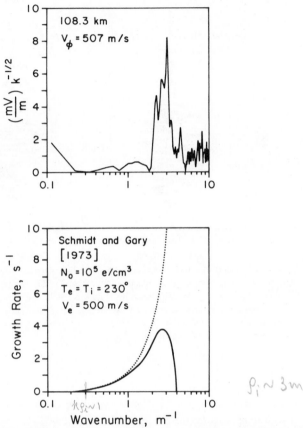

Fig. 4.33. Comparisons of the measured electric field wave spectrum and two-stream growth rate calculations for daytime equatorial conditions by Schmidt and Gary (1973). The dotted line in the growth rate calculation is the result of the fluid theory and the solid line represents the kinetic results. [After Pfaff *et al*. (1988b). Reproduced with permission of the American Geophysical Union.]

threshold velocity for a secondary two-stream instability given by $C_s(1 + \Psi)$. The first thing to note is that between 100 and 105 km the observed perturbation electric field is sufficiently strong to generate a vertical two-stream instability. The region where these waves can be generated agrees remarkably well with the region where vertical up- and downgoing two-stream waves were detected si-

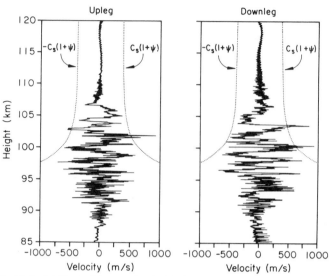

Fig. 4.34. Vertical drift velocities associated with the horizontal electric fields measured during the upleg and downleg traversals of the electrojet by rocket 33.027. The boundaries corresponding to $C_s(1 + \psi)$, which represent the two-stream instability threshold, are also shown for a fixed value of T_e. The figure demonstrates why vertical two-stream waves were only detected in the narrow region of roughly 100–107 km for this day. [After Pfaff *et al.* (1988b). Reproduced with permission of the American Geophysical Union.]

multaneously by the Jicamarca radar. It is safe to say that vertically propagating two-stream waves are created by large-amplitude horizontal electric fields associated with long-wavelength gradient drift waves.

Quantitatively, one can evaluate the required amplitude of the primary waves as follows. As summarized by Fejer and Kelley (1980), the oscillation frequency and growth rate of vertically propagating secondary waves are given by

$$\omega_r(k_s) = -k_s(\Omega_e/\nu_e)(\Psi V_D/(1 + \Psi)^2)A \sin \delta \tag{4.47}$$

$$\gamma(k_s) = (\Psi/1 + \Psi)\{-(\Omega_e^2/\nu_e^2)(k_p V_D/2)[\Psi A^2/(1 + \Psi)^2] \sin(2\delta)$$
$$+ (1/\nu_i)[\omega(k_s)^2 - k_s^2 C_s^2]\}. \tag{4.48}$$

where k_s and k_p are the wave numbers of the secondary vertically propagating and primary horizontally propagating waves, respectively. In this expression, A is the relative amplitude $(\delta n/n)$ of the primary wave and δ is its phase $(\omega_p t -$

$k_p y$). The necessary condition for the generation of vertically propagating type 1 irregularities is given by the requirement (see Section 4.7) that

$$(\delta E/B) > (1 + \Psi)C_s \qquad (4.49a)$$

which corresponds to the dotted lines in Fig. 4.34. Using (4.40) this can also be written

$$A(\nu_i/\Omega_i)[V_D/(1 + \Psi)^2] > C_s \qquad (4.49b)$$

where again we have used A to represent the amplitude of the $(\delta n/n)$ variation as defined above. Neglecting ion inertia, secondary gradient drift irregularities are generated if

$$A^2(M\nu_i V_D k_p)/[2m\nu_e(1 + \Psi)^2] > k_s^2 C_s^2/\nu_i \qquad (4.50)$$

The requirement (4.49b) seems to be met only in the 100–105-km height range, but (4.50) is less stringent and it seems quite possible for secondary gradient drift waves to be generated below 100 km. Such waves were detected during the CONDOR rocket flight with radar and *in situ* probes.

4.8.2   Nonlinear Gradient Drift Theories

Extensive analytical and simulation studies have been performed on the intermediate-scale (≤ 100 m) type 2 or gradient drift irregularities. These studies permit a detailed comparison between theoretical results and the radar and rocket data. McDonald *et al.* (1974, 1975) studied numerically the nonlinear evolution of small-scale ($\lambda \simeq 10$ m) type 2 irregularities by using a grid of 50×50 points with a mesh spacing of 1.5 m in both horizontal and vertical directions. They observed that intermediate wavelengths ($\lambda < 28$ m) are excited only after the large-scale primary horizontally propagating waves grow to an amplitude of 4%, consistent with the "two-step" theory outlined above. The quasi-final state is highly turbulent and almost two-dimensionally isotropic, in agreement with the radar observations. Figure 4.35 shows the saturated nonlinear development of the gradient drift irregularities found in these simulations. In the turbulent state the small-scale irregularities have upward and downward motions with speeds comparable to the background horizontal drift. It is important to note that the two-dimensional power spectra of $(\delta n/n)^2$ and $(\delta E)^2$ are proportional to $k^{-3.5}$. This implies a one-dimensional spectrum varying as $k^{-2.5}$.

More recent detailed quantitative studies of the nonlinear development of the primary gradient drift theory have been reported by Sudan and Keskinen (1977), Keskinen *et al.* (1979), and Sudan (1983). The approach is best suited to zero-order conditions such that $V_D < C_s$, and they therefore were able to apply Kadomtsev's (1965) strong turbulence–weak coupling equations to the gradient drift turbulence. These equations are, essentially, the Fourier-transformed ver-

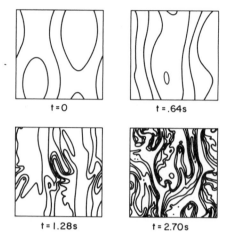

Fig. 4.35. Plasma density contours showing the development of the equatorial irregularities at four selected times. The contour spacing is 2.5% of the ambient density, and the grid spacing is 1.5 m. [After McDonald *et al.* (1975). Reproduced with permission of the American Geophysical Union.]

sion of the direct interaction approximation developed by Kraichnan (1967). One of the crucial features is that the nonlinear damping due to wave emission and cascade is much larger than the linear damping rate because these waves interact for a long time. This in turn relies on the fact that the waves are almost dispersionless and therefore all the waves have nearly equal group velocities. Farley (1985) has reviewed this theory and we follow his approach here. The object is to predict the two-dimensional spectrum of density fluctuation turbulence $I_k(\mathbf{k})$ which is such that the total density fluctuation strength (dimensionless) is given by

$$\langle (\Delta n/n)^2 \rangle = \iint I_k(\mathbf{k}) \, d_\perp \mathbf{k}$$

For isotropic turbulence in the plane perpendicular to $\mathbf{B} = B\hat{a}_y$, $I_k(\mathbf{k}) = I_k(k)$ where $k = |\mathbf{k}| = (k_x^2 + k_z^2)^{1/2}$.

The goal is to derive a differential equation for $I_k(k)$ which describes the flow of energy in the system as a function of k. There are two main effects: the linear growth rate, which includes both linear growth and damping, and the nonlinear flow of energy from eddy to eddy through the spectrum. In a steady state the input of energy at long wavelengths equals the dissipation at short wavelengths. The eddy process forms a mechanism to transfer energy from the growth portion of the spectrum to the region where damping is strongest. The argument goes as follows. In a range of wave numbers Δk the total density fluctuation strength is

$$(\Delta n/n)^2_{\Delta k} = \int_{\Delta k} I_k 2\pi k \, dk \simeq 2\pi I_k k \, \Delta k$$

If we consider the range $\Delta k = k/2\pi$, that is, a bandwidth in k space equal to $k/2\pi$, then

$$(\Delta n/n)^2_{\Delta k} \simeq k^2 I_k$$

Since from (4.40) $\delta V/V_0 = [\nu_i/\Omega_i(1 + \Psi_0)](\delta n/n)$, we also see that in this range of Δk the velocity fluctuation strength is given by

$$\delta V^2_{\Delta k} \propto V_0^2 k^2 I_k$$

Now the classical turbulence argument is that the eddy growth rate (Γ_k) is given by the inverse of the eddy turnover time τ_k, where

$$\Gamma_k = (\tau_k)^{-1} = k \, \delta V_{\Delta k} \propto V_0 k^2 (I_k)^{1/2}$$

This is the time it takes for the material in the eddy to move one eddy scale size (k^{-1}). The total energy ε_k in a given eddy is proportional to the $\delta V^2_{\Delta k}$ given above and so the rate of energy loss is given by

$$(\varepsilon_k/\tau_k)_{\Delta k} = \varepsilon_k \Gamma_k \propto V_0^3 k^4 I_k^{3/2} \tag{4.51}$$

The rate of gain or loss of energy in the same wavelength band from the linear growth and damping processes is determined by the linear growth/damping rate γ_k averaged over all angles at wavelength k. In a time-stationary steady state it must be the case that the spectrum $\varepsilon_k(k)$ has the property that

$$k \, d/dk(\varepsilon_k \Gamma_k) = \gamma_k \varepsilon_k \tag{4.52}$$

so that the steady-state energy spectral density remains the same in any k interval. From linear theory for the primary gradient drift process γ_k is of the form

$$\gamma_k = A - Bk^2 \tag{4.53}$$

where

$$A = \frac{\nu_i V_D}{2(1 + \Psi_0)^2 \Omega_i L} - 2\alpha n_0$$

$$B = \frac{\Psi_0}{\nu_i(1 + \Psi_0)}[C_s^2 - V_D^2(1 + \Psi_0)^{-2}/2]$$

Substituting (4.51) and (4.53) into (4.52) yields

$$d/dk(k^4 I_k^{3/2}) \propto V_0^{-1} k I_k(A - Bk^2)$$

This differential equation may readily be solved to yield

$$I(x) = x^{-8/3}[1 - x^{-2/3} - (x^{4/3} - 1)/2S]^2 \tag{4.54}$$

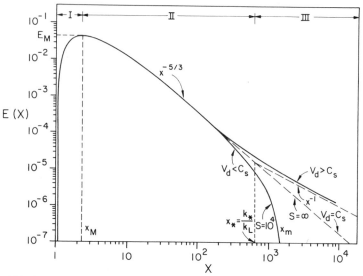

Fig. 4.36. Theoretical one-dimensional power spectrum of the electrojet turbulence. [After Sudan (1983). Reproduced with permission of the American Geophysical Union.]

where $x = k/k_c$, k_c is some long wavelength cutoff, $I(x) = (V_0^2 k_c^4/A^2)I_k$, and $S = A/(Bk_c^2)$. From the definition of γ_k we see that A represents growth and B damping, so that S is therefore a "strength parameter" for the process.

A plot of the quantity $xI(x)$ is given in Fig. 4.36 for values of S equal to 10^4 and ∞ and for negative values. The latter correspond to drifts greater than the primary two-stream value. For positive large values of S the curve rises to a peak value where $x \approx 3$ and then follows a power law for larger x. The value of the power law index for $E(x) = xI(x)$ is the same as the value that would be measured by a one-dimensional cut through the turbulent plasma using devices that measure either $(\delta n/n)^2$ or $(\delta E)^2$, since, as noted above, the one-dimensional measurement spectrum integrates out one power of k and has the same shape as $xI(x)$ (proportional to kI_k). The power law regime occurs only for large S and occurs in regime II in Fig. 4.36, where the eddy growth rate dominates the linear growth. This corresponds to the inertial subrange in neutral turbulence theory (Kolmogorov, 1941). It is interesting that the same one-dimensional spectral form is predicted for this plasma case as for the three-dimensional neutral fluid turbulence in the inertial subrange. For finite S, the spectrum becomes very steep at large k. This corresponds to the viscous subrange in neutral turbulence and in the present case is due to diffusive damping. For S infinite or negative the fluid theory breaks down due to the excitation of short-wavelength waves, which require a kinetic description.

There are a number of ways to check this theory. First, the numerical simulations may be tested against the analytic expression (4.54). The numerical results of McDonald *et al.* (1974, 1975) and Keskinen *et al.* (1979) seem to disagree with the prediction since they report $I_k \propto k^{-3.5}$ rather than $k^{-\gamma/3}$. This problem can be reconciled as shown in Fig. 4.37. The data points in this figure come from the simulation work reported by Keskinen *et al.* (1979). The curved line is a fit to the analytic calculation of Sudan (1983) described above for $k_d = 15k_s$, where k_s is the wave number corresponding to the physical size of the grid. The fit is quite good but it must be realized that k_c, the outer scale for the process, must be considerably smaller than k_s to yield such a steep slope in the range of k space plotted. For smaller values of k, which are not plotted, Sudan's calculations yield a power spectrum of $k^{-2.67}$. Presumably *if* the simulations and theory are in agreement and *if* the simulation occurred in a larger "box," the calculated power spectra would yield a $k^{-2.67}$ two-dimensional power law at small k. However, such a result has not yet been obtained in any simulation and the comparison shown in Fig. 4.37 must remain somewhat suspect.

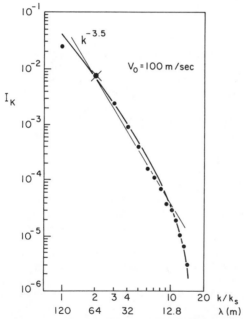

Fig. 4.37. Spectra (dots) from the numerical simulation of Keskinen *et al.* (1979) compared with the theory (solid curve). The curve was normalized at the point marked with the cross and the dissipative cutoff wave number k_d was chosen for the best fit to be $15k_s$, where k_s is determined by the size of the numerical grid and is somewhat analogous to (but in practice considerably larger than) the long-wavelength k_c discussed in the text. [After Sudan (1983). Reproduced with permission of the American Geophysical Union.]

4.9 Future Directions

The physics of equatorial plasma instabilities involves a complex interplay between aeronomic, electrodynamic and plasma physical processes. At the present juncture we have a good understanding of the sources of free energy which lead to the generation of nonthermal fluctuations in the equatorial plasma. The present research thrusts involve (a) understanding the seasonal, local time, and solar cycle control of the onset of instabilities and (b) understanding the nonlinear evolution of the plasma processes themselves. In both of these areas numerical modeling is becoming a very common research tool. Such models allow the researcher to vary different physical parameters and see the effect of the variation. In a geophysical context such control is not available to the experimenter and hence the modeling effort is very important. Major future strides in the space research area will come from direct comparison between experimental results and numerical simulations.

References

Balsley, B. B. (1969). Some characteristics of non-two-stream irregularities in the equatorial electrojet. *J. Geophys. Res.* **74**, 2333.

Basu, Su., Basu, Sa., Aarons, J., McClure, J. P., and Cousins, M. D. (1978). On the coexistence of kilometer- and meter-scale irregularities in the nighttime equatorial F-region. *JGR, J. Geophys. Res.* **83**, 4219.

Beer, T. (1973). Spatial resonance in the ionosphere. *Planet. Space Sci.* **21**, 297.

Buneman, O. (1963). Excitation of field aligned sound waves by electron streams. *Phys. Rev. Lett.* **10**, 285.

Chen, F. F. (1974). "Introduction to Plasma Physics." Plenum, New York.

Cohen, R., and Bowles, K. L. (1967). Secondary irregularities in the equatorial electrojet. *J. Geophys. Res.* **72**, 885.

Costa, E., and Kelley, M. C. (1978a). On the role of steepened structures and drift waves in equatorial spread F. *JGR, J. Geophys. Res.* **83**, 4359.

Costa, E., and Kelley, M. C. (1978b). Linear theory for the collisionless drift wave instability with wavelengths near the ion gyroradius. *JGR, J. Geophys. Res.* **83**, 4365.

Dungey, J. W. (1956). Convective diffusion in the equatorial F-region. *J. Atmos. Terr. Phys.* **9**, 304.

Farley, D. T. (1963). A plasma instability resulting in field-aligned irregularities in the ionosphere. *J. Geophys. Res.* **68**, 6083.

Farley, D. T. (1985). Theory of equatorial electrojet plasma waves: New developments and current status. *J. Atmos. Terr. Phys.* **47**, 729.

Farley, D. T., Balsley, B. B., Woodman, R. F., and McClure, J. P. (1970). Equatorial spread F: Implications of VHF radar observations. *J. Geophys. Res.* **75**, 7199.

Fejer, B. G., and Kelley, M. C. (1980). Ionospheric irregularities. *Rev. Geophys. Space Phys.* **18**, 401.

Fejer, B. G., Farley, D. T., Balsley, B. B., and Woodman, R. F. (1975). Vertical structure of the VHF backscattering region in the equatorial electrojet and the gradient drift instability. *JGR, J. Geophys. Res.* **80**, 1313.

Hoh, F. C. (1963). Instability of Penning-type discharge. *Phys. Fluids* **6**, 1184.

Huba, J. D., and Ossakow, S. L. (1981a). Lower hybrid drift waves in equatorial spread F. *JGR, J. Geophys. Res.* **86**, 829.

Huba, J. D., and Ossakow, S. L. (1981b). Diffusion of small scale irregularities during equatorial spread F. *JGR, J. Geophys. Res.* **86**, 9107.

Hudson, M. K., and Kennel, C. F. (1975). Linear theory of equatorial spread F. *JGR, J. Geophys. Res.* **80**, 4581.

Kadomtsev, B. B. (1965). "Plasma Turbulence." Academic Press, New York.

Kelley, M. C., Baker, K. D., and Ulwick, J. C. (1979). Late time barium cloud striations and their possible relationship to equatorial spread F. *JGR, J. Geophys. Res.* **84**, 1898.

Kelley, M. C., Larsen, M. F., LaHoz, C. A., and McClure, J. P. (1981). Gravity wave initiation of equatorial spread F: A case study. *JGR, J. Geophys. Res.* **86**, 9087.

Kelley, M. C., Livingston, R. C., Rino, C. L., and Tsunoda, R. T. (1982). The vertical wave number spectrum of topside equatorial spread F: Estimates of backscatter levels and implications for a unified theory. *JGR, J. Geophys. Res.* **87**, 5217.

Kelley, M. C., LaBelle, J., Kudeki, E., Fejer, B. G., Basu, Sa., Basu, Su., Baker, K. D., Hanuise, C., Argo, P., Woodman, R. F., Swartz, W. E., Farley, D. T., and Meriwether, J. W., Jr. (1986). The Condor equatorial spread F campaign: Overview and results of the large-scale measurements. *JGR, J. Geophys. Res.* **91**, 5487.

Kelley, M. C., Seyler, C. E., and Zargham, S. (1987). Collisional interchange instability 2. A comparison of the numerical simulations with the *in situ* experimental data. *JGR, J. Geophys. Res.* **92**, 10073.

Keskinen, M. J., Sudan, R. N., and Ferch, R. L. (1979). Temporal and spatial power spectrum studies of numerical simulations of type 2 gradient drift irregularities in the equatorial electrojet. *JGR, J. Geophys. Res.* **84**, 1419.

Keskinen, M. J., Ossakow, S. L., and Chaturvedi, P. K. (1980a). Preliminary report of numerical simulations of intermediate wavelength collisional Rayleigh–Taylor instability in equatorial spread F. *JGR, J. Geophys. Res.* **85**, 1775.

Keskinen, M. J., Ossakow, S. L., and Chaturvedi, P. K. (1980b). Preliminary report of numerical simulations of intermediate wavelength $\mathbf{E} \times \mathbf{B}$ gradient drift instability in equatorial spread F. *JGR, J. Geophys. Res.* **85**, 3485.

Klostermeyer, J. (1978). Nonlinear investigation of the spatial resonance effect in the nighttime equatorial F-region. *JGR, J. Geophys. Res.* **83**, 3753.

Kolmogorov, A. N. (1941). The local structure of turbulence in incompressible viscous fluids for very high Reynolds numbers. *Dokl. Akad. Nauk SSSR* **30**, 301.

Kraichnan, R. H. (1967). Inertial ranges in two-dimensional turbulence. *Phys. Fluids* **11**, 671.

Kudeki, E., Farley, D. T., and Fejer, B. G. (1982). Long wavelength irregularities in the equatorial electrojet. *Geophys. Res. Lett.* **9**, 684.

Kudeki, E., Fejer, B. G., Farley, D. T., and Hannise, C. (1987). The CONDOR Equatorial Electrojet Campaign: radar results. *JGR, J. Geophys. Res.* **92**, 13561.

LaBelle, J. (1985). Mapping of electric field structures from the equatorial F-region to the underlying E-region. *JGR, J. Geophys. Res.* **90**, 4341.

LaBelle, J., and Kelley, M. C. (1986). The generation of kilometer scale irregularities in equatorial spread F. *JGR, J. Geophys. Res.* **91**, 5504.

LaBelle, J., Kelley, M. C., and Seyler, C. E. (1986). An analysis of the role of drift waves in equatorial spread F. *JGR, J. Geophys. Res.* **91**, 5513.

Livingston, R. C., Rino, C. L., McClure, J. P., and Hanson, W. B. (1981). Spectral characteristics of medium-scale equatorial F-region irregularities. *JGR, J. Geophys. Res.* **86**, 2421.

McDonald, B. E., Coffey, T. P., Ossakow, S., and Sudan, R. N. (1974). Preliminary report of numerical simulation of type 2 irregularities in the equatorial electrojet. *JGR, J. Geophys. Res.* **79**, 2551.

McDonald, B. E., Coffey, T. P., Ossakow, S. L., and Sudan, R. N. (1975). Numerical studies of type 2 equatorial electrojet irregularity development. *Radio Sci.* **10**, 247.

McDonald, B. E., Keskinen, M. J., Ossakow, S. L., and Zalesak, S. T. (1980). Computer simulation of gradient drift instability processes in operation Avefria. *JGR, J. Geophys. Res.* **85**, 2143.

Ossakow, S. L. (1981). Spread F theories— a review. *J. Atmos. Terr. Phys.* **43**, 437.

Perkins, F. W., and Doles, J. H. (1975). Velocity shear and the $\mathbf{E} \times \mathbf{B}$ instability. *JGR, J. Geophys. Res.* **80**, 211.

Pfaff, R. F., Kelley, M. C., Fejer, B. G., Maynard, N. C., and Baker, K. D. (1982). *In-situ* measurements of wave electric fields in the equatorial electrojet. *Geophys. Res. Lett.* **9**, 688.

Pfaff, R. F., Kelley, M. C., Fejer, B. G., Maynard, N. C., Brace, L. M., Ledley, B. G., Smith, L. G., and Woodman, R. F. (1985). Comparative *in-situ* studies of the unstable daytime equatorial E region. *J. Atmos. Terr. Phys.* **47**, 791.

Pfaff, R. F., Kelley, M. C., Kudeki, E., Fejer, B. G., and Baker, K. D. (1988a). Electric field and plasma density measurements in the strongly-driven daytime equatorial electrojet. 1. The unstable layer and gradient drift waves. *JGR, J. Geophys. Res.* **92**, 13578.

Pfaff, R. F., Kelley, M. C., Kudeki, E., Fejer, B. G., and Baker, K. D. (1988b). Electric field and plasma density measurements in the strongly-driven daytime equatorial electrojet. 2. Two-stream waves. *JGR, J. Geophys. Res.* **92**, 13597.

Prakash, S., and Pandy, R. (1980). On the production of large scale irregularities in the equatorial F region. *Int. Symp. Equatorial Aeronomy, 6th* pp. 3–7.

Prakash, S., Subbaraya, B. H., and Gupta, S. P. (1972). Rocket measurements of ionization irregularities in the equatorial ionosphere at Thumba and identification of plasma irregularities. *Indian J. Radio Space Phys.* **1**, 72.

Rino, C. L., Tsunoda, R. T., Petriceks, J., Livingston, R. C., Kelley, M. C., and Baker, K. D. (1981). Simultaneous rocket-borne beacon and *in situ* measurements of equatorial spread F— intermediate wavelength results. *JGR, J. Geophys. Res.* **86**, 2411.

Röttger, J. (1973). Wave like structures of large scale equatorial spread F irregularities. *J. Atmos. Terr. Phys.* **35**, 1195.

Satyanarayana, P., Guzdar, P. N., Huba, J. D., and Ossakow, S. L. (1984). Rayleigh–Taylor instability in the presence of a stratified shear layer. *JGR, J. Geophys. Res.* **89**, 2945.

Schmidt, M. J., and Gary, S. P. (1973). Density gradients and the Farley–Buneman instability. *J. Geophys. Res.* **78**, 8261.

Simon, A. (1963). Instability of a partially ionized plasma in crossed electric and magnetic fields. *Phys. Fluids* **6**, 382.

Sudan, R. N. (1983). Unified theory of type 1 and type 2 irregularities in the equatorial electrojet. *JGR, J. Geophys. Res.* **88**, 4853.

Sudan, R. N., and Keskinen, M. (1977). Theory of strongly turbulent two dimensional convention of low pressure plasma. *Phys. Rev. Lett.* **38**, 966.

Sudan, R. N., and Keskinen, M. J. (1984). Unified theory of the power spectrum of intermediate wavelength ionospheric electron density fluctuations. *JGR, J. Geophys. Res.* **89**, 9840.

Sudan, R. N., Akinrimisi, J., and Farley, D. T. (1973). Generation of small-scale irregularities in the equatorial electrojet. *J. Geophys. Res.* **78**, 240.

Tsunoda, R. T. (1980a). Backscatter measurements of 11 cm equatorial spread F irregularities. *Geophys. Res. Lett.* **7**, 848.

Tsunoda, R. T. (1980b). Magnetic-field-aligned characteristics of plasma bubbles in the nighttime equatorial ionosphere. *J. Atmos. Terr. Phys.* **42**, 743.

Tsunoda, R. T. (1981). Time evolution and dynamics of equatorial backscatter plumes. 1. Growth phase. *JGR, J. Geophys. Res.* **86**, 139.

Tsunoda, R. T. (1983). On the generation and growth of equatorial backscatter plumes. 2. Structuring of the west walls of upwellings. *JGR, J. Geophys. Res.* **88**, 4869.

Tsunoda, R. T., Livingston, R. C., McClure, J. P., and Hanson, W. B. (1982). Equatorial plasma bubbles: Vertically elongated wedges from the bottomside F layer. *JGR, J. Geophys. Res.* **87,** 9171.

Valladares, C. E. (1983). Bottomside sinusoidal waves—a new class of equatorial plasma irregularities. Ph.D. Dissertation, University of Texas at Dallas (University Microfilm International, Ann Arbor, Michigan).

Valladares, C. E., Hanson, W. B., McClure, J. P., and Cragin, B. L. (1983). Bottomside sinusoidal irregularities in the equatorial F region. *JGR, J. Geophys. Res.* **88,** 8025.

Vickrey, J. F., and Kelley, M. C. (1982). The effects of a conducting E layer on classical F-region cross-field plasma diffusion. *JGR, J. Geophys. Res.* **87,** 4461.

Vickrey, J. F., Kelley, M. C., Pfaff, R., and Goldman, S. R. (1984). Low-altitude image striations associated with bottomside equatorial spread F: Observations and theory. *JGR, J. Geophys. Res.* **89,** 2955.

Whitehead, J. D. (1971). Ionization disturbances caused by gravity waves in the presence of an electrostatic field and background wind. *J. Geophys. Res.* **76,** 238.

Woodman, R. F., and LaHoz, C. (1976). Radar observations of F-region equatorial irregularities. *JGR, J. Geophys. Res.* **81,** 5447.

Woodman, R. F., Pingree, J. E., and Swartz, W. E. (1985). Spread-F-like irregularities observed by the Jicamarca radar during the daytime. *Atmos. Terr. Phys.* **47,** 867.

Zalesak, S. T., and Ossakow, S. L. (1980). Nonlinear equatorial spread F: Spatially large bubbles resulting from large horizontal scale initial perturbations. *JGR, J. Geophys. Res.* **85,** 2131.

Zalesak, S. T., Ossakow, S. L., and Chaturvedi, P. K. (1982). Nonlinear equatorial spread F: The effect of neutral winds and background Pedersen conductivity. *JGR, J. Geophys. Res.* **87,** 151.

Zargham, S., and Seyler, C. E. (1987). Collisional interchange instability. 1. Numerical simulations of the intermediate scale irregularities. *JGR, J. Geophys. Res.* **92,** 10089.

Chapter 5 | The Mid-Latitude Ionosphere

To a first approximation, study of the mid-latitude ionosphere is based on the principles discussed in the book by Rishbeth and Garriott (1969). Since our primary purpose is to extend rather than repeat the former work, this chapter will necessarily be incomplete. Our goal is merely to treat what we feel are some of the more interesting problems which have arisen in mid-latitude ionospheric physics. In both a geographic and dynamical sense, the mid-latitude zone is a buffer between the low-latitude processes discussed in Chapters 3 and 4 and the high-latitude phenomena presented in subsequent chapters. Both electric fields and perturbed neutral winds penetrate from high-latitude sources, while equatorial plasma streams into the region along the magnetic field lines. Mid-latitude auroras occur, as do airglow depletions which reflect the plasma bubbles discussed in Chapter 4. Lightning electric fields easily penetrate the ionosphere, creating intense whistler mode waves and ionospheric plasma instabilities, while gravity waves continually roll in from high latitudes as well as from stratospheric and tropospheric sources. These are just a few of the dynamical interactions which continue to make the study of mid-latitude ionospheric physics challenging and interesting as well as deserving of a more detailed treatise than we can include here.

5.1 Competing Influences on the Tropical and Mid-Latitude Ionospheres

5.1.1 Background Material

In this context we might define the tropical zone as that region where the magnetic field has a significant dip angle yet cannot be considered to be nearly vertical, as is the case in the auroral zone and polar cap. The latitudes of the Arecibo and St. Santin Radar Observatories are ideally suited for study of this zone (see Appendix A for position data on several sites discussed in the text). The dip angle at Arecibo, for example, is 50°, which allows approximate geometric equality between forces parallel and perpendicular to **B.** Important information also comes from the Millstone Observatory, which is equatorward of St. Santin in geographic coordinates but poleward of St. Santin geomagnetically. Millstone is thus in the transition zone between high and low latitudes. We term this region mid-latitude as opposed to tropical. Until recently most of our dynamical information has come from vector F-region plasma drift measurements made with the incoherent scatter method at these sites. As discussed in Chapter 3, drifts measured perpendicular to **B** in the F region can be interpreted unambiguously and yield the ambient electric field via the relationship $\mathbf{E} = -\mathbf{V} \times \mathbf{B}$. The parallel drift component is much more complex, however, since gravity, neutral winds, and pressure gradients all contribute. The recent availability of neutral wind measurements at Arecibo, Fritz Peak, and other mid-latitude sites has greatly helped interpretation of the incoherent scatter data. Satellite data on neutral and plasma densities and temperatures have been very useful but the dynamics of the mid-latitude region are difficult to study at orbital speeds.

The interrelationship of the forces which act on the ionospheric plasma in the direction parallel to **B** is clear from (2.36a), which we reproduce here:

$$V'_{j\parallel} = [b_j\mathbf{E}' - D_j\nabla n/n + (D_j/H_j)\hat{g}] \cdot \hat{B} \tag{5.1}$$

The primes indicate that the quantities are measured in the neutral frame in which the neutral wind velocity vanishes. Transforming to the earth-fixed frame where the neutral wind has a value **U**, the parallel component of the velocity $\mathbf{V}_j$ is given by

$$V_{j\parallel} = \mathbf{U} \cdot \hat{B} + V'_{j\parallel} \tag{5.2}$$

The plasma is thus closely coupled to the neutral gas motion along **B** but its velocity is modified by the term $V'_{j\parallel}$. For simplicity, in (5.1) we have continued to ignore temperature gradients, which must be included in a complete treatment of the total pressure gradient.

During the day, plasma production or loss by photoionization and recombination dominates in the E and lower F regions, creating the horizontally stratified, slowly varying plasma content of the ionosphere. Rishbeth and Garriott (1969) describe the relevant physics and chemistry in great detail and we refer the reader to their discussion. Below the F peak, production and loss determine the ionization profile, which increases sharply in the 90–100-km range and then more slowly up to a peak, which is typically near 300 km.

Above the F peak we expect the plasma to be in a state something akin to diffusive equilibrium in the gravitational field. Because of its light mass the electron gas diffuses much faster than the ions down a pressure gradient, a process which would tend rapidly to destroy any gradient in electron pressure. However, the resultant charge separation is accompanied by an electric field which restrains the electrons and enhances the ion diffusion. Quantitatively, we can argue as follows. In diffusive equilibrium there should be no net flow of the electron gas or of the ion gas. Our calculations are in the neutral gas frame (or, equivalently, we can assume $\mathbf{U} = 0$) and we take the zero-order perpendicular electric field to be zero for now. Parallel to $\mathbf{B}$ the velocity for each species is given by (5.1). Taking $T_i = T_e = T$ for algebraic ease, $n_i = n_e = n$, and setting both velocities equal to zero gives for the northern hemisphere

$$b_e \left[E_\parallel + \frac{k_B T}{ne} \nabla_\parallel n - \frac{mg_\parallel}{e} \right] = 0 \tag{5.3a}$$

$$b_i \left[E_\parallel - \frac{k_B T}{ne} \nabla_\parallel n + \frac{Mg_\parallel}{e} \right] = 0 \tag{5.3b}$$

remembering that the mobility parameter b_j carries the sign of the charge and that e is a positive number. Dividing each equation by the appropriate mobility, taking $m \ll M$, and adding yields

$$E_\parallel = -(M/2e)g_\parallel$$

Likewise, setting the perpendicular components of velocity equal to zero in (2.36b) and using the large-κ case, which is suitable for the upper F region ($\kappa_j = \Omega_j/\nu_j$), we have

$$0 = \frac{b_e}{\kappa_e} \left[\mathbf{E} + \frac{k_B T}{ne} \nabla n - \frac{m\mathbf{g}}{e} \right] \times \hat{B} \tag{5.4a}$$

$$0 = \frac{b_i}{\kappa_i} \left[\mathbf{E} + \frac{k_B T}{ne} \nabla n + \frac{M\mathbf{g}}{e} \right] \times \hat{B} \tag{5.4b}$$

We can solve these two equations for the perpendicular electric field to find $\mathbf{E}_\perp = -M\mathbf{g}_\perp/2e$, and thus combining both components,

$$\mathbf{E} = -(M/2e)\mathbf{g}$$

Finally, substituting this result into (5.4b) yields

$$\nabla n/n = (M/2k_B T)\mathbf{g}$$

This shows that the equilibrium density gradient is vertically downward even though the magnetic field is inclined at an arbitrary angle. This particular case is an example of a general theorem from statistical mechanics which states that in thermal equilibrium a magnetic field cannot affect the distribution of any fluid, ionized or not. The plasma scale height is equal to $2k_B T/Mg$ or more generally $k_B T_i(1 + T_e/T_i)/Mg$ if $T_e \neq T_i$. The plasma acts like a neutral gas in a gravitational field with mean mass equal to the average of m and M. The electric field is $Mg/2e$, and this has the value 0.8 μV/m in an O^+ plasma. This value is quite important when projected parallel to **B**, due to the high conductivity in that direction, which keeps other sources of $E_\parallel$ small. The force associated with this electric field is countered by the pressure gradient.

Since the assumed dynamic equilibrium is between electrons and the dominant species (O^+), light minor ions such as He^+ and H^+ can be accelerated outward from the ionosphere by this electric field. A considerable literature has developed concerning the effect of this $E_\parallel$ on minor light ions and a complete theory of the topside ionosphere requires its consideration. We take up this topic in some detail in the polar case treated in Chapter 7.

Since hydrogen atoms and molecules can escape the earth's gravitational field, the earth is surrounded with a hydrogen gas "geocorona" which interacts with oxygen ions via the charge exchange reaction

$$O^+ + H \rightleftarrows H^+ + O$$

which is a very rapid process. The result is that a transition occurs between an oxygen and a hydrogen plasma between 500 and 1000 km, depending on seasonal and other effects (Vickrey *et al.*, 1979). Since light ions such as hydrogen can escape gravity, our simple assumption of diffusive equilibrium breaks down and a net upward flux of plasma is possible during the day at the "top" of the ionosphere. The closed dipole magnetic flux tubes at tropical and mid-latitudes act as a reservoir for plasma created during the day by photoionization and carried upward by escaping light ions. At night the plasma can flow back down, tending to maintain the density in the ionosphere. The result is a complex interaction between the ionosphere and a region of hydrogen plasma trapped by the dipole magnetic field.

To summarize, without horizontal transport photoionization coupled with diffusion, recombination, and charge exchange with the geocorona would completely determine the properties of the ionosphere. Rishbeth and Garriott (1969) discuss this situation in great detail, deriving the so-called Chapman layer form for the ionosphere and discussing the various ionospheric layers, E, F_1, F, etc. A diurnal variation in peak plasma density would be expected and some balance would arise between the low-altitude production and outflow of plasma along magnetic field lines during the day and low-altitude recombination and inflow of plasma at night.

The situation is much more complicated than this, however, since neutral winds and electric fields also move the plasma. These forces greatly affect the altitude of ionospheric plasma, particularly in the F region. Since altitude determines the speed with which recombination occurs, these dynamical processes also affect the plasma content. As an example of the variability of the mid-latitude ionosphere, data showing the change in altitude of the F-region peak density ($h_m F2$) relative to average hourly values are presented in Fig. 5.1 for an

HEIGHT and CRITICAL FREQUENCY at 45 deg

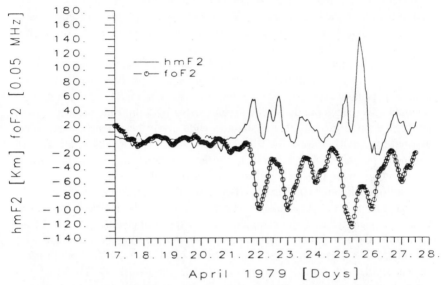

April 1979 [Days]

Fig. 5.1. A plot of the height of the F peak and the value of the peak plasma frequency relative to the long-term hourly average for April as a function of time for eleven days in April, 1979. [After Forbes *et al.* (1988). Reproduced with permission of the American Geophysical Union.]

11-day interval. The first four days were very quiet geomagnetically while the next six days were very active. Also plotted is the maximum value of the plasma frequency in the F region of the ionosphere (f_0F2). The electron density n is related to the plasma frequency f_p by $n = [f_p/8900]^2$ (see Appendix B). Both parameters vary drastically in a pattern referred to as an ionospheric storm. The general trend is for a decrease in plasma density during the storm. Superimposed on the decrease are variations of f_0F2 which seem to correlate positively with the height variations.

The plasma content also affects the electrical conductivity of the ionosphere (and hence the electric field) as well as ion drag (and hence the neutral wind). Throwing in the conjugate hemisphere, which in general will have a different neutral wind and conductivity (but presumably the same electric field), one has a very complex coupled electrodynamic/dynamic system (without even mentioning external influences such as high-latitude electric fields and aurorally driven winds!). We consider some of these complexities in the remainder of this chapter.

5.1.2   The Equatorial Anomaly

Before discussing mid-latitude electrodynamics *per se,* there is an interesting and important tropical ionospheric effect arising from *equatorial* electrodynamics. As mentioned in Chapter 3, the zonal electric field at the magnetic equator is eastward during the day, which creates a steady upward $\mathbf{E} \times \mathbf{B}/B^2$ plasma drift. Just after sunset this eastward electric field is enhanced and the F-region plasma can drift to very high altitudes where recombination is slow, while the low-altitude plasma decays quickly once the sun sets. The result is something of a fountain effect since the dense equatorial plasma rises until the pressure forces are high enough in (5.1) that it starts to slide down the magnetic field lines, assisted by gravity, toward the tropical ionosphere. This results in a region of enhanced plasma density referred to as the equatorial anomaly. The various forces acting on the plasma are illustrated schematically in Fig. 5.2a.

A full global model capable of self-consistently generating electric fields and plasma content is not yet available. However, if the zonal electric field is taken as a given quantity based, for example, on the experimental data shown in Chapter 3, then the actions of production and recombination can be combined with the vertical $\mathbf{E} \times \mathbf{B}$ motion of the ionosphere to yield a reasonable model for the diurnal and latitude variation of the ionospheric plasma density. The effect has been illustrated in model calculations by D. Andersen (private communication, 1979). In the upper contour plot of Fig. 5.2b the zonal electric field was taken to be zero, while in the lower plot it had the typical variation with local time measured at Jicamarca. A considerable difference in the global ionospheric density is predicted in the two different electrodynamic states. In the latter case the zonal electric field drove the plasma to heights where recombination was small. The plasma thus survived longer and had time to flow down the magnetic field lines to higher-latitude regions, forming two symmetrical enhancements on each side of the equator through the combined action of electrodynamic uplift, pressure gradients, and gravity. In either model, a low-density trough occurs just before sunrise. The results, which include the zonal electric field, fit ionospheric observations of the equatorial anomaly quite well.

A visual indication of the equatorial anomaly obtained from an image of the earth taken from the DE-1 satellite located nearly 8000 km above the earth is reproduced here in Fig. 5.3. The image corresponds to an emission line which is produced by O^+ ions when they recombine with electrons. In the photograph, the dayside of the earth is very bright and a ringlike halo surrounds the polar regions. The latter is the auroral oval. Of interest here is the band of light just off the magnetic equator, a remarkable visual representation of the equatorial anomaly region.

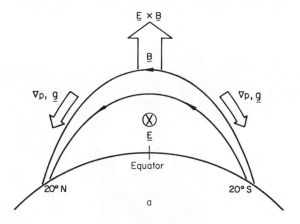

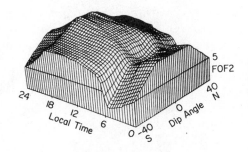

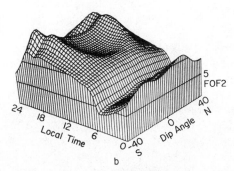

Fig. 5.2. (a) Schematic diagram of how plasma uplift via electric fields transports plasma from equatorial to tropical zones. (b) Contour plots of FOF2 in megahertz (the peak F-region electron plasma frequency, which is proportional to the square of the electron density) for zero zonal electric field (upper plot) and for a typical diurnal variation of the zonal electric field. (Courtesy of D. Anderson.)

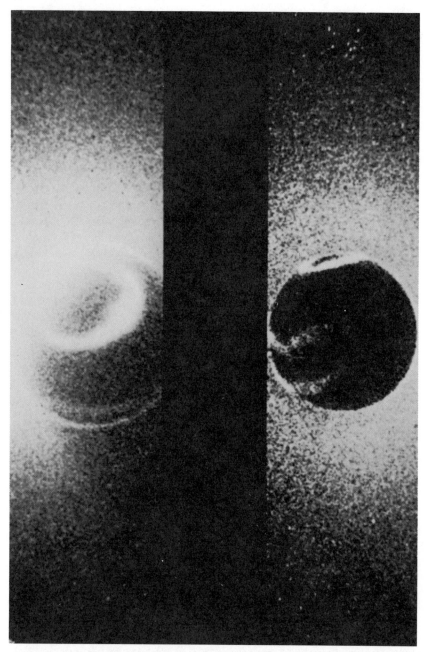

Fig. 5.3. Photographs of the earth's disk with images due primarily to neutral oxygen emissions at 130.4 and 135.6 nm made from the imager on the Dynamics Explorer 1 satellite. The ring of airglow emissions seen at polar latitudes is due to the aurora. Of interest here are the emission bands at mid-latitudes. (Courtesy of J. D. Craven, L. A. Frank, and R. L. Rairden.)

5.1.3   The Corotation Electric Field and Formation of the Plasmasphere

An important dynamical effect involves the nontrivial fact that the plasma on low-latitude flux tubes to first order corotates with the earth. That this must occur can be proved as follows. If the ionospheric plasma did not corotate, in the nonrotating plasma frame there would be a very large neutral wind U_R and a current $J = \sigma \cdot (U_R \times B)$. Since σ depends on altitude this current is not divergence-free and an electric field will build up in the nonrotating frame until $J = \sigma \cdot (E_R + U_R \times B) = 0$. The current vanishes when the plasma $E \times B$ drifts at exactly the same velocity as the earth rotates. Transforming back to the rotating frame, we find $E = 0$. Now it is a remarkable fact (due to the frozen-in condition discussed in Chapter 2) that the electric field in the nonrotating frame is transmitted along the magnetic field to magnetospheric heights and causes the entire inner magnetosphere to corotate with the earth, even though there is virtually no neutral gas at high altitudes along those flux tubes which thread the mid-latitude ionosphere.

Turning this argument around, it is exactly on those flux tubes which corotate with the earth that cold ionospheric plasma can build up to the extent that is observed in the dense plasma-filled region termed the plasmasphere. A crude estimate of the latitude below which corotation dominates can be formed by equating the corotation electric field to the electric field of magnetospheric and solar wind origin. The comparison is properly done in the nonrotating frame. At latitudes where the former dominates, flux tubes and the plasma attached to them make a complete circuit of the earth in 1 day, allowing the trapping and buildup of a cold hydrogen plasma of ionospheric origin (taking into account the charge exchange process which converts O^+ into H^+, of course). At higher latitudes the flux tubes follow trajectories in which they are at times connected to the solar wind or extend to very great distances down the magnetic tail. In either case, the flux tube volume is so large that plasma almost continuously flows outward, never building up a high density such as occurs in the plasmasphere.

Quantitatively, we can proceed as follows. The corotational electric field at ionospheric heights is in the meridional direction and is given by Mozer (1973) as

$$E_c = 14 \cos \theta (1 + 3 \sin^2 \theta)^{1/2} \quad \text{mV/m} \tag{5.5}$$

where θ is the latitude and a centered dipole field has been assumed. The coefficient 14 in this expression is determined by the rotation speed of the earth, its radius, and the magnitude of the magnetic dipole moment. For a planet such as Jupiter, these parameters are all larger and the latitude region of corotation dominance is quite a bit higher. Notice that for $\theta = 0$ and $B_{eq} = 3 \times 10^{-5}$ T, $E_c/B_{eq} = 470$ m/s, which equals the zonal rotation speed of the earth at the equator. To relate this field to the magnitude of the magnetospheric electric field, we treat the magnetic field lines as equipotentials. Following the discussion in

Mozer (1970), the meridional electric field component maps from the ionosphere (I) to the magnetosphere (M) (and vice versa) as

$$E_I/E_M = 2L(L - \tfrac{3}{4})^{1/2} \qquad (5.6)$$

where L is the equatorial crossing distance of a dipole field line measured in earth radii. In Fig. 5.4 the corotation field from (5.5) is plotted along with dashed lines giving the ionospheric electric field associated with two different magnitude magnetospheric source fields of 1 and 0.4 mV/m, mapped to the ionosphere using (5.6). In comparing the magnitudes and effects of these two electric field sources (rotation versus solar wind) it is important to note that the relevant reference frame is the one fixed with respect to the sun. In that frame the solar wind blows by and interacts with the magnetosphere, creating the electric field of magnetospheric origin discussed above. The earth rotates in that reference frame, generating the corotation electric field we refer to. The plasma at the equatorial plane $\mathbf{E} \times \mathbf{B}$ then drifts at the whim of whichever source of electric field dominates.

The lower-value magnetospheric curve (0.4 mV/m) crosses the corotation field at an L value near 7. The value corresponding to the higher magnetospheric field matches the corotation field at about $L = 3$. For this range of magneto-

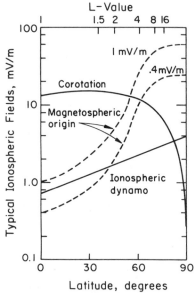

Fig. 5.4. Typical electric fields observed in a nonrotating frame of reference that arise from corotation, the interaction of tidal neutral winds with the ionosphere, and the interaction of the solar wind with the terrestrial magnetic field. [After Mozer (1973). Reproduced with permission of the American Geophysical Union.]

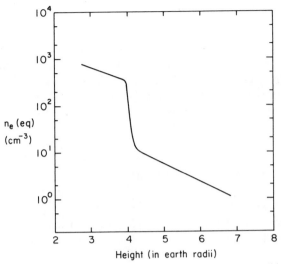

Fig. 5.5. Average equatorial profile of electron density. The magnetic condition represented is one of steady, moderate agitation, with K_p in the range 2–4. [After Angerami and Carpenter (1966). Reproduced with permission of the American Geophysical Union.]

spheric sources and the properties of the earth's dipole field and rotation rate, we see that planetary rotation dominates the physics at the equator within, say, 3–7 earth radii. The "extraterrestrial" source dominates at higher L values.

The region referred to as the plasmasphere corresponds to latitudes where the flux tubes corotate, since such tubes can fill with plasma on the dayside. Sunlight copiously creates F-region plasma, which in turn flows out along the field lines. At night there is not sufficient time to deplete these flux tubes and a dense cold hydrogen plasma fills the plasmasphere. The average profile of plasma density in the magnetospheric equatorial plane reproduced in Fig. 5.5 shows how drastic the decrease in electron density can be near the plasmapause (at $L = 4$ in this case). Since the magnetospheric electric field is quite variable, Fig. 5.4 shows that the position of the plasmasphere and its effect on the mid-latitude ionosphere as a plasma reservoir are strong functions of magnetic activity.

5.2 Electrodynamics of the Tropical and Mid-Latitude Zone

We turn now to consideration of the electric field which exists in the earth-fixed reference frame, where, of course, the corotation field vanishes. In some sense this corresponds to the "electrodynamic weather" just as neutral atmospheric weather-related winds are measured in the rotating frame.

5.2.1   Electric Field Measurements

The lowest line in Fig. 5.4 is a plot of the average value of the Sq dynamo electric field as a function of latitude. This electric field source has been discussed briefly in Chapter 3 and originates from the divergence in the electric currents driven by tidal motions of the neutral atmosphere in the highly conducting E layer (Matsushita, 1967, 1971). The solar heating source dominates the tides, although lunar gravitational effects are also important. Chapman and Lindzen (1970) have discussed tidal theory in great detail and several applications of this theory involving electrodynamic calculations have been made (e.g., Matsushita, 1967; Richmond *et al.*, 1976). The mathematics in these models is complicated but the physics is relatively straightforward and we will not discuss these models extensively here. To first order they yield reasonable agreement with quiet-time ionospheric electric fields, particularly in the daytime low-latitude ionosphere. The results are not at all good at night and also deteriorate with increasing latitude. (A comparison of tidal theory with Jicamarca measurements which shows some of these features was given in Chapter 3.) In this text we emphasize the processes that are less well understood and hence concentrate on the nighttime period at tropical latitudes and the influence of auroral zone effects at the interface between the high-latitude and mid-latitude zones.

To organize our study we use incoherent scatter data from the northern hemisphere sites Arecibo (18.5°; 31°), St. Santin (44.1°; 40°), and Millstone Hill (42.6°; 57°). The first number in the parentheses is the geographic latitude and the second the geomagnetic latitude. Other relevant details concerning these sites are included in Appendix A. In the discussion we distinguish between the tropical and the mid-latitude ionosphere. The former is characterized by an intermediate dip angle and a sufficient distance from the auroral zone that most of the data are usually unaffected by high-latitude electrodynamics. The Arecibo data set is clearly tropical (31° geomagnetic latitude; 50° dip angle). St. Santin and Millstone Hill have similar mid-latitude geographic positions but the latter is much farther north geomagnetically (57° versus 40°). These two sites will thus be the prototypes for our study of mid-latitudes and will be closely investigated for evidence of auroral effects.

Our starting point is the empirical study by Richmond *et al.* (1980), some results of which are presented here in Figs. 5.6a–d. Magnetically quiet periods have been chosen such that, roughly speaking, the 3-hour K_p value is less than or of the order of 3. Overhead data from each radar site have then been averaged in a seasonal manner and plotted. Also plotted as the solid line is an empirical "pseudopotential" fit which uses the data and also satisfies the requirement that $\nabla^2 \phi = 0$ on a spherical shell at F-region heights. The data are plotted in terms of the drift velocity and the electric field. In each panel the left side corresponds to the zonal eastward electric field component and the right side to the equator-

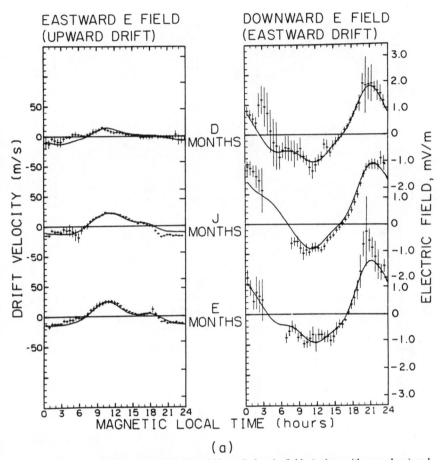

Fig. 5.6. Seasonally averaged quiet-day drifts and electric fields (points with error bars) and model drifts (solid lines) perpendicular to the geomagnetic field at 300 km for (a) Jicamarca, (b) Arecibo, (c) St. Santin, and (d) Millstone Hill. [After Richmond *et al.* (1980). Reproduced with permission of the American Geophysical Union.] (*Figure continues.*)

ward component. The Jicamarca data discussed in detail in Chapter 3 are included for reference.

Considering first the zonal electric field component on the left side of each plot, there are remarkable similarities among all four sites during the period 0300–1400 and among the nonequatorial sites at all times. The latter three data sets seem to exhibit a particularly strong semidiurnal behavior. The Jicamarca zonal component seems to be a mixture of diurnal and semidiurnal oscillations.

ARECIBO

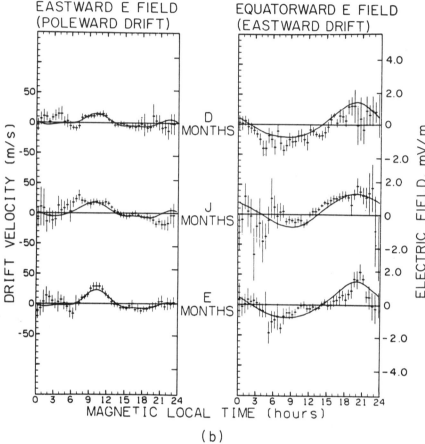

(b)

Fig. 5.6 (*Continued*)

This result is quite reasonable since it is fairly well established that the semi-diurnal atmospheric tide is not very important at equatorial latitudes but that it dominates in the mid-latitude region (Chapman and Lindzen, 1970). Thus, as noted above, these results are reasonably consistent with a tidal E-region source, in particular for the daytime zonal electric field component.

The meridional electric field component in the right-hand panels is generally larger and seems dominated by a diurnal variation at all four sites. It seems curious that one component exhibits a semidiurnal modulation and the other a

ST. SANTIN

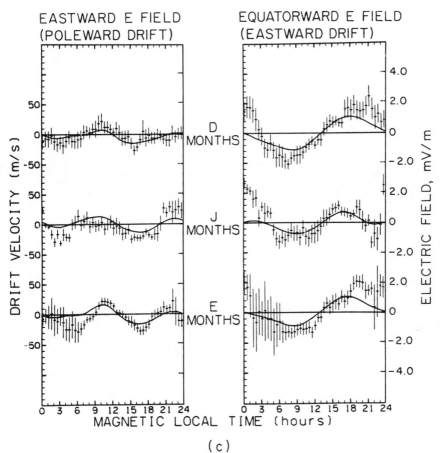

(c)

Fig. 5.6 (*Continued*)

diurnal one! It may be that the two dynamos (E- and F-region) conspire to yield a diurnal pattern for the meridional electric field (zonal drift) at mid- and equatorial latitudes.

Before delving into tropical and mid-latitude electrodynamics in detail, there is an important difference to note between data from the three lowest latitude sites (Jicamarca, Arecibo, and St. Santin) and the Millstone Hill data. In the 1600–2400 local time period even the algebraic *sign* of the Millstone Hill meridional component is different from the other three sites. That is, Millstone

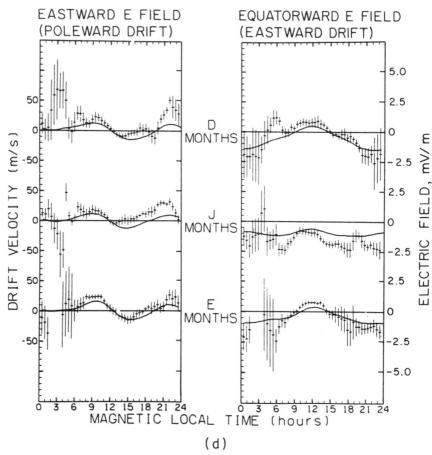

Fig. 5.6 *(Continued)*

registers a poleward field while all three other facilities register an equatorward field ("downward" at Jicamarca). We return to this point later when auroral zone effects are discussed.

The electric field vectors deduced from barium cloud drift data shown in Fig. 5.7 bear out these last comments as well. Notice that in the evening twilight periods the releases all show equatorward electric fields and that the releases were all at latitudes less than 35° geomagnetically. These data agree with St. Santin and Arecibo quite nicely but disagree with Millstone Hill. Mozer (1973) and Gonzales *et al.* (1978) have previously pointed out this discrepancy.

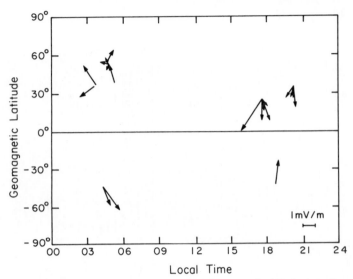

Fig. 5.7. Examples of electric field vectors measured by several mid-latitude barium releases. [After Mozer (1973). Reproduced with permission of the American Geophysical Union.]

5.2.2   The Nighttime Tropical Ionosphere

The detailed dynamics of the earth's ionosphere is governed by a complex interplay between motions of the neutral atmosphere, gravity, electromagnetic forces, pressure gradients, and plasma production and loss. In the nocturnal low- to middle-latitude F layer the latter two processes are relatively unimportant and can be ignored for our purposes once the molecular ions, which have a high recombination coefficient compared to O^+, are removed from the system. Since molecular ions are found at low altitudes, the highly conducting daytime E region virtually disappears, which, as in the equatorial zone, allows the F-layer plasma greater control of the electrodynamics of the region. This can be seen in Fig. 5.8, where the ratios of F-region to E-region height-integrated nighttime conductivities over Arecibo are plotted for several months (Burnside, 1984). Each plot corresponds to an average of 5 days. The ratio $\gamma = \Sigma_{PF}/\Sigma_{PE}$ considerably exceeds unity for all but a few data points. On average the 1981–1982 data (near solar maximum) show nighttime $\gamma \simeq 5$ while in 1983 data $\gamma \simeq 3$. Earlier, Harper and Walker (1977) found γ much closer to unity at solar minimum. This result most likely reflects the lower F-layer plasma densities which occur at solar minimum.

The Arecibo Observatory is particularly well suited to the study of interrelationships between these forces and much of our knowledge concerning the tropical off-equatorial ionosphere comes from Arecibo data. The near-45° dip angle

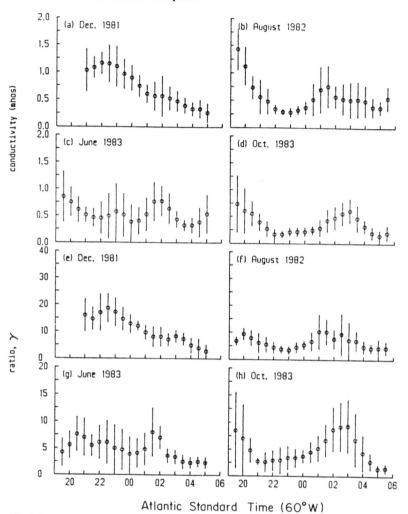

Fig. 5.8. Monthly average values of (a–d) the height-integrated F-region Pedersen conductivity and (e–h) the ratio between the F- and E-region integrated Pedersen conductivities, γ, for four different months. For each season shown, five nights of radar data were used to calculate the mean values and the standard deviation about the mean. (Reproduced with permission of R. Burnside.)

allows each of the important forces an "equal vote" in the control of the F-region plasma dynamics. This is not entirely a geometric effect but stems also from the near equality of the following four velocities at altitudes near the F-layer peak:

$$E/B \; : \; U \; : \; g/\nu_{in} \; : \; V_i^{th}(l/L) \tag{5.7}$$

where V_i^{th} is the ion thermal speed, l the ion mean free path, and L the plasma pressure gradient scale length. The latter term is a measure of the term $\nabla p_i/\rho_i \nu_i$

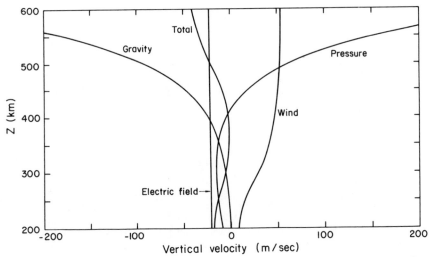

Fig. 5.9. Magnitudes of the four contributions to the vertical plasma flow velocity over Arecibo for a typical plasma density profile, neutral wind, and zonal electric field component. (Data supplied by R. Burnside and R. Behnke.)

in the ion equation of motion, which is the velocity at which a pressure gradient would drive the ion flow velocity against friction with the neutral gas (ignoring ambipolar effects). Representative values and their sum are plotted in Fig. 5.9 for a particular (observed) electron density profile over Arecibo. If any of these terms dominated the dynamics, the F layer as we know it could not be limited to the relatively modest altitude excursions found experimentally. Turning this argument around, we will see that the F layer seeks out an altitude where a balance between these factors is reached.

Numerous measurements of vector plasma velocities have been conducted at Arecibo. The "natural" coordinate system for display of ionospheric F-region drift data is, of course, geomagnetic, so the Arecibo data were displayed in this manner. The remarkable result first reported by Behnke and Harper (1973), an example of which is shown in Fig. 5.10, is the existence of a strong anticorrelation between the components of the drift velocity in the geomagnetic meridian plane, that is, between the components parallel and perpendicular to **B**. The resulting vector plasma motion was therefore nearly horizontal! The parallel and perpendicular dynamics, which seem so nicely separated by the dominance of the magnetic field effect, are thus not at all decoupled. Just as with the neutral atmosphere, nature seems to abhor a vertical velocity, even for the plasma. At least three explanations have been put forward to explain this observation: ion drag (Dougherty, 1961), the F-region dynamo (Rishbeth, 1971), and parallel ion diffusion (Stubbe and Chandra, 1970).

In the first of these theories, the atmosphere is assumed initially to be at rest

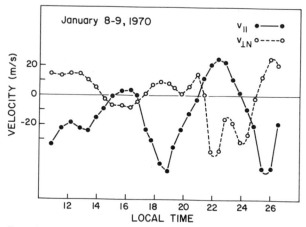

Fig. 5.10. Example of the observed anticorrelation between components of plasma drift perpendicular to **B** (positive is geomagnetic northward) and parallel to **B** (positive is upward along −**B**). [After Behnke and Harper (1973). Reproduced with permission of the American Geophysical Union.]

and an external zonal electric field is applied which puts the plasma in motion via the **E** × **B** drift. For example, a zonally eastward electric field would cause the plasma to move northward and upward at a 40° angle over Arecibo with velocity $-(E/B)\hat{a}_{y'}$ where $\hat{a}_{y'}$ is a unit vector perpendicular to **B** in the magnetic meridian (see Fig. 5.11). We ignore the declination of the magnetic field. The moving ions would, after a while, set the neutral gas in motion via the ion drag effect. Scaling arguments (Holton, 1979) show that the neutrals cannot have a very large vertical velocity so they would eventually attain a poleward horizontal velocity $\mathbf{U} = v\hat{a}_y$ due to momentum transfer from the ions. The plasma carries a zonal current $\mathbf{J}_\perp = \sigma_P \mathbf{E}_\perp$, where we can use a scalar Pedersen conductivity since we are considering perpendicular motion in the F region where σ is diagonal. The **J** × **B** force due to this zonal current is, in fact, the origin of the poleward ion drag force. Once the neutrals begin to move with some velocity **U** the current is given by $\mathbf{J} = \sigma_P(\mathbf{E} + \mathbf{U} \times \mathbf{B})$. In the final equilibrium force-free state, however, **J** × **B** = 0 and no net electric current can flow. In this state the plasma and neutrals must therefore move horizontally together. This is illustrated in Fig. 5.11a and can be seen mathematically as follows. Once the neutrals begin to move with velocity **U**, they produce field-aligned motion of the ions such that $V_{i\parallel} = U_\parallel$. The final vector plasma velocity in geomagnetic (primed) coordinates is given by

$$\mathbf{V}_i = -(E/B)\,\hat{a}_{y'} + v \cos I\,\hat{a}_{z'} \qquad (5.8)$$

where $\hat{a}_{z'}$ is a unit vector parallel to **B**. Since no current can flow in the force-free state (**J** × **B** = 0) and since as noted above the neutral atmospheric velocity

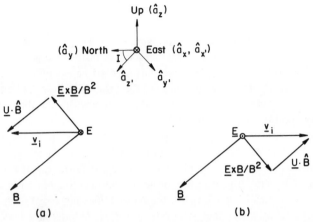

Fig. 5.11. Schematic diagram of motor (a) and dynamo (b) electromagnetic phenomena in the tropical F-region ionosphere.

U must always be nearly perfectly horizontal, we can set $J = \sigma_P(E + U \times B) = 0$ in the final state and solve for **U**. This implies in the F region that the neutrals will be driven geographically northward with the velocity $v = E(B \sin I)^{-1}$. Finally, substituting this result into (5.8) yields the plasma *geographic* velocity $V_y = E(B \sin I)^{-1}$, and $V_z = 0$. The plasma thus does not end up with a vertical velocity component at all, but moves horizontally at the same speed as the neutrals. The time constant for horizontal (meridional) acceleration of the neutrals by a zonal electric field can be estimated from the initial $J \times B$ force via

$$\partial v/\partial t = (J \times B) \cdot \hat{a}_y/n_n M = (\sigma_P EB/n_n M) \sin I$$

In the F region $\sigma_P = ne^2 \nu_{in}/M\Omega_i^2$, which is proportional to the neutral density due to the ν_{in} term. Thus the n_n terms cancel out and the acceleration depends only on the plasma density. The acceleration time is thus of the order of

$$\delta t = (2 \times 10^{15}/n)(\delta vB/E) \quad s$$

and for a density of 3×10^{11} m^{-3}, a time of the order of 1 h is required to accelerate the neutral atmosphere to a velocity v comparable to E/B. Thus, although this mechanism does lead to a net horizontal motion for the plasma, a considerable time is required to establish the effect. In addition to the long time constant, this explanation ignores the problem of the origin of the electric field, which presumably requires an external magnetospheric source or a field applied from the other hemisphere which maps to the local ionosphere along **B**. This process is thus not likely to explain the common observation of horizontal ion motion. Notice that $J \cdot E \geq 0$ initially, as required when electrical energy is converted to the mechanical energy in the neutral atmospheric flow. This process can therefore be termed a "motor."

A dynamo explanation is more promising since the process self-consistently generates an electric field and operates on a much faster time scale, which is of the order of $(\Omega_i)^{-1}$. For example, if we start with zero electric field but a neutral meridional wind $-v\hat{a}_y$ in the equatorward direction (see Fig. 5.11b), a zonally eastward current given by $\mathbf{J} = -\sigma_P v\hat{a}_y \times \mathbf{B}$ will flow with magnitude $\sigma_P vB \sin I$. If boundary conditions are applied which force the net zonal current to be zero, a westward electric field must build up such that $E = vB \sin I$. The net plasma velocity will again be due to the combined effect of the electric field and the neutral velocity. In this case the plasma velocity due to the $\mathbf{E} \times \mathbf{B}$ force and the component of $\mathbf{V}$ parallel to $\mathbf{B}$, when projected into geographic coordinates, becomes $V_y = -v$ and $V_z = 0$. Again, the plasma does not move vertically at all and matches the horizontal neutral speed. Here, $\mathbf{J} \cdot \mathbf{E} \leq 0$ initially since electrical energy is created from a mechanical source as in a dynamo.

Finally, the diffusion effect can also act to create horizontal ion motion. Since the downward velocity due to gravity is proportional to $(\nu_{in})^{-1}$, which increases rapidly with height, there is a natural limitation on the altitude to which either an equatorward wind (which pushes plasma up along $\mathbf{B}$) or an eastward electric field (which $\mathbf{E} \times \mathbf{B}$ drifts the plasma upward and northward) can drive the plasma. Eventually the material falls along $\mathbf{B}$ as fast as it is pushed up and no net vertical motion occurs. We return to this case again in Section 5.4.3 when the so-called Perkins instability is discussed.

We now turn to some experimental data aimed at trying to sort out these various processes. All these effects (and more!) seem to compete for control of the mid-latitude F-layer plasma. An example in which the local dynamo effect seems to explain everything is illustrated in Figs. 5.12a and b. Two components of the plasma drift, the height of the F peak (h_{max}) and the maximum value of the electron density, are plotted in Fig. 5.12a. Unfortunately, the standard coordinate system in which the Arecibo data are presented is different from ours. Here and in subsequent plots $V_\parallel$ is positive for drifts along $-\hat{a}_{z'}$ ($V_\parallel$ is positive in the $-\mathbf{B}$ direction) and $V_{\perp N}$ is positive along the $-\hat{a}_{y'}$ direction. The two middle curves are anticorrelated until about 0340 LT, when sunrise occurred in the *conjugate* hemisphere, which is a nonlocal effect. The plasma motion is therefore nearly horizontal all night long. Indeed, h_{max} only slowly changes from 2300 until 0300, which also shows that the vertical velocity was very small, averaging less than 1 m/s (downward) during this time. Fabry–Perot wind measurements are available for this night, and the measured northward neutral wind (v) and the northward horizontal component of the ion velocity are compared in Fig. 5.12b. The agreement between the meridional neutral wind *direction* and that of the corresponding horizontal ion wind is clear. The ion velocity is much smaller, however, suggesting that there may be a partial shorting out of the electric field in the local E region or in the conjugate hemisphere.

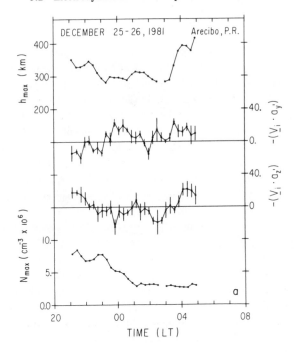

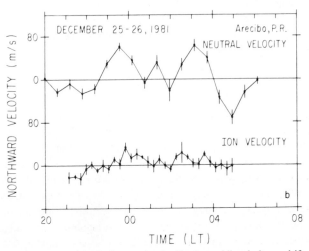

Fig. 5.12. (a) Height of the peak electron density (h_{max}), meridional plasma drift velocity components ($V_{\perp N}$ and $V_{\parallel}$), and peak in electron density (n_{max}) measured over Arecibo. [After Behnke *et al.* (1985). Reproduced with permission of the American Geophysical Union.] (b) Horizontal neutral wind and ion drift velocities (positive northward) for the same night as the data shown in (a). [After Behnke *et al.* (1985). Wind data supplied by R. Burnside. Reproduced with permission of the American Geophysical Union.]

Although consistent with an F-layer dynamo process, even this event raises some interesting questions. The relatively constant large ratio between local neutral wind and ion velocity implies that the electrical loading effect was larger than suggested in Fig. 5.8e for a local E-region load. Also the loading was relatively constant with time. This seems surprising considering the length of time involved. However, we note that the local height-integrated conductivity, which is proportional to $n\nu_{in}$, may have been relatively constant since the layer descended (larger ν_{in}) at the same time as the peak density was decreasing (2100 to 0100 LT). The density and height of the layer were subsequently relatively constant until 0330. If the conjugate F layer had a smaller neutral wind and a relatively constant height-integrated conductivity which was slightly higher than the local F-region conductivity, then it would act as a load and the data could be explained.

Turning to the effects of an applied electric field, the magnetosphere supplied such an event on October 10–11, 1980, as illustrated in Fig. 5.13a. The large eastward perturbation in **E** seen at Arecibo was also detected at Jicamarca and was a clear magnetospheric effect (see the earlier discussion in Chapter 3). Unfortunately, no neutral wind data were taken on this night. The anticorrelation between $V_{\perp N}$ and $V_{\parallel}$ was occurring as usual until about 0300, when the tropical and low-latitude ionospheric dynamo was interrupted by the auroral event. The F-layer height increased dramatically due to the large eastward electric field. It seems that the uplift "perpendicular to **B**" was compensated by a strong downward gravity-driven flow. To test this, the large observed downward $V_{\parallel}$ has been superimposed on a calculated value of $(g/\nu_{in}) \sin I$ in Fig. 5.13b. The two curves match very well, illustrating the tendency of gravity to counter extreme vertical electrodynamic forcing. The F layer did display a net upward velocity. This is the "test" for an external electrodynamic source (Walker, 1980) for the plasma motion, since the main driver is the upward **E** × **B** drift with gravity-driven flow being a secondary effect resulting from this uplift.

Although complex, these two days at least can be more or less explained in a straightforward manner. So as not to leave the reader with the impression that this is always the case, we discuss an event recorded on June 15–16, 1980. Data from this night are presented in Figs. 5.14a and b in the same format as Figs. 5.12a and b. Comparison of Figs. 5.14a and b shows that the field-aligned ion motion, $V_{\parallel}$, was directed *oppositely* to the parallel component of the local neutral wind until 0100. Although counterintuitive, this situation can occur if a strong downward gravity-driven velocity is simultaneously present due to the high altitude of the layer. In fact, Burnside *et al.* (1983) have calculated the gravity term for this data set and have found that indeed the downward flow is sufficient to overpower the oppositely directed neutral wind during this time. In this same period (from 2130 LT until 0100), the perpendicular ion flow shown in Fig. 5.14a was upward, which corresponds to an eastward electric field. But,

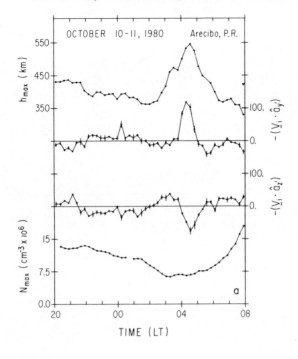

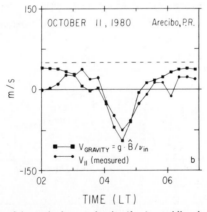

Fig. 5.13. (a) Height of the peak electron density (h_{max}), meridional plasma drift velocity components ($V_{\perp N}$ and $V_{\parallel}$), and peak in electron density (n_{max}) measured over Arecibo. (b) Calculated diffusion velocity and observed ion motion for the event illustrated in (a). [After Behnke *et al.* (1985). Reproduced with permission of the American Geophysical Union.]

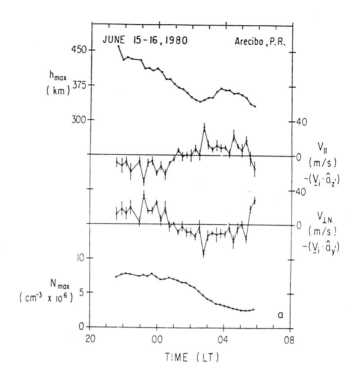

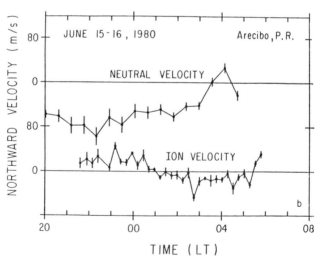

Fig. 5.14. (a) Height of the peak electron density (h_{max}), meridional plasma drift velocity components ($V_{\perp N}$ and $V_{\parallel}$), and peak in electron density (n_{max}) measured over Arecibo for June 15–16, 1980. (b) Horizontal neutral winds and ion drift velocities (positive northward) for the same night as the data shown in (a). [After Behnke *et al.* (1985). Reproduced with permission of the American Geophysical Union.]

the neutral wind was equatorward at this time, which would tend to generate a westward electric field component by the dynamo process. This rules out a local F-region dynamo process as a source of the electric field. Magnetic activity was low, so there is no reason to expect a magnetospheric origin for the eastward zonal electric field. A more likely explanation is an electric field which mapped from the other hemisphere. In fact, if the neutral wind extended into the southern hemisphere with the same direction (toward the south pole) and generated an F-layer dynamo field at the conjugate point, the zonal component in the northern hemisphere would be eastward as observed. This postulated south-poleward wind in the southern hemisphere would also tend to keep the F layer low, which in turn would enhance the southern hemisphere Pedersen conductivity. If two ends of a magnetic field line with differing dynamo wind fields compete for the electric field, the region with the higher internal conductivity (lower internal resistance) will determine the electric field.

Comparison of $V_{\parallel}$ and $V_{\perp N}$ shows a remarkable anticorrelation with a net poleward horizontal ion flow prior to 0030 LT and an equatorward flow thereafter. Although the individual behavior of $V_{\parallel}$ and $V_{\perp N}$ can be accounted for by invoking nonlocal effects, the strong anticorrelation of the velocity components observed on this night cannot be explained using any of the conventional mechanisms discussed above. Certainly the local F-layer dynamo mechanism is not operating, as demonstrated by the ions moving oppositely to the local wind. The external electric field/gravity mechanism is also not operating, as witnessed by the lack of a net upward layer motion. (In fact, the F layer falls throughout the period 2100 to 0230 LT.) What is responsible for the close coupling of the $V_{\parallel}$ and $V_{\perp N}$ velocity components on this night? We do not know. A great deal of careful comparison between optical data and radar data is likely to be necessary before we arrive at an understanding of this apparently new mechanism in F-layer dynamics.

5.2.3   The Transition Zone between Mid- and High Latitudes

The magnetic dip angle increases very rapidly with increasing latitude. One effect is to decouple parallel and perpendicular dynamics. To some extent this simplifies the physics, since gravity and pressure gradients are most important in the vertical direction, while electric fields and neutral winds are most associated with horizontal motions of the plasma and neutral constituents. Atmospheric motions driven by solar heating still create E- and F-region dynamo electric fields, of course, but these sources are in competition with other processes. We have already pointed out the marked difference between the St. Santin/Arecibo nighttime observations and those at Millstone Hill, even during relatively quiet times. The strong implication is that some dominating high-latitude factor is present at $L = 3.2$ even during very modestly active times when the plasmasphere very likely extends to large L shells.

Millstone Hill

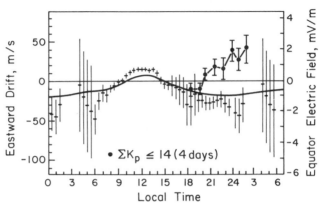

Fig. 5.15. Average Millstone Hill electric field measurements along with 1-h averages of the same component during times of very low activity.

This result is further emphasized in work by Gonzales *et al.* (1978), who studied the evening local time period using Millstone Hill data during "super-quiet" periods. Their results are summarized in Fig. 5.15, where hourly averages of the electric field/zonal drift obtained on days with $\Sigma\, K_p < 14$ are plotted for the evening period (filled circles) along with the "low" K_p Millstone Hill values published by Richmond *et al.* (1980) which were presented earlier in Fig. 5.6. The "superlow" K_p electric field is found to be equatorward in the evening pe-riod, as is the normal evening case for both St. Santin and Arecibo. It seems that even in what one might consider magnetically quiet times, electrodynamics in the Millstone Hill area are strongly affected by high-latitude process.

Two possible explanations were advanced by Gonzales *et al.* (1978). One is the direct penetration of the high-latitude electric field into the subauroral region. As we shall see in Chapter 6, this field has the proper sign since the auroral zone electric field is poleward in the evening-to-midnight period, in agreement with the smooth curve in Fig. 5.15. However, it is not so simple since the high-latitude field changes sign near midnight, switching from poleward to equator-ward, which is not observed at Millstone Hill. Blanc (1983) has studied the effect of the high-latitude electric field on the drift patterns over St. Santin both ex-perimentally and theoretically. He has studied magnetically active times and finds experimentally the *disturbance* drift vectors plotted here in Fig. 5.16a as a function of local time (for $K_p \geq 3$). These vectors are determined by subtracting the quiet-time drift from the disturbed-day measurements. The magnetospheric electric field perturbations are primarily westward and peak near 2220 local time. If these perturbations are added to the quiet-time St. Santin curves, they match the average Millstone Hill data reasonably well. One of the simulations of the

ST.-SANTIN AVERAGE DISTURBANCE DRIFTS

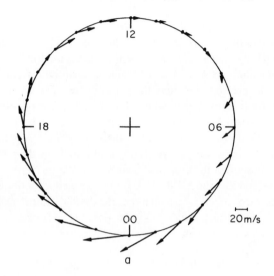

a

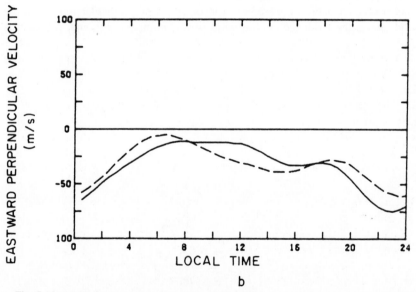

b

Fig. 5.16. (a) St. Santin average disturbance drift pattern, presented as a vector plot of a latitude–local time diagram. The data were obtained by a harmonic fit to half-hourly values of the
difference between the average drifts during disturbed periods ($K_p > 3$) and quiet periods ($K_p <$
$2 +$). The drifts, westward clockwise and northward toward the center, are westward at all local
times and maximize in the late evening. (b) Comparison between the average disturbance drifts above
St. Santin (continuous curve) and the drifts produced by a magnetospheric source. [After Blanc
(1983). Reproduced with permission of the American Geophysical Union.]

magnetospheric effect on the zonal drift component published by Blanc (1983) is plotted as the dashed curve in Fig. 5.16b. The solid curve is the experimental disturbance field. The agreement is quite good.

A second possibility is that the electric field is, after all, created by an F-region dynamo but that the local solar-driven winds are modified by high-latitude energy and momentum sources. The heat sources stem both from Joule heating and from particle precipitation into the atmosphere. The momentum source arises from the $\mathbf{J} \times \mathbf{B}$ forcing term where $\mathbf{J} = \boldsymbol{\sigma} \cdot \mathbf{E}$. In our previous discussion of this term we showed that at low latitudes it yields the ion drag effect. In the neutral wind dynamo region $\mathbf{J} \cdot \mathbf{E} < 0$ and the electric field is generated by the winds. At high latitudes the electric field is imposed on the ionosphere and $\mathbf{J} \cdot \mathbf{E} > 0$. This shows that the electrical energy is available for the Joule heating mentioned above and that the $\mathbf{J} \times \mathbf{B}$ force is not so much a frictional drag on the neutrals as it is a mechanism for accelerating the thermospheric neutrals. Results of very simple calculations which show the possible importance of this effect even at low latitudes are presented in Fig. 5.17. Here the "applied electric field" event observed at Arecibo on October 11, 1980 was used as input to a numerical atmospheric model which included $\mathbf{J} \times \mathbf{B}$ forcing (Behnke *et al.*, 1985). The local conductivity at Arecibo was also measured and used in the model. Data points show the measured electric field and perpendicular ion drift. The solid line is the predicted neutral wind perturbation. Even though the electric

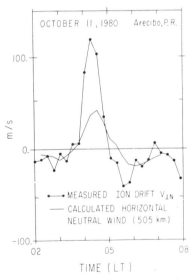

Fig. 5.17. Calculation of the perturbed neutral wind subject to the enhanced ion drift pattern, which is also plotted. [After Behnke *et al.* (1985). Reproduced with permission of the American Geophysical Union.]

field event was impulsive, the perturbed wind still reached a respectable value of 40 m/s.

At high latitudes the applied electric fields to first order form a two-celled plasma flow pattern which is virtually always present. It is not surprising then that the high-latitude neutral atmospheric motions are greatly affected by electrodynamic forcing. Suppose, as discussed in Chapter 7, that winds in the thermosphere are driven across the polar cap by the plasma flow. The plasma turns to follow the auroral oval but the wind, to first order, has no such constraint. A flywheel-like effect may then occur with disturbance winds blowing out of the auroral oval even during relatively quiet times. Such winds would be reinforced by the equatorward pressure gradient due to Joule and particle heating in the auroral oval. Once equatorward of the oval the Coriolis force might subsequently deflect the wind toward the west. This wind would then create a disturbance dynamo electric field in the poleward direction.

During very active times there is good evidence for such a wind pattern. Quiet-time neutral wind measurements at Fritz Peak, Colorado (39.9°N, 105.5°W, $L = 3$) for six nights are gathered in Fig. 5.18a, along with the predictions of a thermospheric global circulation model (Hernandez and Roble, 1984). The data and model both show eastward winds in the evening sector. On the other hand, measurements made on an active day and shown in Fig. 5.18b display strong westward winds until 0200 LT and an equally strong equatorward wind from 2300 until 0400 LT. the unusual wind was detected only north of the station.

There is a classical "chicken or egg" problem in the relationship between these mid-latitude neutral wind and electric field patterns which will probably not be resolved until more simultaneous neutral wind and electric field data are available.

5.2.4   E-Region Mid-Latitude Dynamics

When the E-region conductivity is high, the electrodynamics are driven by tidal modes in the E region. These have already been discussed in Chapter 3. As noted there, the semidiurnal tidal mode becomes important at mid-latitudes and we expect that the daytime electric field will be dominated by semidiurnal tides at Millstone Hill and St. Santin. Both diurnal and semidiurnal tides should contribute at Arecibo. However, since the F-region dynamo and the high-latitude electric field sources are both primarily diurnal in form, we might expect the composite picture to be quite complex. This is borne out by the electric field data in Fig. 5.6. Arecibo, St. Santin, and Millstone Hill all display semidiurnal variations in the zonal electric field component and diurnal behavior in the meridional field.

Some of the richness of E-region ionospheric dynamics may be visualized using motion of the layers which form in the ionosphere as natural tracers. Un-

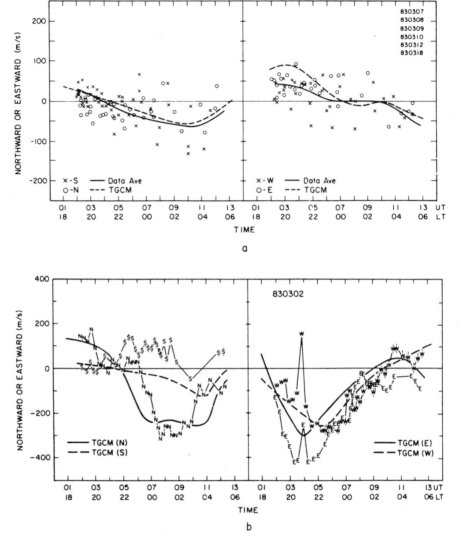

Fig. 5.18. (a) Nighttime variation of thermospheric winds measured during six geomagnetic quiet days in early March 1983 with the year, month, and day given in the upper right corner. The meridional and zonal wind measurements, positive northward and eastward, are given at the left and right, respectively, with the solid line being an average of the data points and the dashed line representing TGCM predictions for geomagnetic quiet conditions. (b) Nighttime variation of thermospheric winds measured on March 2, 1983: (left) meridional winds (positive northward) measured to the north (N) and south (S); (right) zonal winds (positive eastward) measured to the east (E) and west (W) of Fritz Peak Observatory. The solid and dashed curves represent TGCM predictions for a constant-pressure surface near 300 km and for grid points north and south (left) and east and west (right) from Fritz Peak. [After Hernandez and Roble (1984). Reproduced with permission of the American Geophysical Union.]

fortunately, these layers are much more obvious at night than during the day since plasma production by sunlight tends to wash them out during the daytime. However, long-period oscillations such as diurnal and semidiurnal tides can be extrapolated from the nighttime data if some care is taken. Several examples of plasma density profiles over Arecibo for the period including sunset, sunrise, and the nighttime hours are presented in Figs. 5.19a–c. Nighttime layers are very common over Arecibo and the tidal modes can be visualized quite well. Presumably the neutral wind fields which create the nighttime layers are present during the day as well. In the next section we discuss how the layers are formed, but for now we take their existence as an experimental fact and study their motion.

The altitude of the peak in the plasma density for a number of layers of this type has been determined by a careful study which includes even the more difficult daytime hours, and the results are plotted in Fig. 5.19d. Seventy-eight consecutive hours of data are shown. A well-defined motion is seen in the central part of the time period in which a layer starts descending from 150 km at 0600 LT, reaching 90 km at about midnight. Parts of the same pattern can be discerned on the previous day as well as on the following day, which suggests that a diurnal mode exists. Higher-frequency motions are also clearly indicated in the figure. For example, there is a semidiurnal daytime pattern around 110 km which seems to be masked during the nighttime period by noiselike fluctuations. When discussing atmospheric waves, as the frequency increases above the tidal range it is no longer necessary to discuss the atmospheric winds in terms of tidal modes since the horizontal wavelengths become much smaller than the radius of the earth. In this case the appropriate modes are termed internal gravity waves, a topic we take up in some detail in the next section. The noiselike features of the data in Fig. 5.19d are thus geophysical in origin and are probably due to short-period gravity waves.

In summary, the electrodynamics of the mid-latitude ionosphere is controlled by tidal and higher-frequency atmospheric wave modes in the E region during the day. At night a very complex combination of the thermospheric wind dynamo, high-latitude electric field penetration, conjugate hemisphere effects, and gravity waves all play a role. These effects are modulated by a strong diurnal variation in the E-region conductivity which allows the F-layer dynamo to gain control at night.

5.3   Irregularities in the Mid-Latitude Ionosphere

Although dynamical considerations do affect the dayside mid-latitude ionospheric content and altitude, production and recombination are sufficiently rapid to mask much of the structure in the medium other than large-scale vertical and horizontal gradients. At night, however, a number of processes can and do con-

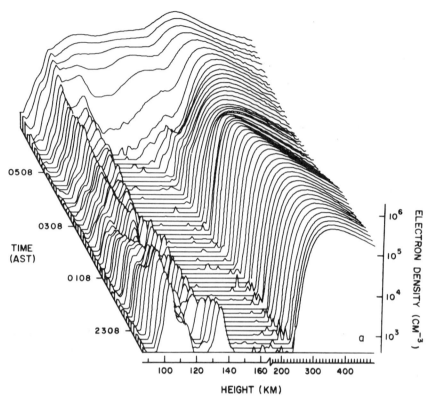

Fig. 5.19. (a) Electron density profiles for the night of April 16–17, 1974. This night was extremely quiet magnetically. (b) Electron density profiles for the night of April 17–18, 1974. This night was somewhat disturbed, and an intermediate layer can be seen. Note also the undulations of the F layer. (c) Profiles for the night of April 5–6, 1974. The intermediate layer on this night was particularly well defined and hence suitable for detailed study. The K_p index on this night ranged from 3 to 6. [Parts (a)–(c) after Shen *et al.* (1976). Reproduced with permission of the American Geophysical Union.] (d) Altitude variation of a number of layers detected during a 78-h period over Arecibo. (Courtesy of J. D. Mathews.)

tribute to the formation of irregularities. In this section we discuss a few of the more important sources of ionospheric structure.

5.3.1   Large-Scale Organization of the Mid-Latitude Nighttime Ionospheric Plasma

Three graphic examples of consecutive plasma density profiles detected in the nighttime Arecibo ionosphere were presented in Figs. 519a–c. The April 16–17, 1974 night shown in Fig. 5.19a was very quiet magnetically, while the other

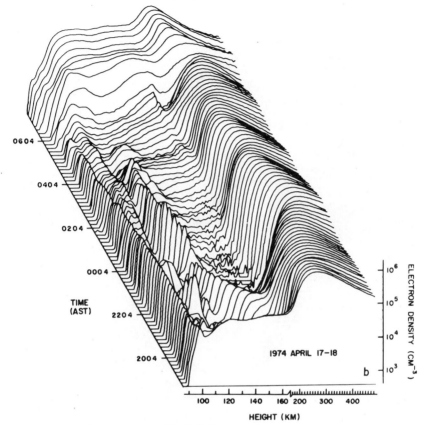

Fig. 5.19 (*Continued*)

nights were moderately disturbed. The solar influence can be clearly seen at sunrise, when the deeply depressed plasma in the F-layer "valley" between 160- and 240-km altitude fills in and causes even the E-region structure to merge into the fairly featureless daytime ionospheric profiles at the top of each figure.

Some features are common to all nights. The high-density F layer itself displays undulations with a typical period of 2 h. The F layer rose and fell by many tens of kilometers during these long-period oscillations. In the E region between 90 and 120 km very intense layers developed on each night and lasted from sunset to sunrise.

On the more magnetically active nights the plasma density in the valley region is very much elevated over the quiet night. Whenever there are electrons to scatter from, the radar can detect motion and organization of the plasma. For example, on the nights when plasma is present in the valley a piece of the F region

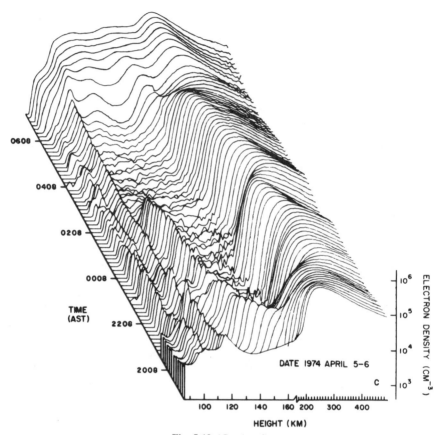

Fig. 5.19 (*Continued*)

seems to "peel off" the bottomside of the layer at sunset and to propagate downward into the F-layer valley region. Such a structure has been termed an intermediate layer by Shen *et al.* (1976) since it occurs between the F layer and the
more classical sporadic E layers below. Since ionosonde signals are often reflected by the intense lower-altitude layers, the intermediate structures are not
visible with an ionosonde and can be studied only via incoherent scatter radars
such as Arecibo. The sporadic nature of these various layers, which gives them
the name sporadic E, is evident in these profiles. The strongest intermediate layer
lasted almost all night on April 17–18, while it died out at 0230 LT on April
5–6. Other more sporadic and weaker intermediate layers came and went on
April 17–18.

In the remainder of this section we discuss the formation and dynamics of
these layers, which are primarily due to oscillatory behavior of the neutral at-

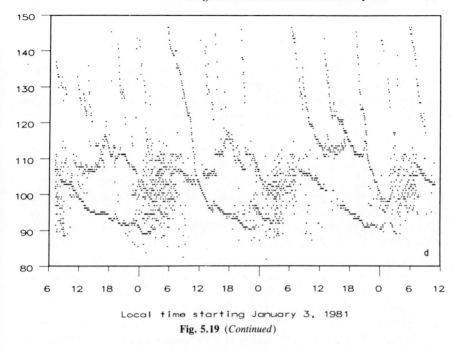

Fig. 5.19 (*Continued*)

mosphere. Such motions are classified as either tides or gravity waves depending on their frequency. Although, as we shall see, these neutral atmospheric motions are capable of organizing the plasma into layered structures, this is not the entire story. The intermediate-layer ionization in particular cannot be explained by photoionization processes, and we briefly discuss the observation and effect of ionizing energetic particle fluxes in this regard.

Finally, once the layers are formed they may be subject to plasma instability processes similar to those discussed for the equatorial zone, and we end the section with a brief discussion of such processes.

5.3.2   Oscillations of the Neutral Atmosphere

From the discussion in Section 5.1 it is clear that when $\kappa_i = \Omega_i/\nu_i$ is large and any motion of the neutral atmosphere has a component parallel to the magnetic field, the plasma will be carried along **B** with the same velocity as the parallel component of the neutral atmosphere. Any neutral velocity with a component in the magnetic meridian will therefore create a similar velocity of F-region plasma projected in the direction of the magnetic field. We previously discussed the effect of large-scale meridional neutral winds in causing F-region plasma motion upward or downward along **B**. Here we expand this discussion to include tides and gravity waves. Traditionally, *upper* thermospheric dynamics are discussed in terms of wind patterns rather than tidal modes, even though they display clear diurnal variations. On the other hand, long-period *lower* thermospheric dynamics are discussed in terms of diurnal and semidiurnal atmospheric tides. This notation is partly historical but is also related to the origin of the atmospheric forcing. The upper thermospheric winds are driven *in situ* by solar heating, Joule heating, and momentum transfer with the plasma, whereas the lower E-region winds are usually ascribed to upward-propagating tides generated at tropospheric and stratospheric heights. The oscillation periods of the semidiurnal and diurnal tides are, of course, 12 and 24 h, respectively. Higher-order tides have also been considered but as the oscillation period nears several hours the motions are usually referred to as gravity waves. Upper thermospheric forcing via solar UV heating has a strong diurnal component and could be referred to as *in situ* tide, but this usage is not very common.

The upper atmosphere is continuously bombarded with gravity waves from a number of sources. These include tropospheric weather fronts, tornadoes and thunderstorms, impulsive auroral zone momentum injection and heating events, and even earthquakes and volcanic eruptions. The famous monograph entitled "The Upper Atmosphere in Motion" by Hines (1974) is an excellent annotated collection of gravity wave studies published by Hines and co-workers over about a 10-year period. The reader is referred to that work for details about gravity waves and tidal oscillations as well as to the excellent review of tidal theory by Chapman and Lindzen (1970) mentioned earlier. Here our approach is much more modest in scope, aiming at physical intuition rather than detailed analysis.

We study gravity waves first, in effect finding the normal modes of a flat nonrotating inviscid atmosphere. These results will be valid as long as the periods do not approach the tidal range and the wavelengths are not long enough that the curvature of the earth matters. We assume an isothermal, inviscid atmosphere initially in hydrostatic equilibrium, so that if ρ_0 and p_0 are the zero-order mass density and pressure, the relation

$$\rho_0 \mathbf{g} = \nabla p_0$$

applies. In addition, it can be shown that ρ_0 and p_0, which vary only in the vertical direction, are of the form

$$\rho_0, \ p_0 \propto e^{-z/H}$$

where H is the scale height of the atmosphere; that is, $1/H = -(1/\rho_0)(d\rho_0/dz)$. Here we again choose our coordinates using the meteorological convention and take x eastward, y northward, and z vertically upward. We assume there are no neutral winds in the unperturbed atmosphere. The equations governing the behavior of the atmosphere are the mass continuity equation (2.2), the equation of motion (2.20), and the adiabatic condition (see Yeh and Liu, 1974). In the equation of motion, only terms due to gravity, pressure gradients, and inertia are retained.

Now consider atmospheric oscillations in the presence of gravity. We assume there are small perturbations in the mass density, pressure, and wind velocity denoted by $\delta\rho$, δp, and $\mathbf{U} = (u, v, w)$. Without the Coriolis or viscous forces there is no coupling between oscillations in the y–z plane and those in the x direction, so we can ignore the x component of velocity making the problem two-dimensional. We define a column vector $\mathbf{F}$ by

$$\mathbf{F} = \begin{vmatrix} \delta\rho/\rho_0 \\ \delta p/p_0 \\ v \\ w \end{vmatrix}$$

and assume that atmospheric perturbations can be described by plane waves of the form

$$F \propto e^{i(\omega t - k_y y - k_z z)} \tag{5.9}$$

Substituting $\rho = \rho_0 + \delta\rho$, $p = p_0 + \delta p$, $\mathbf{U} = (0, v, w)$ into the equations describing the atmosphere (see Chapter 2) and retaining terms up to first order in $\delta\rho$, δp, and $\mathbf{U}$ gives the linearized forms of the mass continuity, motion, and adiabatic state equations, that is,

$$\partial(\delta\rho)/\partial t + \mathbf{U}\cdot\nabla\rho_0 + \rho_0\nabla\cdot\mathbf{U} = 0 \tag{5.10a}$$

$$\rho_0\,\partial v/\partial t + \partial(\delta p)/\partial y = 0 \tag{5.10b}$$

$$\rho_0\,\partial w/\partial t + \partial(\delta p)/\partial z + \delta\rho g = 0 \tag{5.10c}$$

$$\partial(\delta p)/\partial t + \mathbf{U}\cdot\nabla p_0 - C_0^2\,\partial(\delta\rho)/\partial t - C_0^2\mathbf{U}\cdot\nabla\rho_0 = 0 \tag{5.10d}$$

In (5.10d), C_0 is the "speed of sound," given by

$$C_0^2 = \gamma p_0/\rho_0 = \gamma g H$$

where γ is the ratio of specific heats at constant pressure and constant volume. Using (5.9), the condition for hydrostatic equilibrium, and the expression for the scale height H, (5.10) can be rewritten as a matrix equation:

$$\begin{vmatrix} i\omega & 0 & -ik_y & -1/H - ik_z \\ 0 & -ik_y C_0^2/\gamma & i\omega & 0 \\ g & -C_0^2(1/H + ik_z)/\gamma & 0 & i\omega \\ -i\omega C_0^2 & i\omega C_0^2/\gamma & 0 & (\gamma - 1)g \end{vmatrix} \cdot \mathbf{F} = 0$$

In deriving (5.10c) we used the fact that from (5.9), $\delta p \propto p_0 \exp(i\omega t - ik_y y - ik_z z)$. This leads to $\partial(\delta p)/\partial z = \delta p[(1/p_0)(dp_0/dz) - ik_z]$ and eventually to the corresponding entry in row 3. Setting the determinant of the 4×4 matrix equal to zero yields the dispersion relation for linear modes of a nonrotating neutral atmosphere on a flat earth,

$$\omega^4 - \omega^2 C_0^2(k_y^2 + k_z^2) + (\gamma - 1)g^2 k_y^2 + i\gamma g\omega^2 k_z = 0 \tag{5.11}$$

A variety of possible wave modes are buried in this dispersion relation. Suppose we take the limit that $g = 0$. Then (5.11) reduces to

$$\omega^2 = C_0^2(k_y^2 + k_z^2)$$

which is the dispersion relation for sound waves propagating without attenuation, growth (pure real ω and $\mathbf{k}$), or dispersion ($\omega/k = $ constant).

We now turn to the gravity wave case. If there are no sources of energy or dissipation (viscosity was ignored), waves will not grow or decay in time at a fixed point in space so we can assume ω is real. If we are including gravity, however, it can be shown that there are no solutions of (5.11) with both k_y and k_z purely real. Anticipating the final result, let us assume k_y is purely real and investigate k_z. This corresponds to a wave propagating in an unattenuated fashion with a component in the horizontal direction. Then we can write (5.11) as

$$\omega^4 - \omega^2 C_0^2 k_y^2 + (\gamma - 1)g^2 k_y^2 = -i\gamma g\omega^2 k_z + \omega^2 C_0^2 k_z^2 \qquad (5.12)$$

where the left-hand side is purely real. Now if we let k_z be a complex number,

$$k_z = k_z' + ik_z''$$

it is straightforward to show that the right-hand side of (5.12) is purely real if and only if

$$k_z'' = (1/2H)$$

Dropping the superscript (prime) notation, we can now see that the solutions for the quantities in the column vector $\mathbf{F}$ are of the form

$$e^{i(\omega t - k_y y - k_z z)} e^{z/2H} \qquad (5.13)$$

In (5.13) both k_y and k_z are real.

Atmospheric waves which propagate in the manner described by (5.13) are termed internal gravity waves. Some of the complexity of the wind patterns which arise in the 90–120-km height range due to such waves can be gauged from the photograph of a trimethyl aluminum (TMA) vapor trail deployed by a sounding rocket which is shown in Fig. 5.20a. This photograph yields only one perspective on the distortion of the trail by the ambient winds but numerous reversals and shears are evident. From (5.13) the theoretical prediction is that the wave amplitude should grow as it propagates upward (positive z). The physical explanation for this somewhat bizarre result is that to conserve the wave perturbation energy (e.g., terms of the form $\rho_0 v^2$) as ρ_0 decreases with z, v^2 must increase. The factor of 2 in the exponential form occurs since, in order to keep $\rho_0 v^2$ constant, v need only e-fold over a height interval $2H$ when ρ_0 decreases by a factor of e in the height interval H. The classical observation supporting this result is shown in Fig. 5.20b. The dashed curve shows the mean wind. The actual wind fluctuates about this mean with an amplitude which increases with

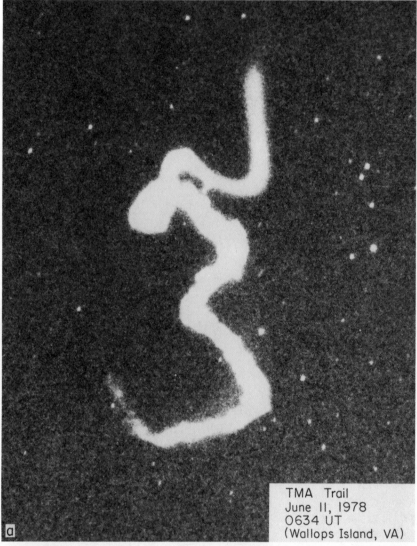

TMA Trail
June 11, 1978
0634 UT
(Wallops Island, VA)

Fig. 5.20. (a) A TMA trail deployed from Wallops Island, Virginia on June 11, 1978 at 0634 UT. The trail was photographed from the NASA C54 airplane. (Courtesy of I. S. Mikkelsen.) (b) Wind components at meteor levels in a vertical plane in one representative case, derived by Liller and Whipple (1954) from the distortion of a long-enduring meteor trail. (c) Normalized wind profile at meteor heights, measured to the right and to the left from the "0" position, deduced from (b) by removal of the general shear and reduction of the residual by a factor proportional to $\rho_0^{1/2}$. (d) Pictorial representation of internal atmospheric gravity waves. Instantaneous velocity vectors are shown, together with their instantaneous and overall envelopes. Density variations are depicted by a background of parallel lines lying in surfaces of constant phase. Phase progression is essentially downward in this case, and energy propagation obliquely upward; gravity is directed vertically downward. [Parts (b)–(d) after Hines (1974). Reproduced with permission of the American Geophysical Union.] (*Figure continues.*)

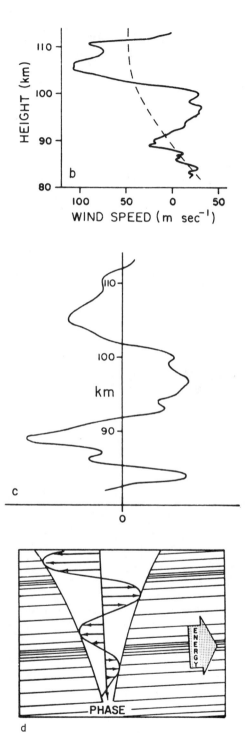

Fig. 5.20 (*Continued*)

height. In Fig. 5.20c a detrended version of the same data is given after subtracting the mean and multiplying by $\exp[(112 - z)/2H]$ where z and H are in kilometers. A schematic representation of the effect is shown in Fig. 5.20d. Rocket data such as these usually show that the fluctuating component of the neutral wind velocity increases with increasing height. The perturbations about the average profile are often comparable to the mean wind.

An important implication of this exponential growth effect is that waves of little or no importance to tropospheric or stratospheric dynamics grow to monumental proportions in the E and F regions. Eventually, either these waves break down nonlinearly due to their large amplitude or the kinematic viscosity (μ/ρ_0) gets so large (as ρ_0 decreases) that viscous dissipation balances growth.

Equation (5.11) can also be written in the form

$$\omega^4 - \omega^2 C_0^2(k_y^2 + k_z^2) + \omega_b^2 C_0^2 k_y^2 + \omega_a^2 \omega^2 = 0$$

where $\omega_b^2 = (\gamma - 1)g^2/C_0^2$ is the square of the Brunt–Väisälä frequency and $\omega_a^2 = C_0^2/H^2$. At a given value of k_y, if ω is large enough (the high-frequency branch with $\omega > \omega_a$) the first and second terms dominate and we recover the sound wave dispersion relation found above for $g = 0$. The low-frequency branch corresponds to gravity waves which propagate only for $\omega < \omega_b$. Physically, the Brunt–Väisälä frequency is the frequency at which a parcel of air oscillates about its equilibrium position when it is initially displaced from that position. For a nonisothermal atmosphere (one in which C_0^2 varies with height),

$$\omega_b^2 = (\gamma - 1)g^2/C_0^2 + (g/C_0^2)(dC_0^2/dz)$$

Representative values for the buoyancy period $T_b = (2\pi/\omega_b)$ derived from this expression are plotted in Fig. 5.21 as a function of height. Clearly the 1–3-h oscillations in the ionospheric parameters discussed above fall in the gravity wave branch $\tau = (2\pi/\omega) > T_b$.

Tides can be considered to be low-frequency gravity waves. Complications arise in the analysis because of the requirement that the solutions satisfy boundary conditions on the spherical earth as well as the fact that the rotation frequency of the earth cannot be ignored. The solutions must exhibit certain altitudinal and latitudinal forms referred to as Hough functions. Which tidal modes are generated depends on how well the forcing function, for instance, solar or lunar forcing, matches the mode structure. As alluded to earlier in the discussion of electrodynamics, the diurnal tide due to atmospheric heating is important at low latitudes but the mode response is small above 45° latitude. The semidiurnal forcing is smaller, but the altitude profile of ozone and water vapor content fit the so-called (2, 2) semidiurnal tidal mode quite well. The local heating due to these minor constituents thus couples well to the semidiurnal tide, explaining its importance at the higher latitudes.

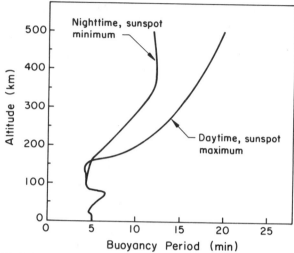

Fig. 5.21. Vertical structure of the buoyancy period computed for a standard (nonisothermal) atmosphere. [After Yeh and Liu (1974). Reproduced with permission of the American Geophysical Union.]

5.3.3 Role of Gravity Waves and Tides in Creating Ionospheric Structure

Given that gravity waves and tides can be generated by a variety of sources and that they grow to respectable amplitudes by the time they reach ionospheric heights, we can now investigate their effect on the ionization.

For ω much less than ω_b and λ_z, the vertical wavelength, much less than the scale height $2\pi H$, (5.10a) becomes

$$k_y v + k_z w = 0 \tag{5.14}$$

This equation is equivalent to the assumption of incompressible flow $(\nabla \cdot \mathbf{U}) = 0$ when expressed in linearized form. The so-called Boussinesq approximation, for example, uses $(\nabla \cdot \mathbf{U}) = 0$ for perturbations when $\omega < \omega_b$ but allows for compressibility in the zero-order equations. Equation (5.14) can also be written

$$|v/w| = |\lambda_y/\lambda_z|$$

Since $\lambda_z < 2\pi H$ in our approximation, while λ_y can be shown to be quite large for the 1–3-h periods of interest here ($\simeq$ hundreds of kilometers), we conclude that $v \gg w$. That is, the horizontal components of the perturbation wind are much stronger than the vertical.

As a first application consider the F layer and F-layer valley in which κ_i is large. For a wave having a small vertical wavelength, we have sketched the wind fluctuations for a typical gravity wave as a function of height in Fig. 5.22. Con-

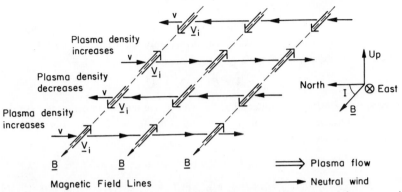

Fig. 5.22. Neutral wind (horizontal vector) and ion flow vectors (large arrows) due to a gravity wave with a large wavelength in the y direction and a smaller wavelength in the z direction.

sider the case in which the perturbation wind lies in the magnetic meridian and, as also shown, the magnetic field has a finite dip angle. Because the plasma is constrained to move with the component of the neutral wind motion parallel to **B** when κ_i is large, in regions where the perturbation wind changes sign, the plasma either converges to or diverges from the altitude where the shear occurs. A layer of ionization will arise in the convergence zone. This wind shear theory for layer formation was first suggested by Dungey (1959), was extended by Whitehead (1961), and seems to explain many of the observations quite well. The case illustrated in Fig. 5.22 is most effective in the upper E and lower F regions. At higher altitudes (F layer), diffusion parallel to **B** keeps sharp layers from forming. This can be seen by comparing the rate at which plasma converges, which is of the order of $k_z v \cos I$, with the rate at which it diffuses, which is of the order of $k_z^2 D$. For

$$D \gtrsim v \cos I / k_z$$

diffusion is too strong for layers to form. Since $\lambda_z < 2\pi H$ and $v \cos I \simeq 10$ m/s, we can estimate the critical value for D to be roughly 10^4 m²/s. The appropriate value for D is the ambipolar diffusion rate (twice the ion parallel diffusion) and we find that gravity waves are not very efficient in producing layers above about 200 km. In addition, at low altitudes where $\kappa_i \lesssim 1$ (<120 km) collisions become so frequent that the plasma velocity follows the neutral gas velocity *everywhere*. The plasma merely sloshes back and forth in the horizontal north–south direction with no convergence or divergence occurring along **B** and this version of wind shear theory breaks down.

However, shears in the zonal wind component can take over the production of layers below about 130 km. The mechanism stems from the Lorentz force felt by the ions when subject to a wind field. For example, referring to (2.22b), which gives a general relation for the ion velocity, we are interested in the equilibrium case $d\mathbf{V}_i/dt = 0$. In addition, we ignore the effects of $\mathbf{E}$, ∇n, and $\mathbf{g}$ and consider collisions between ions and neutrals only. Expressing (2.22b) for ions in the earth-fixed frame we have

$$\kappa_i(\mathbf{V}_i \times \hat{B}) = \mathbf{V}_i - \mathbf{U} \tag{5.15}$$

where $\kappa_i = eB/M\nu_{in}$. To understand this result, we use the coordinate system in Fig. 5.11 and take $\mathbf{U}$ along the $\hat{a}_x$ axis and $\hat{B}$ along the $\hat{a}_{z'}$ axis. Then the x, y, and z components of (5.15) are

$$\kappa_i(V_{iy} \sin I + V_{iz} \cos I) = u - V_{ix} \tag{5.16a}$$

$$\kappa_i V_{ix} \sin I = V_{iy} \tag{5.16b}$$

$$\kappa_i V_{ix} \cos I = V_{iz} \tag{5.16c}$$

At the magnetic equator and at a height where $\kappa_i = 1$, which occurs at about 130 km, the solutions to (5.16) are

$$V_{ix} = V_{iz} = u/2, \qquad V_{iy} = 0$$

The ions therefore have a component parallel to $\mathbf{U}$ but are also deflected in the direction of the Lorentz force $q(\mathbf{U} \times \mathbf{B})$ giving a net motion at a 45° angle to the neutral wind $\mathbf{U}$, the deflection being upward in the coordinate system shown. This net velocity is illustrated in Fig. 5.23. For larger κ_i the deflection angle is larger and for smaller κ_i the ion motion is nearly parallel to $\mathbf{U}$. The mechanism which creates the plasma layers is illustrated in Fig. 5.24 for the case of a vertical shear in the zonal wind and $\kappa_i = 1$. The ions above the shear point drift down at a 45° angle, while those below drift upward. Plasma accumulates at the point of maximum wind shear.

Note that in the last sentence we shifted from a discussion of *ion motion* to a statement about *plasma accumulation*. In the idealized geometry of Fig. 5.24, with the magnetic field perfectly horizontal, the highly magnetized, high-κ_e electrons could not move perpendicular to $\mathbf{B}$ to join the converging ions. A huge space charge electric field would build up and the whole process would grind to a halt. However, if there is even a slight dip angle, electrons can move along the

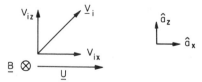

Fig. 5.23. Ion velocity vector ($\mathbf{V}_i$) and its components subject to a neutral wind $\mathbf{U}$ perpendicular to $\mathbf{B}$ when $\kappa_i = 1$.

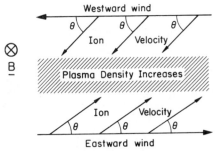

Fig. 5.24. Illustration of the wind shear mechanism which operates at E-region and lower F-region heights. [After Hines (1974). Reproduced with permission of the American Geophysical Union.]

magnetic field from a region of ion divergence to one of ion convergence in response to a very slight initial charge imbalance. This keeps the plasma charge neutral and allows the plasma buildup into layers cited above. This wind shear process can be considered as an example of the compressibility of the ionospheric plasma in the direction perpendicular to **B** when $\kappa_i \approx 1$. As noted earlier, F-region plasma flow is virtually incompressible $(\nabla \cdot \mathbf{V}) = 0$ but in the lower F and E regions this is not the case.

The theory outlined here yields a very nice dynamical explanation for ionospheric layering. Even the downward progression of layers such as seen in Figs. 5.19a–c is accounted for since gravity wave and tidal theories both predict downward phase propagation for an upward group velocity (expected for lower atmospheric sources). The theory initially ran into quantitative difficulty, however, since the standard recombination rate at the heights of interest was too large to support the observed layer densities (up to 10^6 cm^{-3}). Metallic ions such as Mg^+, Si^+, and Fe^+ due to meteoric sources proved to be the dominant ions in three layers. This fact removed objections to the wind shear theory since such ions have very long lifetimes. An example of ion composition measured in a sporadic E layer during a rocket flight is presented in Fig.5.25 (Hermann *et al.*, 1978). The ΣM^+ curve shows all the metal ions and tracks the peaks in the electron density quite well. Such data strongly support the notion that metallic ions are responsible for long-lived intense sporadic E layers. The intermediate layers remain a problem, however, since they do not contain metallic ions. The resolution of this problem, discussed in some detail in the next section, involves additional ionizing radiation at mid-latitudes over and above the usual photoionization sources.

Studies by Bowman (1981, 1985) seem to show clearly that much of mid-latitude spread F as registered on ionosondes is due to gravity wave modulation of the electron density in the E and F layers. Some of the more violent disturbances which include small-scale structure most likely also involve plasma instabilities discussed below.

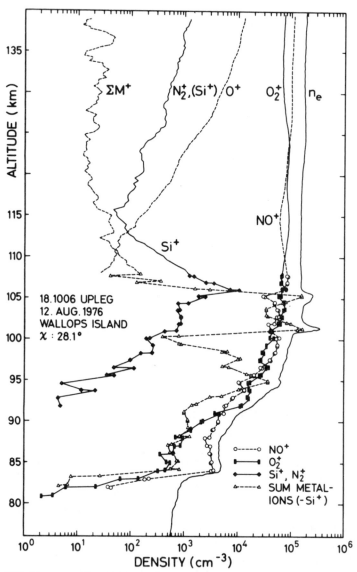

Fig. 5.25. Electron and ion density profiles from a rocket flight. Below 108 km the individual measurements are shown. Above this altitude the curves represent running averages from four consecutive mass spectrometer sweeps. Localized peaks in the electron density profile are associated with metallic ions. (Courtesy of L. Smith.)

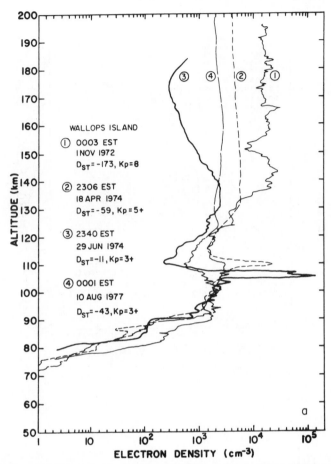

Fig. 5.26. (a) Nighttime electron density profiles from rockets launched at Wallops Island. (b) Variation with magnetic activity of the nighttime ionization rate in the upper E region and of the particle energy flux obtained from rocket experiments. [Parts (a) and (b) after Voss and Smith (1979). Reproduced with permission of the American Geophysical Union.] (c) Total energetic particle count rate and count rate of electrons only as a function of altitude. (Courtesy of L. Smith.) (*Figure continues.*)

5.3.4   Particle Precipitation at the Mid-Latitudes

Shen *et al.* (1976) concluded that both the intermediate layers in Figs. 5.19b and c and the general enhancement of F-layer valley ionization were due to particle precipitation. To test this idea, extensive studies involving a number of rocket flights at mid-latitudes have been conducted by Voss and Smith (1979, 1980a, b). Summary plots are reproduced here in Figs. 5.26a–c. Figure 5.26a shows plasma density profiles for four different values of the magnetic activity index K_p. The highest K_p profile shows F-layer valley densities an order of magnitude

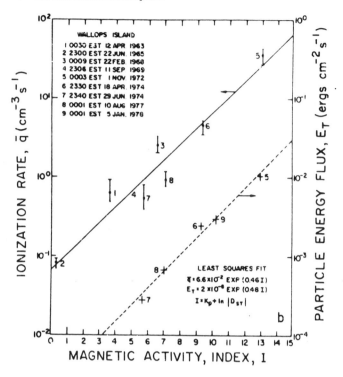

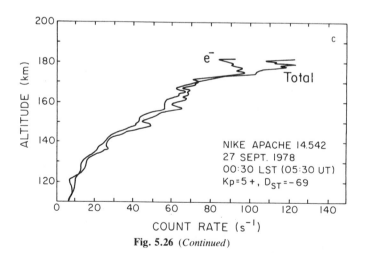

Fig. 5.26 (*Continued*)

higher than the lowest K_p profile. A layer with peak density at 135 km is also evident in the high-K_p profile. Figure 5.26b shows both the particle flux and the ionization rate as functions of a specialized magnetic index $I = K_p + \ln(DST)$ where DST is the global mid-latitude average magnetic deflection in units of γ.

It is clear that such photon sources as galactic x-rays and scattered Lyman α (from the geocorona) cannot produce much ionization above 105 km. A consensus is growing that particle ionization sources are important at mid-latitudes and that the primary component involves positively charged particles rather than electrons, which produce most of the auroral ionization. Evidence in support of this hypothesis is given in Fig. 5.26c for a rocket flight over Wallops Island, Virginia ($L = 2.6$). In comparing the two curves, electrons contribute only about 10% of the count rate in the energy range surveyed. A schematic latitude profile of precipitating particles based on satellite observations is reproduced in Fig. 5.27. Positive particles dominate in three bands: equatorial, mid-latitude, and the subauroral zone. Electrons dominate in the auroral zone proper and in a "low" latitude region between 20° and 30°. The schematic picture is valid when averaged over longitude. However, in the northern hemisphere the mid-latitude zone disappears almost entirely between 45°W and 75°E. Conversely, the low-latitude electron zone disappears between 135°W and 150°W. These latter effects are presumably due to the South Atlantic anomaly, a region of enhanced mag-

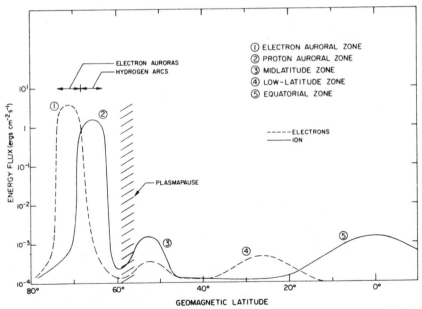

Fig. 5.27. Particle precipitation as a function of latitude. [Adapted from Voss and Smith (1980b). Reproduced with permission of Pergamon Press.]

netic field strength which modifies the mirror heights of energetic particles quite drastically.

In summary, particle precipitation plays an important role in creating structure in the mid- and low-latitude sector by providing an ionizing source for the F-layer valley. The production rate is very low, but neutral atmospheric waves gather the ions and produce easily observable features in the plasma profile. To this is added meteoric sources of metallic ions at lower altitudes. There, long-lived ions are also gathered together by wind and wave patterns to form the ionization patterns observed.

5.4 Mid-Latitude Plasma Instabilities

In this section the plasma physics of the mid-latitude ionosphere is discussed. Many of the processes are similar to those discussed already in Chapter 4 and hence need only be briefly reintroduced here. Once again, F-region and E-region processes are presented separately.

5.4.1 F-Region Plasma Instabilities in the Equatorial Anomaly Region

The plasma bubbles discussed in Chapter 4 are electrodynamically produced and hence involve uplift of the entire flux tube. Indeed, the radar map in Fig. 4.4 shows that the depletions are field aligned. These flux tubes reach sufficiently high altitudes that they map into the equatorial anomaly airglow bands shown in the satellite photograph earlier in this chapter. Since the airglow is caused by enhanced plasma density and the associated recombination emission at 6300 Å, a depleted flux tube should have a depressed airglow.

Just such an effect is shown in Fig. 5.28a, an all-sky camera airglow photograph taken in the anomaly region over Ascension Island (geographic position 8°S, 14°W). The dark bands correspond to plasma depletions near 300-km altitude, where the airglow originates. They extend from horizon to horizon in the north–south direction and are 50–100 km in east–west horizontal size. When mapped to the equatorial plane along magnetic field lines these structures take on the vertical wedgelike shapes shown in Fig. 5.28b. These wedges tilt to the west as noted in Chapter 4 and seem to give definitive proof of the wedge versus bubble geometry for the equatorial depletions.

These depletion regions owe their origin to an equatorial phenomenon but in turn create the seed for very strong local off-equatorial F-region instabilities. This stems from their associated east–west horizontal density gradients. Such gradients are unstable to the generalized Rayleigh–Taylor instability discussed in Chapter 4. To first order gravity does not play a role since $\mathbf{g}$ is perpendicular

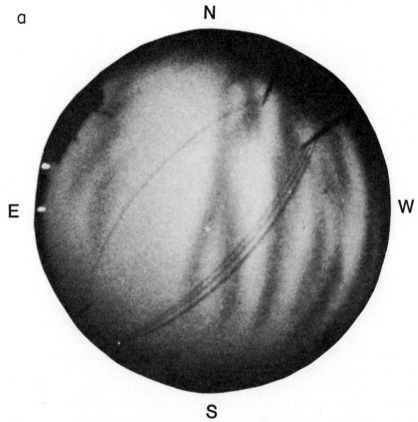

Fig. 5.28. (a) All-sky photograph at 6300 Å taken from Ascension Island at 2120 LT on February 7, 1981. (b) Linear and bifurcated airglow features recorded in the all-sky camera photograph shown in (a) after mapping to the equatorial geomagnetic plane. [After Mendillo and Tyler (1983). Reproduced with permission of the American Geophysical Union.] (*Figure continues.*)

to ∇n for a horizontal gradient. For $g = 0$, the criterion for instability discussed in Chapter 4 is

$$(\mathbf{E}' \times \mathbf{B}) \cdot \nabla n > 0 \tag{5.17a}$$

where

$$\mathbf{E}' = \mathbf{E} + \mathbf{U} \times \mathbf{B} \tag{5.17b}$$

is the electric field which would be measured in the neutral frame of reference, while $\mathbf{E}$ and $\mathbf{U}$ are the electric field and neutral wind vectors in the earth-fixed

6300 Å AIRGLOW DEPLETIONS FROM ASCENSION ISLAND

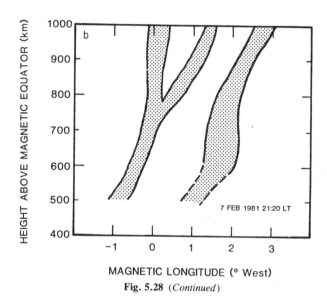

Fig. 5.28 (*Continued*)

frame. As noted earlier, **B** is unchanged in such a transformation if $U \ll c$, the speed of light. At mid-latitudes it is usually the case that $|U \times B| > |E|$. In the F-region dynamo case this is a necessary condition and an example is shown in Fig. 5.12b. For $E = 0$, (5.17a) implies that an eastward wind is destabilizing to a westward density gradient and vice versa. The depletions should thus structure on one side or the other, depending on the zonal neutral wind direction.

Since the equatorial anomaly region is one of very high electron density, possibly the highest average electron density anywhere in the earth's ionosphere, the effect of such structuring on transionospheric propagation is very pronounced. The scintillation technique discussed in Appendix A has been used extensively in the anomaly region to characterize the structure there and to better understand the scintillation process itself. Even at a transmission frequency of 6 GHz, up to 6 dB of scintillation-induced amplitude variation has been detected in this region. An example of data from such a study is shown in Figs. 5.29a and b. In Fig. 5.29a, a schematic diagram shows the relative motion of an airplane conducting scintillation measurements using satellite transmission through an uplifted plasma wedge. Data from five consecutive traversals are shown in Fig. 5.29b. The line-of-sight total electron content (TEC) and the scintillation level at 249 MHz are depicted for each pass. The westward edge is clearly more structured than the eastward edge, which is consistent with the discussion above

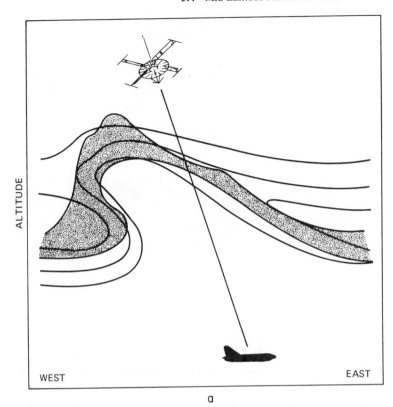

WEST   EAST

a

Fig. 5.29. (a) Schematic diagram of an airplane flight path and a satellite transmission link through a plasma uplift. (b) Data from five consecutive passes similar to the one illustrated in (a). For each leg the scintillation intensity and total electron content are plotted. The structure is clearly enhanced on one side of the bubble and decays with time. (Courtesy of R. Livingston.) (*Figure continues.*)

and the fact that the wind is usually eastward at this local time and latitude. Structures on the top and eastward regions are probably due to instabilities caused either by gravity or by large-scale secondary perturbation electric fields δE, which can also create $\delta E \times B$ instabilities on the gradients.

Many "active experiments" have been conducted at mid-latitudes to model this process by creating artificial plasma density gradients perpendicular to B and observing the resulting structure. The active technique is to inject large amounts of barium gas into the ionosphere from a rocket, where the barium is vaporized from the metal state by the intensely energetic thermite chemical reaction. The barium is ionized by sunlight and, if released at sunset or sunrise, results in a visible long-lived plasma (see Appendix A for more details). A photograph of such a release made over the Gulf of Mexico is given in Fig. 5.30. Notice

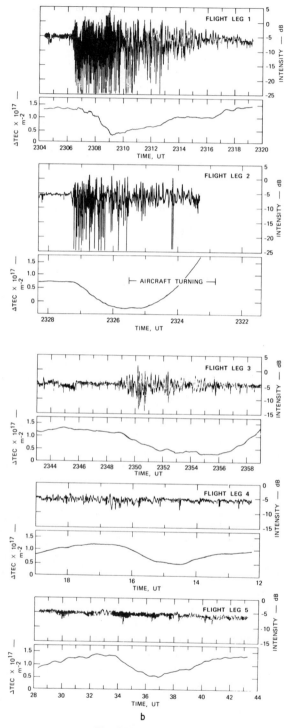

Fig. 5.29 (*Continued*)

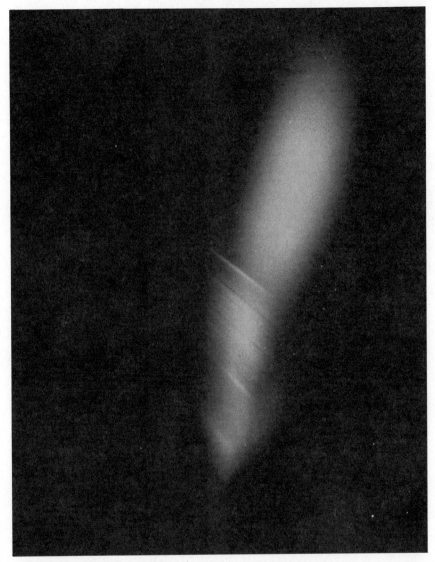

Fig. 5.30. Photograph of a striated barium cloud taken at right angles to the magnetic field. (Courtesy of W. Boquist.)

that striated regions form in the plasma and that they occur in only a portion of the cloud.

Whether a cloud is large or small depends on its electrical properties. A small cloud is required whenever the barium technique is used for tracer experiments such as those discussed in relation to Fig. 5.7. The circuit diagram analogy in

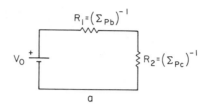

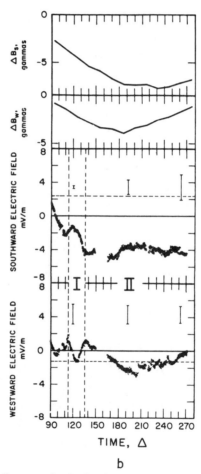

Fig. 5.31. (a) Circuit diagram analog for the effect of a barium cloud on the electrodynamics of the ionosphere. Here, V_0 is the applied voltage due to the ambient electric field, $R_1 = (\Sigma_{Pb})^{-1}$ is the resistivity of the ionosphere and $R_2 = (\Sigma_{Pc})^{-1}$ is the resistivity of the cloud. (b) Electric field signatures during an ionospheric rocket flight through a mid-latitude barium cloud. The measurements were made *in situ* and show that **E** is considerably smaller inside the barium cloud (between the dashed lines) than outside. The magnetic field instrument was located on the ground. [Part (b) after Schutz *et al.* (1973). Reproduced with permission of the American Geophysical Union.]

Fig. 5.31a applies if we associate R_1 with the reciprocal of the height-integrated Pedersen conductivity in the cloud [$R_1 = (\Sigma_{Pc})^{-1}$], while R_2 is the inverse of the height-integrated conductivity in the background ionosphere [$R_2 = (\Sigma_{Pb})^{-1}$]. If $R_1 \ll R_2$, the full ambient potential appears across the cloud and it drifts with the full velocity $\mathbf{E} \times \mathbf{B}/B^2$. If the cloud is large ($\Sigma_{Pc} > \Sigma_{Pb}$), the ambient electric field is shorted out and the cloud moves more slowly than the background. This has in fact been observed during a rocket flight which traversed a dense barium plasma cloud. The data in Fig. 5.31b show that the electric field was lower inside the barium cloud than outside (Schutz *et al.*, 1973).

We have already discussed in Chapter 4 why an electric field produces an instability in the plasma. Another way to understand the instability of a barium cloud and formation of the striations which are apparent in the photograph in Fig. 5.30 involves the shorting effect discussed above. Consider alternating regions of enhanced and depleted plasma density. That is, consider a small initial sinusoidal perturbation which occurs on the particular side of a cylindrically symmetric plasma cloud where $(\mathbf{E}' \times \mathbf{B}) \cdot \nabla n > 0$. If we work in the neutral frame of reference where $\mathbf{U} = 0$, then $\mathbf{E}'$ is the electric field and $\mathbf{E}' \times \mathbf{B}/B^2$ is the background plasma flow velocity in that reference frame. A region of enhanced plasma density will have an elevated height-integrated Pedersen conductivity and, due to the shorting effect, will drift slower than the adjacent depleted region for which Σ_P is lower. Now since $(\mathbf{E}' \times \mathbf{B})$ is parallel to ∇n, the lower-density region will move up the gradient to regions of higher surrounding density while the high-density region will move down the gradient. Thus, relative to the surroundings, both enhancements and depletions seem to grow in intensity and we say the configuration is unstable. This argument is equivalent to that associated with Fig. 4.8.

Mid-latitude barium cloud striations have been penetrated by sounding-rocket payloads. Some data from one such experiment are reproduced in Fig. 5.32a. The smooth, enhanced density profile due to the plasma cloud is interrupted on one edge by fingers of alternating high and low plasma density. These are the unstable electrostatic perturbations discussed above. Such a process is compared to the naturally occurring phenomenon of bottomside equatorial spread F in Fig. 5.32b by comparing the power spectrum of the barium fluctuations with spread F data, which is driven by the gravitational term in the generalized Rayleigh–Taylor instability. The similarity is remarkable and gives further evidence that the active barium cloud experiment indeed mirrors natural phenomena quite well.

The growth rate for the $\mathbf{E} \times \mathbf{B}$ instability for a "local" calculation, ignoring, among other things, the plasma effects due to electrical coupling along the magnetic field lines, is from Chapter 4

$$\gamma = E'/BL \qquad (5.18a)$$

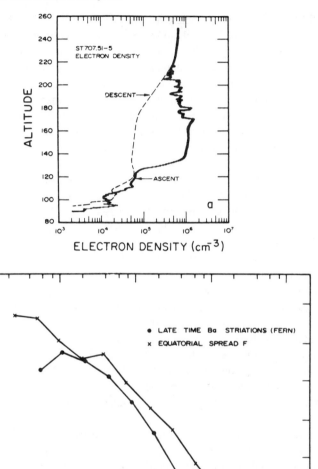

Fig. 5.32. (a) Electron density profiles from a probe rocket flight through a barium cloud. The dashed curves are the measurements of the undisturbed F-region profile on rocket descent. (b) Filled circles indicate the wave number power spectrum of the data shown in (a). Also shown (crosses) is a spectrum for a rocket flown into equatorial spread F conditions. [After Kelley *et al.* (1979). Reproduced with permission of the American Geophysical Union.]

where L is the inverse gradient scale length of the cloud. For typical mid-latitude conditions $E'/B = 50$ m/s while typical barium clouds have $L \sim 6000$ m, and hence $\gamma \simeq (2 \text{ min})^{-1}$.

A first-order correction to (5.18a) (see Francis and Perkins, 1975) is given by

$$\gamma(k) = [\Sigma_{Pb}/(\Sigma_{Pb} + \Sigma_{Pc})](E'/BL)[k^2/(k^2 + k_0^2)] \qquad (5.18b)$$

which takes into account the shorting effect of a conducting background E or F region and the finite size of the cloud, since $k_0 = 2\pi/L$. Considerable effort has gone into the barium cloud striation problem and a vast literature exists. Much of this research is applicable to naturally occurring spread F phenomena as well. Some subtleties which arise involve the effect of velocity shear on the instability, which is a stabilizing factor (Perkins and Doles, 1975), and the production of image striations in the background ionosphere (Goldman *et al.*, 1976). Both of these processes introduce additional scale size factors into the linear growth rate. Images are discussed further in Chapter 8 as is the high-latitude $\mathbf{E} \times \mathbf{B}$ instability.

5.4.2 Mid-Latitude F-Region Plasma Instabilities

When one looks at the effect of the $\mathbf{E} \times \mathbf{B}$ process on the plasma, there is no fundamental difference between the equatorial anomaly region and higher latitudes except for the different origins of unstable density gradients perpendicular to $\mathbf{B}$. We have separately discussed the anomaly region in Section 5.4.1 since a prime source of gradients in that region is upwelling wedges due to equatorial spread F. Likewise, we treat the polar latitude case separately in Chapter 8 since the plasma gradients there come from particle precipitation, large electric fields, and/or other strictly high-latitude sources. In this section we treat the region poleward of bubble penetration and equatorward of the auroral zone.

Of course, any zonal density gradient will still form plasma structures if there is a zonal wind or a poleward electric field such that $(\mathbf{E}' \times \mathbf{B}) \cdot \nabla n > 0$. However, another wrinkle to the mid-latitude problem was pointed out by Perkins (1973). He noted that the typical mid-latitude ionospheric conditions will adjust such that there is a force balance (see Section 5.2.2). In fact, although the process we are discussing is called the Perkins instability, the mid-latitude ionospheric state could just as well be termed the Perkins "stability" since we have already noted the remarkable tendency for the plasma to move horizontally. Perkins chose to treat the case in which an eastward electric field supports the plasma against gravity (he points out that an equally valid approach is to assume there is a southward neutral wind which also supports the plasma against gravity). He assumes a vertically stratified ionosphere; that is, ∇n is vertical even though $\mathbf{B}$ has a significant dip angle. This is quite consistent with a slowly varying photoionization source for the mid-latitude F-region plasma since, as shown in Section 5.1.1, in such a case the plasma must be horizontally stratified. This need not, and often is not, the case at high latitudes, where convection and particle precipitation can create magnetic field-aligned plasma structures (see Chapter 8).

First, we note that without an eastward electric field, equilibrium can exist only exactly at the magnetic equator, where the $\mathbf{J}_g \times \mathbf{B}$ force due to gravitational currents can balance the vertical gravity term in the force balance equation

$$\mathbf{J}_g \times \mathbf{B} + \rho\mathbf{g} = 0$$

At the equator $\mathbf{J}_g$ has a magnitude $\rho g/B$ and is directed toward the east. In general, using (2.36b) for $m \ll M$, $\mathbf{J}_g = (\rho/B^2)\mathbf{g} \times \mathbf{B}$, and if the magnetic field has a dip angle I, then the $\mathbf{J}_g \times \mathbf{B}$ force is

$$\mathbf{J}_g \times \mathbf{B} = -\rho g \cos I \hat{a}_{y'} \tag{5.19}$$

where the $\hat{a}_{y'}$ axis is defined in Fig. 5.11. This force cannot balance the gravitational force $\rho\mathbf{g} = -\rho g \hat{a}_z$ unless we are at the equator. The Rayleigh–Taylor theory given in Chapter 4 shows that this equilibrium is not stable.

For the off-equatorial case, Perkins showed that if a purely zonal electric field (or a southward wind) is added, equilibrium can again be achieved but the resulting equilibrium is stable. In essence, in equilibrium the ionosphere rises due to the $\mathbf{E} \times \mathbf{B}$ drift as fast as it falls due to gravity. This, of course, is exactly the typical ionospheric state measured so often at Arecibo and discussed in detail above, in which there is a nearly horizontal plasma flow either poleward (eastward electric field driven) or equatorward (equatorward wind driven). For the zonal electric field case, the velocity due to E_x is $-(E_x/B)\hat{a}_{y'}$. The velocity due to gravity is in the $\hat{a}_{z'}$ direction and has a magnitude $g \sin I/\nu_{in}$. For the vertical components of these two velocities to be equal and opposite, we require

$$(E_x/B) \cos I = (g/\nu_{in}) \sin^2 I \tag{5.20a}$$

or

$$E = (Bg/\nu_{in})(\sin^2 I/\cos I) \tag{5.20b}$$

The only free parameter in this equation is ν_{in} since g and the dipole angle I are fixed. The plasma must therefore adjust in *height* to satisfy (5.20a) since ν_{in} is height dependent. A similar equilibrium holds if

$$(g/\nu_{in}) \sin^2 I = V_E \cos I \sin I \tag{5.20c}$$

or

$$V_E = (g/\nu_{in}) \tan I \tag{5.20d}$$

where V_E is the equatorward neutral wind component.

A rigorous analysis requires that height-dependent quantities be used. Perkins introduced Σ_P and N, the height-integrated Pedersen conductivity and electron density, into the analysis as the two perturbation parameters. In the course of his analysis, Perkins used the fact that if there are no horizontal gradients in N or Σ_P, then there also are no gradients perpendicular ($\nabla_\perp$) to **B**. This seems to be a contradiction since **B** makes a finite angle with the vertical, as illustrated in Fig. 5.33a, and it seems that a purely vertical density gradient would imply that the perpendicular gradient would not vanish. Indeed, the local gradient $\nabla_\perp n$ does *not* vanish. However, N and Σ_P are integrated over the entire F-region ionosphere. Thus, the integral will be the same along every flux tube, $\nabla_\perp N = \nabla_\perp \Sigma_P = 0$ and there is no **E** × **B** instability.

Consider now a zonal electric field E_x supporting the ionospheric plasma against gravity. Perkins showed that it is in fact a stable equilibrium, which can be understood as follows. First, we note that ignoring production and loss (as Perkins does), the integrated plasma content N cannot change on a given flux tube. Referring now to Fig. 5.33b, suppose a sinusoidal perturbation in Σ_P exists with **k** perpendicular to **B** and in the zonal direction. Where Σ_P is high, the flux tube must be at a lower altitude (larger ν_{in}), and where Σ_P is low, the flux tube plasma must be displaced to a higher altitude (smaller ν_{in}). Now, where Σ_P is high the zonal current is high and the **J** × **B** force tends to drive the flux tube back up from its lower-altitude position. Likewise, where Σ_P is reduced the supporting current is decreased and the flux tube falls back down toward its equilibrium position. In other words, the configuration is stable.

The more formal analysis of the linearized equations for N and Σ_P yields, for the case of an eastward electric field E_x and no horizontal gradients (and therefore also no gradients perpendicular to **B**),

$$dN/dt = 0$$

since the height-integrated plasma content cannot change, and

$$\partial \Sigma_P / \partial t = geN \sin^2 I / \Omega_i B L_n - \Sigma_P E_x \cos I / B L_n \qquad (5.21a)$$

where L_n is the neutral scale height. In equilibrium $\partial \Sigma_P / \partial t = 0$ and

$$\Sigma_{P0} E_x = (N_0 eg / \Omega_i)(\sin^2 I / \cos I) \qquad (5.21b)$$

which reduces to $(E_x/B) = (gB/\nu_{in})(\sin^2 I/\cos I)$ if $(N_0 e^2 \nu_{in}/M\Omega_i^2)$ is substituted for Σ_{P0}. This result is identical to 5.20b.

Evidence that this equilibrium can exist in the ionosphere comes from the ubiquitous occurrence of horizontal plasma flows over Arecibo. Evidence for the instability comes from a very remarkable data set published by Behnke (1979). Examples of five nights of data are presented in Fig. 5.33c and show that the ionosphere can display a very banded structure with adjacent regions characterized by abrupt changes in h_{max}, the height of the peak in electron density. These structures were detected on five out of eight nights studied. The extent of the

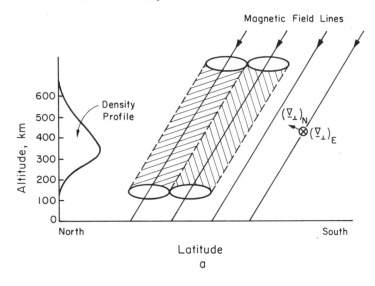

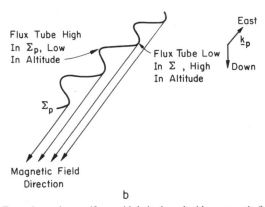

Fig. 5.33. (a) For an ionosphere uniform with latitude and with a magnetic field which does not vary in dip angle with latitude (both approximations) neither the height-integrated content (N) nor the height-integrated conductivity (Σ_P) vary from one flux tube to the other. The two components of $\nabla_\perp$ are illustrated. If, as assumed by Perkins, the ionosphere is also longitudinally invariant, $\nabla_\perp N_0 = \nabla_\perp \Sigma_{P0} = 0$. (b) Perturbations in Σ_P with $\mathbf{k}_\perp$ in the zonal direction and $\mathbf{k}_\parallel = 0$.(c) Variations of h_{max}, the altitude of the F peak over Arecibo, versus time for five nights. The open circles indicate points assumed to be transitions between regions of dissimilar h_{max}. The nights are (a) March 6, 1973; (b) March 5, 1973; (c) March 7, 1973; (d) October 14, 1973; and (e) June 21–22, 1973. [Part (c) after Behnke (1979). Reproduced with permission of the American Geophysical Union.]

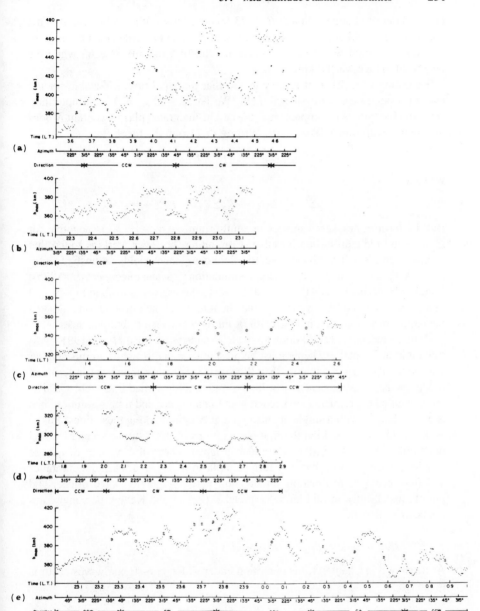

C

Fig. 5.33 (*Continued*)

elevated density band on March 6, 1973 (panel a) was sufficiently large that the zonal electric field could be measured inside (E_{xi}) and outside (E_{xo}) the region. The reported ratio was $E_{xi}E_{xo} = (17.2 \text{ mV/m})/(6.5 \text{ mV/m}) = 2.65$ while the height difference was 60 km.

Returning to (5.21a), it is easy to see that the equilibrium is stable for the case of a purely eastward electric field. We let $\Sigma_P = \Sigma_{P0} + \Sigma'$ and note that, since the F region is incompressible, the height-integrated plasma content N does not change with time $(dN/dt = 0)$. Writing (5.21a) in the form

$$\partial\Sigma_P/\partial t = K_1 N - K_2\Sigma_P$$

we have

$$\gamma\Sigma' = [K_1 N - K_2\Sigma_{P0}] - K_2\Sigma'$$

But the term in brackets vanishes in equilibrium, so $\gamma = -K_2$ is negative for $E_x > 0$ and the equilibrium is stable. Continuing, however, Perkins found that adding even a small northward electric field component yields an unstable situation, a result verified via computer simulations (Scannapieco et al., 1975). Behnke (1979) interpreted the 50–100-km slabs themselves (shown in Fig. 5.33c) as a manifestation of the instability since he found that their orientations satisfied the requirement for maximum growth (**k** midway between **E** and due east).

A more recent study of kilometer-scale structures by Basu et al. (1981) indicated that horizontal gradients must be present to create structure in this smaller scale size. They argued that the **E** × **B** process is the most important in terms of producing scintillations and, most likely, mid-latitude spread F as well. Clearly, more detailed experiments are needed with higher space and time resolution than is available now with a single antenna feed at Arecibo. Along these lines, Fukao et al. (1988) have reported the first observations of mid-latitude spread F using the multiple-beam 46.5-MHz MU radar in Japan. These observations of intense 3-m irregularities show that very turbulent plasmas can occur at middle latitudes and may provide information on the link between gravity wave effects and plasma instabilities in mid-latitude spread F (Fukao and Kelley, personal communication, 1989).

5.4.3   Mid-Latitude E-Region Instabilities

As discussed in Chapter 4, the condition under which two-stream waves occur is given by

$$V_D \simeq E/B > (1 + \Psi_0)C_s \tag{5.22}$$

where $\Psi_0 = \nu_i\nu_e/\Omega_i\Omega_e$ and C_s is the acoustic speed. The magnetic field increases quickly with latitude, while the mid-latitude electric field is actually smaller than the vertical polarization field at the equator. Thus, (5.22) could hold only at times when large electric fields penetrate to low latitudes. To our knowledge, no

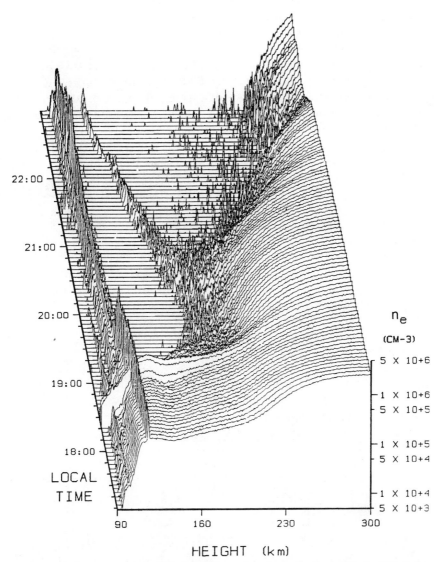

Fig. 5.34. Consecutive electron density profiles measured over Arecibo on May 7, 1983. Notice the intense sporadic E layer which existed even prior to sunset and the intermediate layer which descended through the F-layer valley. [After Riggin *et al.* (1986). Reproduced with permission of the American Geophysical Union.]

clear evidence for pure two-stream waves exists in the E region between the equatorial and polar regions. This is not the case for the gradient drift process, which has been observed in the HF frequency range in Japan (Tanaka and Venkateswaran, 1982) and with 50-MHz radars from the Caribbean islands of Guadeloupe (Ecklund *et al.*, 1981) and St. Croix (Riggin *et al.*, 1986).

As in the equatorial and high-latitude experiments, radar backscatter from plasma waves in the ionosphere occurs only when the radar wave vector is nearly perpendicular to **B.** From St. Croix, this geometry holds in the E region directly over the Arecibo Observatory. Conditions in the ionosphere during one such 50-MHz backscatter event are illustrated with the contour plot in Fig. 5.34, which is similar to the Arecibo plots given in Figs. 5.19a–c. Prior to sunset a very strong sporadic E layer with $n_e > 5 \times 10^5$ cm^{-3} was in evidence near 125 km.

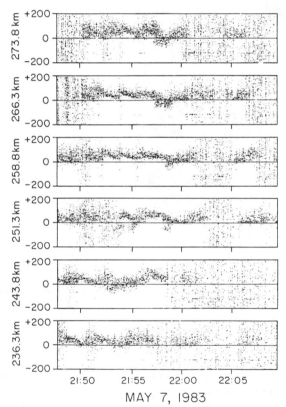

Fig. 5.35. Doppler spectrogram for the event illustrated in Fig. 5.34. The gray scale denotes intensity normalized to the peak intensity in a given spectrum and the vertical axis Doppler shift in meters per second. [After Riggin *et al.* (1986). Reproduced with permission of the American Geophysical Union.]

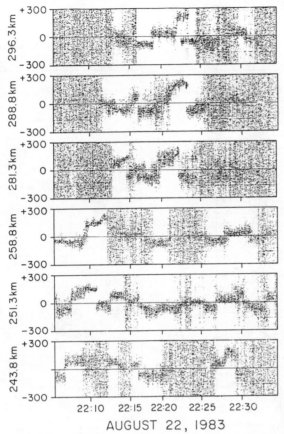

Fig. 5.36. (a) Doppler spectrogram for August 22, 1983. (b) Individual Doppler spectra are shown on the left side for the event in (a). The position across the beam of a localized scatterer is plotted on the right side. The scatterer is moving across the beam at a speed of about 30 m/s. [After Riggin *et al.* (1986). Reproduced with permission of the American Geophysical Union.] (*Figure continues.*)

It drifted downward after sunset, decreasing drastically in peak density near 2000 LT. The layer continued to move downward until 2148 LT, when, at 105 km, the layer density began to rise again. At this time 50-MHz echoes were received from St. Croix which continued for some 20 minutes. The Doppler spectrogram for the event, which is given in Fig. 5.35, is very similar to those associated with "large-scale waves" in the gradient-dominated portion of the equatorial electrojet discussed at length in Chapter 4. The line-of-sight phase velocity of 3-m waves is seen to oscillate about a mean value as the event proceeds. The different panels correspond to different ranges from the radar. The spectacular example from August 22, 1983 is given in Fig. 5.36a and shows that in some cases quite

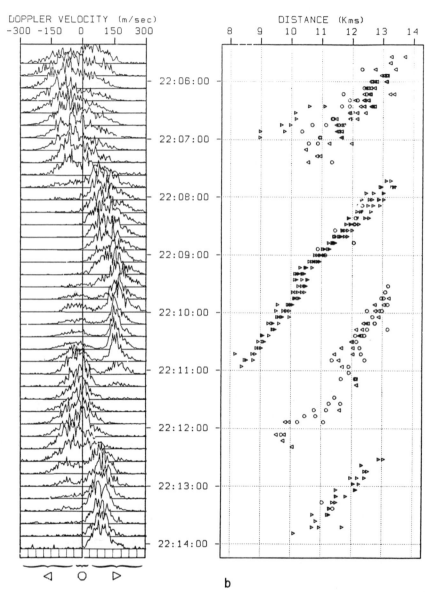

Fig. 5.36 (*Continued*)

substantial Doppler shifts can occur, nearly 300 m/s here. The waveforms are "squared off," suggesting that a limiting phase velocity exists, much as observed at the equator, where the 3-m wave phase velocities seem limited to the acoustic speed. A more quantitative Doppler presentation is given in the left side of Fig. 5.36b for one full large-scale wave cycle. Interferometric measurements in Fig. 5.36b (right side) show that the horizontal wavelength was in the range of 6 km (Riggin *et al.*, 1986). The implication of these large Doppler shifts is that the perturbation electric fields in these large waves are of the order of 10–15 mV/m.

Many questions about these mid-latitude E-region instabilities remain but it seems likely that they arise in much the same way as their equatorial counterparts. The prediction of simple linear theory is that the side of the sporadic E layer where $\mathbf{E} \cdot \nabla n > 0$ should be unstable while the other side should be stable. Since the two-stream condition is seldom satisfied, these waves offer a "pure" gradient drift system against which to test nonlinear theories. They also provide a way to test the effect of electric field mapping along the magnetic field lines. As discussed in Chapter 2, large-scale waves map for long distances along magnetic field lines. This means that the wave potential will map to regions where the density gradient is stable. How this affects the wave growth is not well understood at all.

References

Angerami, J. J., and Carpenter, D. L. (1966). Whistler studies of the plasmapause in the magnetosphere. 2. Electron density and total tube electron content near the knee in magnetospheric ionization. *J. Geophys. Res.* **71**, 711.

Basu, Su., Basu, Sa., Ganguly, S., and Klobuchar, J. A. (1981). Generation of kilometer scale irregularities during the midnight collapse at Arecibo. *JGR, J. Geophys. Res.* **86**, 7607.

Behnke, R. A. (1979). F layer height bands in the nocturnal ionosphere over Arecibo. *JGR, J. Geophys. Res.* **84**, 974.

Behnke, R. A., and Harper, R. M. (1973). Vector measurements of F region ion transport at Arecibo. *J. Geophys. Res.* **78**, 8222.

Behnke, R. A., Kelley, M., Gonzales, G., and Larsen, M. (1985). Dynamics of the Arecibo ionosphere: A case study approach. *JGR, J. Geophys. Res.* **90**, 4448.

Blanc, M. (1983). Magnetospheric convection effects at mid-latitudes. 3. Theoretical derivation of the disturbance convection pattern in the plasmasphere. *JGR, J. Geophys. Res.* **88**, 235.

Bowman, G. G. (1981). The nature of ionospheric spread-F irregularities in mid-latitude regions. *J. Atmos. Terr. Phys.* **43**, 65.

Bowman, G. G. (1985). Some aspects of mid-latitude spread-E_s and its relationship with spread-F. *Planet. Space Sci.* **33**, 1081.

Burnside, R. G. (1984). Dynamics of the low-latitude thermosphere and ionosphere. Ph.D. Thesis, University of Michigan, Ann Arbor.

Burnside, R. G., Behnke, R. A., and Walker, J. C. (1983). Meridional neutral winds in the thermosphere at Arecibo: Simultaneous incoherent scatter and airglow observations. *JGR, J. Geophys. Res.* **88**, 3181.

Chapman, S., and Lindzen, R. S. (1970). "Atmospheric Tides: Thermal and Gravitational." Gordon & Breach, New York.

Dougherty, J. P. (1961). On the influence of horizontal motion of the neutral air on the diffusion equation of the F-region. *J. Atmos. Terr. Phys.* **20**, 167.

Dungey, J. W. (1959). Effect of a neutral field turbulence in an ionized gas. *J. Geophys. Res.* **64**, 2188.

Ecklund, W. L., Carter, D. A., and Balsley, B. B. (1981). Gradient drift irregularities in mid-latitude sporadic E. *JGR, J. Geophys. Res.* **86**, 858.

Forbes, J. M., Codrescu, M., and Hall, T. J. (1988). On the utilization of ionosonde data to analyze the latitudinal penetration of ionospheric storm effects. *Geophys. Res. Lett.* **15**, 249.

Francis, S. H., and Perkins, F. W. (1975). Determination of striation scale sizes for plasma clouds in the ionosphere. *JGR, J. Geophys. Res.* **80**, 3111.

Fukao, S., McClure, J. P., Ito, A., Sato, T., Kimura, I., Tsuda, T., and Kato, S. (1988). First VHF radar observations of mid-latitude F-region field aligned irregularities. *Geophys. Res. Lett.* **15**, 768.

Goldman, S. R., Baker, L., Ossakow, S. L., and Scannapieco, A. J. (1976). Striation formation associated with barium clouds in an inhomogeneous ionosphere. *JGR, J. Geophys. Res.* **81**, 5097.

Gonzales, C. A., Kelley, M. C., Carpenter, L. A., and Holzworth, R. H. (1978). Evidence for a magnetospheric effect on mid-latitude electric fields. *JGR, J. Geophys. Res.* **83**, 4397.

Harper, R. M., and Walker, J. C. G. (1977). Comparison of electrical conductivities in the E and F regions of the nocturnal ionosphere. *Planet. Space Sci.* **25**, 197.

Hernandez, G., and Roble, R. G. (1984). Nighttime variation of thermospheric winds and temperatures over Fritz Peak Observatory during the geomagnetic storm of March 2, 1983. *JGR, J. Geophys. Res.* **89**, 9049.

Herrmann, U., Eberhardt, P., Hidalgo, M. A., Kopp, E., and Smith, L. G. (1978). Metal ions and isotopes in sporadic E-layers during the Perseid meteor shower. *Space Res.* **18**, 249.

Hines, C. O. (1974). "The Upper Atmosphere in Motion: A Selection of Papers with Annotation," Geophys. Monogr. 18. Am. Geophys. Union, Washington, D.C.

Holton, J. R. (1979). "An Introduction to Dynamic Meteorology," 2nd ed. Academic Press, New York.

Kelley, M. C., Baker, K. D., and Ulwick, J. C. (1979). Late time barium cloud striations and their possible relationship to equatorial spread F. *JGR, J. Geophys. Res.* **84**, 1898.

Liller, W., and Whipple, F. L. (1954). *J. Atmos. Terr. Phys.* **1**, Spec. Suppl., 112.

Matsushita, S. (1967). Solar quiet and lunar daily variation fields. *In* "Physics of Geomagnetic Phenomena" (S. Matsushita and W. H. Campbell, eds.), Vol. 1, p. 301. Academic Press, New York.

Matsushita, S. (1971). Interactions between the ionosphere and the magnetosphere for Sq and L variations. *Radio Sci.* **6**, 279.

Mendillo, M., and Tyler, A. (1983). Geometry of depleted plasma regions in the equatorial ionosphere. *JGR, J. Geophys. Res.* **88**, 5778.

Mozer, F. S. (1970). Electric field mapping from the ionosphere to the equatorial plane. *Planet. Space Sci.* **18**, 259.

Mozer, F. S. (1973). Electric fields and plasma convection in the plasmasphere. *Rev. Geophys. Space Phys.* **11**, 755.

Perkins, F. W. (1973). Spread F and ionospheric currents. *J. Geophys. Res.* **78**, 218.

Perkins, F. W., and Doles, J. H., III (1975). Velocity shear and the $E \times B$ instability. *JGR, J. Geophys. Res.* **80**, 211.

Richmond, A. D., Matsushita, S., and Tarpley, J. D. (1976). On the production mechanism of electric currents and fields in the ionosphere. *JGR, J. Geophys. Res.* **81**, 547.

Richmond, A. D., Blanc, M., Emery, B. A., Wand, R. H., Fejer, B. G., Woodman, R. F., Ganguly, S., Amayenc, P., Behnke, R. A., Calderon, C., and Evans, J. V. (1980). An empirical model of quiet-day ionospheric electric fields at middle and low latitudes. *JGR, J. Geophys. Res.* **85**, 4658.

Riggin, D., Swartz, W. E., Providakes, J., and Farley, D. T. (1986). Radar studies of long wavelength waves associated with midlatitude sporadic-E layers. *JGR, J. Geophys. Res.* **91**, 8011.

Rishbeth, H. (1971). The F-layer dynamo. *Planet. Space Sci.* **19**, 263.

Rishbeth, H., and Garriott, O. K. (1969). "Introduction to Ionospheric Physics," Int. Geophys. Ser., Vol. 14. Academic Press, New York.

Scannapieco, A. J., Gordon, S. R., Ossakow, S. L., Book, D. L., and McDonald, B. E. (1975). "Theoretical and Numerical Simulation Studies of Midlatitude F Region Irregularities," NRL Memo. Rep. 3014. Naval Res. Lab., Washington, D.C.

Schutz, S., Adams, G. J., and Mozer, F. S. (1973). Probe electric field measurements near a midlatitude ionospheric barium release. *J. Geophys. Res.* **78**, 6634.

Shen, J. S., Swartz, W. E., Farley, D. T., and Harper, R. M. (1976). Ionization layers in the nighttime E region valley above Arecibo. *JGR, J. Geophys. Res.* **81**, 5517.

Stubbe, P., and Chandra, S. (1970). The effect of electric fields on the F-region behavior as compared with neutral wind effects. *J. Atmos. Terr. Phys.* **32**, 1909.

Tanaka, T., and Venkateswaran, S. V. (1982). Gradient-drift instability of nighttime mid-latitude Es-layers. *J. Atmos. Terr. Phys.* **44**, 939.

Vickrey, J. F., Swartz, W. E., and Farley, D. T. (1979). Postsunset observations of ionospheric–protonospheric coupling at Arecibo. *JGR, J. Geophys. Res.* **84**, 1310.

Voss, H. D., and Smith, L. G. (1979). Nighttime ionization by energetic particles at Wallops Island in the altitude region 120 to 200 km. *Geophys. Res. Lett.* **6**, 93.

Voss, H. D., and Smith, L. G. (1980a). Rocket observations of energetic ions in the nighttime equatorial precipitation zone. *In* "Low Latitude Aeronomic Processes" (A. P. Mitra, ed.), p. 131. Pergamon, Oxford.

Chapter 6 | High-Latitude Electrodynamics

In this chapter we study the macroscopic motion of the high-latitude ionospheric plasma in the plane perpendicular to the magnetic field lines. Since the magnetic field is nearly vertical, this corresponds to the horizontal motion of plasma. At the large scales (> 100 km) considered here the electric force in (2.36b) dominates the pressure gradient and gravitational forces so that only an imposed electric field and a neutral wind need be considered in the perpendicular plasma motion. In order to understand the characteristics and the sources of the imposed electric field, we first deal briefly with the relationships between electric fields and currents that exist in the ionosphere, outer magnetosphere, and solar wind. These fields and currents are coupled along the earth's magnetic field. Following this we discuss the observed characteristics of electric fields and currents in the ionosphere and their relationships to the magnetic field topology throughout the ionosphere, the magnetosphere, and the solar wind system.

6.1 Electrical Coupling between the Ionosphere, Magnetosphere, and Solar Wind

6.1.1 General Relationships

We begin by considering two regions—one in the ionosphere and one in the magnetosphere. Below about 200 km the ionosphere is a resistive medium, and in Chapter 2 we showed that the electric field $\mathbf{E}$ and electric current $\mathbf{J}$ are related by the equation

$$\mathbf{J} = \boldsymbol{\sigma} \cdot (\mathbf{E} + \mathbf{U} \times \mathbf{B}) \qquad (6.1)$$

where $\boldsymbol{\sigma}$ is the ionospheric conductivity tensor and $\mathbf{U}$ and $\mathbf{B}$ are the neutral wind velocity and magnetic field, respectively. In that chapter we also pointed out that above about 2000 km the magnetospheric plasma is essentially collisionless and the equations governing $\mathbf{E}$, $\mathbf{J}$, and the plasma velocity $\mathbf{V}$ are

$$\mathbf{E} + \mathbf{V} \times \mathbf{B} = 0 \tag{6.2}$$

and

$$\rho \, d\mathbf{V}/dt = -\nabla p + \rho \mathbf{g} + \mathbf{J} \times \mathbf{B} \tag{6.3}$$

Taking the cross product of (6.3) with $\mathbf{B}$, the current density perpendicular to $\mathbf{B}$ is given by

$$\mathbf{J}_\perp = (1/B^2)(\rho \mathbf{B} \times d\mathbf{V}/dt + \rho \mathbf{g} \times \mathbf{B} + \mathbf{B} \times \nabla p) \tag{6.4}$$

The gravitational term in (6.4) can usually be neglected, so we note that in the high-altitude magnetosphere the cross-field current is controlled by pressure gradients and space or time-dependent flow and not directly by the electric field. Electrical coupling between the regions is described by the equation for current continuity

$$\nabla \cdot \mathbf{J} = 0 \tag{6.5}$$

and Faraday's law

$$\nabla \times \mathbf{E} + \partial \mathbf{B}/\partial t = 0 \tag{6.6}$$

which must apply throughout the system. The boundaries of these regions can be defined rather loosely for our purposes since only a qualitative treatment of magnetosphere–ionosphere coupling is being undertaken. The region between 200 and 2000 km is a transition zone, but for most purposes the plasma motion there is also given by (6.2). Throughout this chapter we assume that the magnetic field lines that connect these regions are electric equipotentials. This assumption breaks down drastically in the lower magnetosphere on auroral zone field lines, where electrons and ions are accelerated by parallel electric fields. This source region for auroral arcs is not considered in this text.

The space above 2000 km can be divided into two topologically different regions by the magnetic field geometry described in Chapter 1 (see Fig. 1.13). At the highest latitudes the magnetic field lines extend either to the magnetopause and subsequently into the magnetosheath and solar wind, or far down the magnetotail into the boundary layer that lies just inside the magnetopause. Field lines extending to the magnetosheath are "open"; that is, they have one foot on the earth and the other connected to the interplanetary magnetic field (IMF). Field lines extending to the boundary layer and inner magnetosphere are "closed"; that is, they have both feet on the earth even though the field line may be extremely long. In the magnetosheath and in the boundary layer the plasma is flowing rapidly antisunward and is driven by the expanding solar atmosphere. In the inner magnetosphere the plasma flow is dependent on the plasma pressure and upon both internally and externally applied electric fields.

6.1.2   A Qualitative Description for Southward IMF

Some fundamental properties of the coupling of electric fields, currents, and energy in the ionosphere, magnetosphere, and solar wind system can be understood as follows. Consider the case where magnetic field lines in the ionosphere connect to a southward IMF. This defines an area at high latitudes in the ionosphere called the polar cap. The geometry is shown schematically in Fig. 6.1. This interaction is described qualitatively in Chapter 1, which the reader is encouraged to revisit at this time. Since the solar wind plasma is collisionless and expands radially outward from the sun, the electric field in the solar wind is given by $\mathbf{E}_{sw} = -\mathbf{V}_{sw} \times \mathbf{B}_{sw}$ and for a southward IMF the field will have a component pointing from dawn to dusk. The electric potential across the connected field lines will be applied across the magnetosphere and will map down to the polar cap ionosphere, where a dawn-to-dusk directed ionospheric electric field $\mathbf{E}_I$ will also result. This electric field drives the *ionospheric* F-region plasma in the antisunward direction at a speed $\mathbf{V}_I = \mathbf{E}_I \times \mathbf{B}_I / B_I^2$. The magnetic flux density is higher in the ionosphere than in the solar wind and since the equipotential surfaces converge the electric field in the ionosphere will be larger than in the solar wind. Typical numbers are

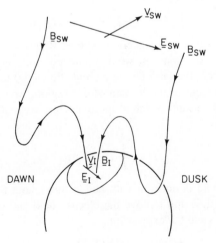

Fig. 6.1. Schematic representation of the magnetic connection between the solar wind dynamo and the ionospheric load.

$$B_1/B_{SW} = 50{,}000 \text{ nT}/5 \text{ nT} = 10^4$$

$$E_1/E_{SW} = 50 \text{ mV m}^{-1}/1 \text{ mV m}^{-1} = 50$$

$$V_1/V_{SW} = 1 \text{ km s}^{-1}/400 \text{ km s}^{-1} = 2.5 \times 10^{-3}$$

Returning to (6.1) and ignoring the neutral wind in the ionosphere for the present, the electric field will drive a current at ionospheric heights given by

$$\mathbf{J} = \boldsymbol{\sigma} \cdot \mathbf{E}_1 \tag{6.7}$$

The Pedersen component of this ionospheric current is parallel to $\mathbf{E}_1$ and hence is such that $\mathbf{J} \cdot \mathbf{E} > 0$. In this interaction then the ionosphere is a load and we must ascertain where the electrical energy originates. Referring to Fig. 6.1, suppose that the solar wind slows down slightly when it is in contact with the region shown and that we can ignore gravity and pressure gradients. Then (6.4) shows that $\mathbf{J}_\perp = (\rho/B^2)\mathbf{B} \times d\mathbf{V}/dt$ is antiparallel to $\mathbf{E}_{SW}$, that is, $\mathbf{J}_{SW} \cdot \mathbf{E}_{SW} < 0$. The solar wind acts as an MHD generator feeding energy to the ionosphere as a load. Note that in this region the $\mathbf{J} \times \mathbf{B}$ force is directed back toward the sun, which shows consistency with the concept that the solar wind is slowing down due to the interaction. These same principles hold in any electromotive generator in which kinetic energy is converted to electrical energy. Finally, since the interaction region where the solar wind slows down is bounded, $\mathbf{J}_{SW}$ has a divergence which provides the field-aligned currents feeding the load.

From the viewpoint of ionospheric physics it is important to understand how the currents from the generator link up to the load currents. Dividing $\mathbf{J}$ into components perpendicular ($\perp$) and parallel ($\parallel$) to the magnetic field and setting the divergence of $\mathbf{J}$ to zero yields

$$\nabla \cdot \mathbf{J} = \nabla_\perp \cdot \mathbf{J}_\perp + \partial J_\parallel/\partial s = 0 \tag{6.8}$$

where s is distance along the magnetic field line. Repeating and slightly extending the discussion in Section 2.4, we may integrate this equation over the field line distance Δs where the ionospheric currents flow (200 to 80 km). Assuming that no current flows out of the bottom of this region, we obtain an expression for the field-aligned current (which is approximately vertical in the polar cap) at the top of the region:

$$J_\parallel = \int_{\Delta s} (\nabla_\perp \cdot \mathbf{J}_\perp) \, ds \tag{6.9}$$

Since the electric field is uniform with height throughout the integral, we can substitute (6.7) in (6.9) and remove the electric field and divergence operator from the integral to yield the expression

$$J_\parallel = \nabla_\perp \cdot (\boldsymbol{\Sigma}_\perp \cdot \mathbf{E}_1) \tag{6.10}$$

where $\Sigma_\perp$ is the height-integrated horizontal conductivity tensor. Breaking this expression into Hall and Pedersen components, we have finally

$$J_\parallel = \Sigma_P(\nabla_\perp \cdot \mathbf{E}_1) + \mathbf{E}_1 \cdot \nabla_\perp \Sigma_P + \Sigma_H[(\nabla_\perp \cdot (\mathbf{E}_1 \times \hat{s})]$$
$$+ (\mathbf{E}_1 \times \hat{s}) \cdot \nabla_\perp \Sigma_H \tag{6.11}$$

We showed in Chapter 3 that for an electrostatic field $\nabla_\perp \cdot (\mathbf{E}_1 \times \hat{s})$ is very small in the F region and above, and thus

$$J_\parallel = \Sigma_P(\nabla_\perp \cdot \mathbf{E}_1) + \mathbf{E}_1 \cdot \nabla_\perp \Sigma_P + (\mathbf{E}_1 \times \hat{s}) \cdot \nabla_\perp \Sigma_H \tag{6.12}$$

This expression shows explicitly that field-aligned currents are intimately related to spatial variations in the ionospheric electric field and conductivity. It is important to realize, however, that the field-aligned current in (6.12) is usually *caused* by some divergence of current in the generator, not in the ionosphere.

Now consider the electric field in the closed field line region of the magnetosphere. (We are still considering the southward IMF case.) In Chapter 1 we saw that the magnetic field lines in the boundary layer and the plasma sheet are distorted from a dipole shape to produce a magnetic tail extending away from the sun. This magnetic geometry has a tension (see Chapter 2) that exerts a force on the plasma. Together with the pressure gradient and the potential difference applied across the magnetosphere by the flowing solar wind, these forces produce motion of the magnetospheric plasma on closed field lines toward the sun and an associated dawn-to-dusk magnetospheric electric field in the tail. Further, since the gradient and tension forces are equivalent to the $\mathbf{J} \times \mathbf{B}$ force, the geometry requires the existence of electric currents. Figure 6.2a shows the configuration of electric fields and currents in the ionosphere and magnetosphere. The tail, or neutral sheet, current $\mathbf{J}_T$ flows across the tail to support the curl implied by the stretched magnetic field geometry. The tail current is closed primarily in the magnetosheath by currents flowing on the magnetopause which are not shown in the figure. The ring current $\mathbf{J}_R$ to first order flows in closed loops around the earth at distances between 2 and 10 earth radii (R_e) and is driven primarily by pressure gradient forces. Any divergences in these currents must be closed by field-aligned currents that enter the ionosphere. The portion of the tail and ring current that is closed through the ionosphere is called the partial ring current, J_{PR}. These currents are labeled R_2 in the figure, the so-called region 2 currents, and link the inner magnetosphere with the auroral oval near its equatorward edge. Region 1 currents, labeled R_1, link the poleward portion of the auroral oval and the polar cap to the magnetosheath, solar wind, or the boundary layer plasma near the magnetopause. Figure 6.2b shows how the magnetospheric electric field $\mathbf{E}_m$ maps to the ionosphere in this system.

The ionosphere is not a passive element in this circuit because some of the hot plasma in the magnetosphere can move along the magnetic field and strike

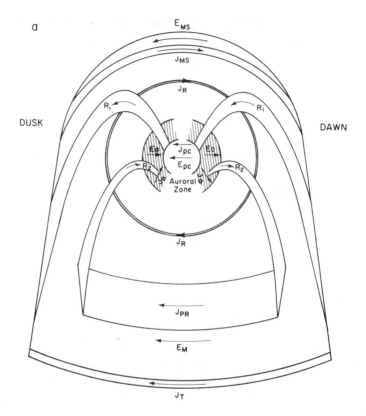

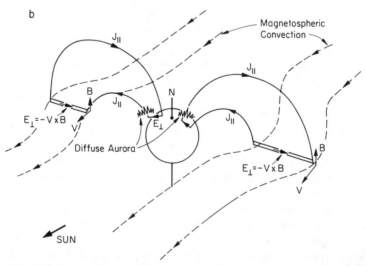

Fig. 6.2. (a) Schematic diagram of the currents and electric fields that exist as a result of the extended magnetic field in the tail of the magnetosphere, and the interaction between the solar wind and the earth's magnetic field. (b) Three-dimensional view of the electric and magnetic field geometry on auroral zone flux tubes.

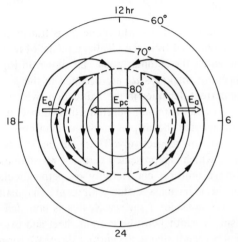

Fig. 6.3. Representation of ionospheric electric fields in the northern hemisphere polar cap and auroral zone, as well as the plasma flow due to those fields.

the atmosphere, producing significant ionization. This particle precipitation produces the discrete and diffuse auroral airglow and can play a dominant role in determining the ionospheric conductivity in the eclipsed ionosphere. Notice that the electric field mapping causes the sign of the electric field to be reversed (e.g., Fig. 6.2b) so that in the ionospheric auroral zone the electric field $\mathbf{E}_a$ is directed from dusk to dawn, which is opposite from the polar cap electric field. The resulting ionospheric plasma flow and electric field patterns are shown in Fig. 6.3. The flow is made up of antisunward flow at the highest latitudes resulting from the connection of the open magnetic field lines to the solar wind electric field discussed earlier. The return sunward flow in the auroral zones results from the electric field $\mathbf{E}_a$, which is in turn determined by the potential difference across the closed field line portion of the magnetosphere.

The electric field reversal at the polar cap boundary is associated with the so-called region 1 field-aligned or Birkeland currents (R_1 in Fig. 6.2a) after the scientist who first postulated their existence. They are closed at one end by currents in the magnetosheath. In the summer polar cap, a large fraction of the region 1 currents close across the conducting polar cap. Note from examination of Fig. 6.2 that the ionospheric and internal magnetospheric currents and associated electric fields all have the same direction at any one place, designating a load, whereas in the magnetosheath the electric field and current have opposite signs, as required for a generator. In fact, the entire system is analogous to a magnetohydrodynamic (MHD) generator, where the solar wind in the magnetosheath is the flowing conductor connected by the region 1 Birkeland currents to the ionosphere–magnetosphere system, which is the load.

6.1.3 Energy Transfer

Further insight into the methods by which energy is transmitted and converted in this system can be obtained from the idealized model of region 1 currents and the closure current across the summer polar cap shown in Fig. 6.4. Here, two parallel current sheets of thickness dx are oriented in the y–z plane and carry equal and opposite current densities in the z direction, which is along the magnetic field. These represent the region 1 currents that connect the MHD generator in the solar wind or magnetosheath plasma to the ionosphere on the dawn and dusk sides of the polar cap, respectively. These current sheets are closed in the polar cap ionosphere, which is represented here as a resistive medium of vertical extent h and having a uniform conductivity σ_P perpendicular to the current sheets. The current sheets have associated with them a magnetic field $\delta\mathbf{B}$ and an electric field $\delta\mathbf{E}$. Now consider a surface S_1 that is bounded by a rectangular loop 1, encompassing a width w in the current sheet and extending a distance $d/2$ in each direction perpendicular to the current sheet. We further take $d \gg dx$. For such a loop the steady-state integral form of Ampère's law can be written

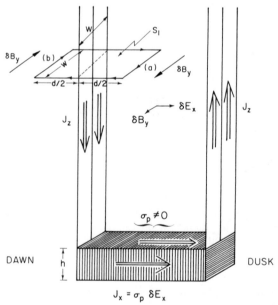

Fig. 6.4. Schematic representation of field-aligned currents and closure currents in the polar cap ionosphere and the electric and magnetic fields associated with them.

$$\oint_l \delta\mathbf{B} \cdot d\mathbf{l} = \mu_0 \iint_{S_1} \mathbf{J} \cdot d\mathbf{a}$$

When the surface S_1 is far from the magnetosheath and ionospheric ends of the current sheet we may assume that the magnetic perturbation δB_x is zero at the surface edges. We may also assume from symmetry that δB_y along part (a) of the loop is equal and opposite to δB_y along part (b) of the loop. Thus, evaluating both sides of the previous equation gives

$$2\,\delta B_y\, w = \mu_0 J_z w\, dx$$

Hence

$$\delta B_y = \mu_0 J_z\, dx/2 \tag{6.13}$$

Note that the magnetic perturbation amplitude is independent of the distance from an infinite current sheet. Thus the magnetic perturbations from the two equal and opposite current sheets shown in Fig. 6.4 will add together in the region between the two sheets and exactly cancel each other in the regions outside. The result will be a magnetic perturbation $\delta B_y = \mu_0 J_z\, dx$ confined completely to the region between the current sheets.

Now consider the ionospheric load. In this resistive medium the horizontal current from dawn to dusk is $J = \sigma_P\, \delta E_x\, h$. In a steady state the current entering the ionosphere in current sheet 1, $J_z\, dx$, must be equal to the total horizontal current in the vertical extent h of the ionosphere. Thus, making use of (6.13) and the fact that $\Sigma_P = \sigma_P h$, we get

$$\delta B_y = \mu_0 \Sigma_P\, \delta E_x \tag{6.14a}$$

This equation may be rewritten in equivalent forms

$$\delta E_x = \delta B_y / \mu_0 \Sigma_P \tag{6.14b}$$

$$\delta B_y / \delta E_x = \mu_0 \Sigma_P \tag{6.14c}$$

or

$$\Sigma_P = \delta B_y / \mu_0\, \delta E_x \tag{6.14d}$$

These expressions show that for a uniform Σ_P, the electric and magnetic fields at high altitudes away from the ionospheric load should be highly correlated. DE-2 satellite observations have shown some events with such correlations with a 99% confidence level (Sugiura *et al.*, 1982), although Smiddy *et al.* (1980) have also shown examples where the correlation breaks down. The latter authors have interpreted their results in terms of the $\mathbf{E} \cdot \nabla_\perp \Sigma_P$ term in (6.11), which must be ignored to derive the relationship above.

One may consider the solar wind MHD generator as a voltage source characterized by the electric field $\delta\mathbf{E}$ which is attached to a resistive load characterized by Σ_P. Then (6.14a) yields the perturbation magnetic field as a function of the load conductivity. This viewpoint may be applicable for the large-scale magnetosphere–ionosphere interaction (Lysak, 1985). Alternatively, the generator may be viewed as a current source. There is some evidence that the latter viewpoint is more accurate at smaller scales (≤ 100 km) since the summer polar cap has much smaller electric fields in this scale size regime than does the winter polar cap (Vickrey *et al.*, 1985). For such a current source, $\delta\mathbf{B}$ would be fixed by the source and from (6.14b) the electric field would then be determined by the ionospheric conductivity. The electric field would then be inversely proportional to Σ_P and hence larger in the winter polar cap, where the lack of solar illumination makes Σ_P small.

Further insight comes from consideration of the Poynting flux. In the geometry of Fig. 6.4, $\delta\mathbf{E} \times \delta\mathbf{H}$ is downward between the current sheets and the energy input is $(\delta E_x\,\delta B_y/\mu_0)$ W/m². This energy must be dissipated as Joule heat in the ionosphere at the rate $W = \mathbf{J} \cdot \mathbf{E} = \sigma_P\,\delta E_x^2$. Integrating over the vertical extent of the ionosphere yields a dissipation rate of $(\Sigma_P\,\delta E_x^2)$ W/m². Since the Poynting flux yields the power flow into the region per unit area we may equate the two expressions and once again we have the result (6.14c)

$$\delta B_y/\delta E_x = \mu_0\Sigma_P$$

The two approaches are therefore self-consistent.

To summarize, mechanical energy is converted into electromagnetic energy in the solar wind generator. It flows down the magnetic field lines to the ionosphere as Poynting flux, where it is converted into heat by Joule dissipation. For typical parameters $\delta E_x = 50$ mV/m and $\delta B_y = 500$ nT, we can estimate the Poynting flux to be $\delta E_x\,\delta B_y/\mu_0 = 0.02$ W/m² $= 20$ ergs/cm²·s. This is a substantial amount of energy, roughly 10^{11} W over the whole region. It is important to notice also that an energy flux of 20 ergs/cm²·s is very large compared to typically observed energy fluxes in auroral particle precipitation, except for extremely intense localized auroral arcs. In fact, the Joule heat input is the primary reason that the thermosphere has a local temperature maximum in the high-latitude region, which competes with the solar photon-driven temperature maximum that occurs near the subsolar point.

6.1.4 Additional Complexities

Before leaving this introductory discussion we emphasize several important considerations to be kept in mind when examining the observations of high-latitude ionospheric plasma motion in the next section. First, the qualitative discussion presented above is centered around a direct connection between the earth's magnetic field and the IMF. The subsequent communication of the interplanetary electric field to the ionosphere and the magnetosphere gives rise to a two-cell convection pattern as shown in Fig. 6.3. This process, which produces antisunward flow on open field lines if the IMF has a component in the *southward* direction, was first described by Dungey (1961). Axford and Hines (1961) showed that a similar motion of the plasma at high latitudes would result if solar wind momentum was transferred across the magnetopause without any direct connection between the ionosphere and the solar wind magnetic fields. This process, called "viscous interaction," produces a relatively narrow region of antisunward-flowing plasma just inside the magnetopause. This region is now called the equatorial magnetospheric boundary layer, and in this region the antisunward plasma flow occurs on closed field lines. The ionospheric flow pattern we described above for the electrical connection model (often called reconnection) does not differ dramatically for a theory in which viscous interaction produces the plasma flow in the ionosphere and hence the electric field pattern. The major difference is that the antisunward convecting plasma is on closed instead of open field lines. Second, when the interplanetary magnetic field is *northward,* the viscous interaction idea does not change very much but connection to the interplanetary magnetic field and the solar wind potential is drastically different. Since B_z is northward roughly one-half of the time, it is not surprising that some of the flow patterns shown below are quite different from the idealized two-cell pattern discussed thus far.

We also need to discuss the physics at the inner boundary of the region dominated by magnetospheric processes. We know from Chapters 3 and 5 that the electric field pattern on field lines within 3 or 4 earth radii is controlled by the atmosphere. How does this transition take place? As the plasma in the magnetosphere flows toward the earth, it encounters an increasing magnetic field strength. Since the first adiabatic invariant is conserved (see Section 2.5.2), the perpendicular energy of the plasma increases. Since the gradient and curvature drifts of these particles depend on both their energy and charge, a zonal charge separation occurs with positive charges at dusk and negative charges at dawn. This creates an electric field pointed from dusk toward dawn in the inner magnetosphere, which tends to cancel out the applied dawn–dusk electric field. The inner magnetosphere is therefore shielded from the magnetospheric electric field and the plasma flows around this region. This shielding only operates on long time scales, and fluctuations of the external field with periods shorter than several hours can penetrate (see Chapter 3).

Since the magnetospheric electric field is reduced to almost zero earthward of the ring current on lower-latitude field lines, the electric field due to the rotation of the earth becomes comparable to the magnetospheric field near this boundary. When this source of the plasma motion is included, the *ionospheric* flow paths look like those shown in Fig. 6.5. At latitudes below about 50° (not shown on this figure) the plasma undergoes circular convection paths around the earth at

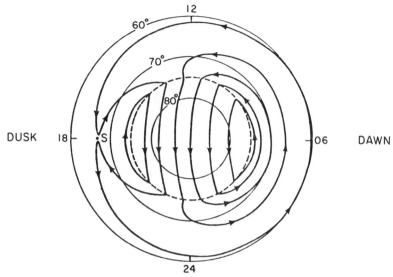

Fig. 6.5. Ionospheric flow paths at high latitudes including the effect of corotation of the plasma with the earth. The diagram is fixed with respect to the sun.

the corotation speed. On the dusk side near 60° latitude the corotation and the magnetospheric electric fields oppose each other, leading to complex flow trajectories that involve a stagnation point, marked S. At still higher latitudes the two-cell convection pattern is preserved. The corresponding plasma flow paths in the equatorial plane of the inner *magnetosphere* might look like those shown in Fig. 6.6. Near the earth in the shaded region, plasma flows in concentric circles. This region is called the plasmasphere since it contains a cool dense plasma. This region has a high plasma content since once per day the associated flux tubes fill from below with ionospheric plasma produced on the dayside of the earth (see Chapter 5). To first order the plasmasphere lies just inside the ring current and the plasma sheet field lines. Outside this region the flow is more or less toward the sun in the equatorial plane. Where a flow line meets the magnetopause it is assumed either to make contact with the interplanetary field and flow back over the top of the magnetopause or to flow back down the flanks of the tail in the boundary layer. Figures 6.5 and 6.6 are in the sun-fixed frame.

On the magnetic field lines in contact with the corotating plasmasphere, the ionosphere to first order corotates with the earth just as the atmosphere does. In Chapters 3 and 5 we worked in this earth-fixed frame and discussed the ionospheric "weather." This involved electric fields and consequent plasma motions in the rotating frame which had typical magnitudes of 1–10 mV/m and velocities of 20–200 m/s. For reference, in the equatorial plane the earth's rotation speed

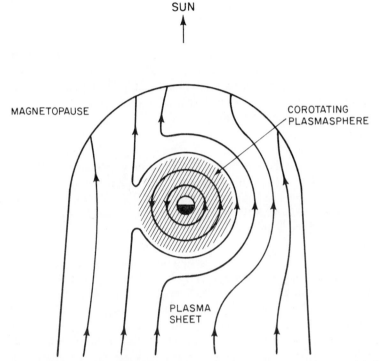

Fig. 6.6. Contours of plasma flow velocity and equipotentials in the magnetic equatorial plane. The concentric circles indicate corotating plasma deep in the magnetosphere. The diagram is fixed with respect to the sun.

at the surface is 434 m/s while the corotation speed at $L = 4$ in the equatorial plane is near 2 km/s.

6.2 Observations of Ionospheric Convection

Our knowledge of the large-scale motion of the high-latitude F-region plasma has come from satellite, rocket, balloon, and ground-based radar and magnetic field measurements. These and other measurement techniques have also been used extensively to study smaller-scale features of the plasma density and plasma motion associated with the aurora. This topic is dealt with in Chapter 8. No single measurement can provide a complete description of the large-scale motion of F-region plasma. Over a 24-h period, for example, a satellite measurement can be repeated 10 to 15 times over the entire high-latitude range in the iono-sphere but only in a very limited local time region. Over the same time period a ground-based radar measurement can be made over the entire local time region but in a limited latitude range. A description of the high-latitude plasma motion is therefore made up from a synthesis of these complementary measurements taken over a period of many years.

For a time-independent system $\nabla \times \mathbf{E} = -\partial \mathbf{B}/\partial t = 0$ and the electric field may be described by an electrostatic potential ϕ such that $\mathbf{E} = -\nabla \phi$. The electric field is perpendicular to a line of constant potential and is also perpendicular to the $\mathbf{E} \times \mathbf{B}$ motion of the plasma. Thus the plasma flows along lines of constant potential and a pattern of equipotentials also represents the plasma flow pattern. Each measurement of electric field or plasma velocity therefore provides a signature of some portion of the convection pattern. These measurements show that the convection pattern is highly dependent on the orientation and the magnitude of the IMF. Recall from Chapter 1 that when discussing the IMF, its direction can be described in a number of ways. The z component can be negative or positive and the IMF is often referred to as southward or northward, respectively. Similarly, the y component can be directed opposite to the earth's revolution about the sun or parallel to it, which is referred to as positive or negative, respectively. Finally, it is worth noting that the tendency of the IMF to assume a garden hose-like spiral in the solar wind means that when the IMF is directed away from the sun the y component is positive and when the field line is directed toward the sun B_y is negative. Thus the terms "toward" and "away" are sometimes used in the literature to describe the sign of the IMF y component. One final note concerns the presentation of observational data at high latitudes. Since the earth's magnetic field lines can be highly stretched on the nightside and slightly compressed on the dayside, a dipole representation is not very informative. Thus it is frequently the practice to describe a point in the high-latitude region with two parameters. One is the L value, which is defined as the equatorial crossing point of the magnetic field line passing through the point. The second is the magnetic local time (MLT), defined using the angle between the earth–sun line and the plane containing the magnetic axis of the earth and a line from the center of the earth to the point. Frequently the L value is expressed in terms of an invariant latitude Λ where

$$\Lambda = \cos^{-1}(1/\sqrt{L})$$

An L value of 4 then corresponds to $\Lambda = 60°$. In a centered dipole field, the latitude at which a magnetic field line passed through the surface of the earth would be equal to the invariant latitude. The MLT is almost always expressed in hours. It is also quite common to express the date as a five-digit number called the Julian day. In this format the first two digits of the number denote the year while the last three digits denote the day number assuming that January 1 is day 1.

The most radical differences in the convection pattern can be seen by comparing signatures when the IMF is southward with signatures when the IMF is northward. The discussion below is divided into these two categories. However, it should be emphasized that substantial variability exists in the convection pattern and our understanding of the nature of the controlling factors is still evolving.

6.2.1 Observations during Southward IMF

Figure 6.7a shows vectors representing the plasma velocity perpendicular to the magnetic field at points along a satellite track as the spacecraft passes through the high-latitude convection pattern during a period of southward IMF. The regions of sunward and antisunward convection are separated by well-defined reversals near dawn and dusk and the signature of the two-cell convection pattern is indicated by the dashed lines. The auroral zone is coincident with the sunward flow regions as described above. If we arbitrarily assign a zero value for the electrostatic potential at low latitudes, data of this nature can be integrated along the satellite track to produce a representative electrostatic potential distribution, which is shown in Fig. 6.7b. By joining points of equal potential deduced from the lower plot with a line that is locally parallel to the observed flow, it is possible to deduce the flow pattern as shown by the dashed lines in the upper plot. Typical electric field values of 10 to 50 mV/m, corresponding to flow velocities of 200 to 1000 m/s, in the sunward flow regions lead to maxima in the potential of the order of plus and minus 30 kV and a total potential drop across the open field line region of the polar cap of about 60 kV. This observed convection pattern is in qualitative agreement with our expectations from earlier consideration of the solar wind–magnetosphere interaction. However, only a fraction of the magnetospheric potential drop across a distance equal to the dimension of the magnetosphere appears across the ionospheric polar cap. For example, using our previous estimate for the solar wind electric field of 1 mV/m, we obtain a total potential of about 200 kV across a magnetosphere of typical width $30R_e$. This indicates that either the transmission of the solar wind and magnetosheath electric field into the magnetosphere is far from 100% efficient or the region of direct connection of the earth's magnetic field to the IMF is much smaller than the width of the magnetosphere.

The next series of figures shows some more examples of the measurements used in the derivation of high-latitude convection patterns. They illustrate the most important variations in the plasma flow pattern when the IMF has a southward component ($B_z < 0$). The following points should be noted.

1. *Average auroral zone pattern.* The average electric field pattern in the auroral oval determined from 500 h of balloon electric field measurements is presented in Fig. 6.8a. The dominant variation is diurnal with the meridional component considerably larger than the zonal component. When collated with respect to high and low K_p values, the pattern does not change much in shape

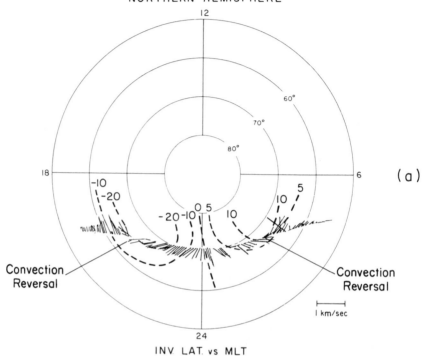

AE-C
ION DRIFT VELOCITIES
DAY 75044   ORBIT 5516
NORTHERN HEMISPHERE

(a)

INV. LAT. vs MLT

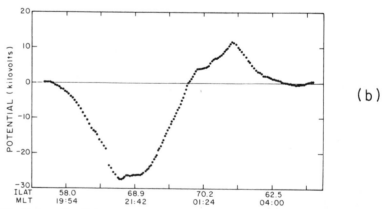

(b)

Fig. 6.7. (a) A satellite flight across the high-latitude convection pattern provides drift velocity profiles, which are shown along with the inferred convection pattern. (b) The potential distribution resulting from this convective flow pattern shows maxima and minima at the polar cap boundaries and a total potential difference of about 60 kV across the polar cap. [After Heelis and Hanson (1980). Reproduced with permission of the American Geophysical Union.]

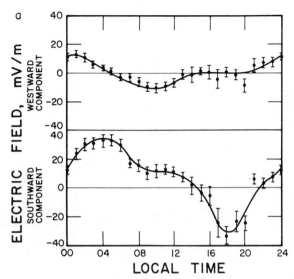

Fig. 6.8. (a) Hourly averages of auroral zone electric field data from balloon flights. The error bars are standard deviations of the means, and the solid curves are empirical fits to the data satisfying the constraint imposed by $\nabla \times \mathbf{E} = 0$ that the 24-h average westward field be zero. [After Mozer and Lucht (1974). Reproduced with permission of the American Geophysical Union.] (b) The antisunward component of the plasma velocity, measured in both the northern and southern hemispheres, is plotted as a satellite moves across the auroral zone and polar cap. (c) The plasma drift velocity perpendicular to the magnetic field measured during two satellite passes through the high-latitude ionosphere is represented here by drift velocity vectors. (d) Measurements of the vector plasma drift velocity in the dayside high-latitude ionosphere show that the antisunward flow can be preferentially biased toward the dawn or dusk directions depending on the sign of B_y. [Parts (b)–(d) courtesy of R. A. Heelis and W. B. Hanson.] *(Figure continues.)*

but the amplitude of the dominant feature decreases by about a factor of 2 as K_p varies from 6 to 0.

2. *Seasonal dependence.* In Fig. 6.8b the plasma motion in a direction almost parallel to the earth–sun line is shown for two dawn–dusk satellite passes. The figure illustrates that the plasma velocity is considerably more structured in the winter (southern) hemisphere (top panel) than in the summer hemisphere. In both cases, however, the data are consistent with a two-cell convection pattern. This is characterized by two large-scale reversals, shown by the heavy arrows, that separate regions of antisunward flow (dawn-to-dusk electric field) from regions of sunward flow (dusk-to-dawn electric field).

3. *Variability.* Figure 6.8c shows the plasma drift velocity perpendicular to the magnetic field measured along a satellite track during two passes through the high-latitude ionosphere using a vector representation. Notice that for these quite similar IMF conditions the latitudinal extent of the sunward flow in the auroral

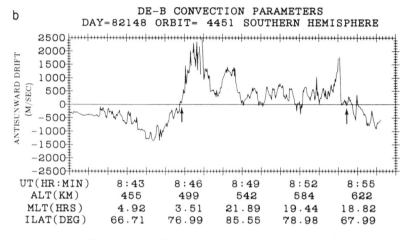

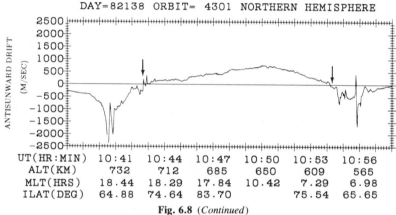

Fig. 6.8 (*Continued*)

zones is quite different. Also the magnitude of the antisunward flow in the polar cap is quite different in these two cases.

4. *Response to* B_y. Figure 6.8d again shows the ion drift velocity vector perpendicular to the magnetic field for two passes through the high-latitude ionosphere. In the first pass (top panel) B_y is negative, and in the bottom panel B_y is positive. Notice that when B_y is positive the sunward flow from the dusk side passes across local noon in the auroral zone before flowing antisunward. When B_y is negative, the sunward flow from the dawn side passes across local noon before flowing antisunward.

These observations show that the ionospheric conductivity, the internal state of the magnetosphere, and the external state of the interplanetary environment

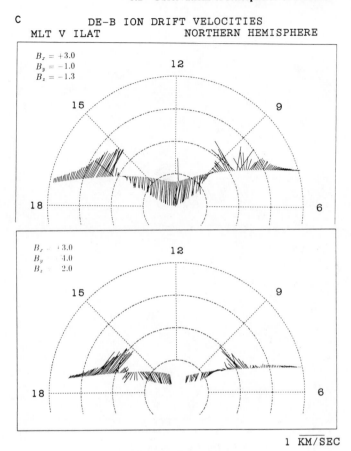

C

DE-B ION DRIFT VELOCITIES

MLT V ILAT   NORTHERN HEMISPHERE

$B_x = +3.0$
$B_y = -1.0$
$B_z = -1.3$

$B_x \quad +3.0$
$B_y \quad 4.0$
$B_z \quad 2.0$

1 $\overline{\text{KM/SEC}}$

Fig. 6.8 (*Continued*)

can all affect the high-latitude ionospheric convection pattern. However, when the IMF has a southward component, a two-cell convection pattern made up of predominantly antisunward convection at the highest latitudes with sunward convection occurring at lower latitudes is almost always observed.

In principle, a number of different radar stations and satellites operating at the same time can provide a simultaneous signature of the convection pattern at different local times. Alternatively, if some degree of stability in the convection pattern is assumed, a single radar site can sample all local times over a substantial latitudinal width of the pattern during a day. These techniques and the previously mentioned statistical syntheses of data bases are currently being used to elucidate details of the seasonal dependence in the convection pattern and the factors determining the convection speed. To date, we have only a qualitative

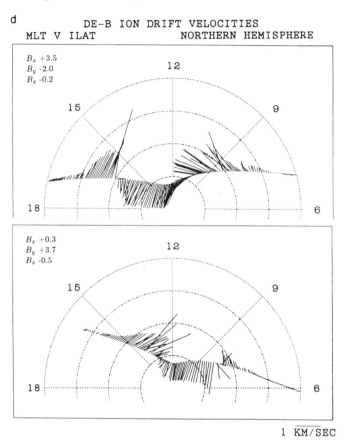

d
DE-B ION DRIFT VELOCITIES
MLT V ILAT   NORTHERN HEMISPHERE

B_x +3.5
B_y -2.0
B_z -0.2

B_x +0.3
B_y +3.7
B_z -0.5

1 $\overline{\text{KM/SEC}}$

Fig. 6.8 (*Continued*)

description of the convection cell geometry and its dependence on the y component of a southward-directed IMF. This dependence is well documented for that portion of the convection pattern on the sunward side of the dawn–dusk meridian. It is shown schematically in Fig. 6.9. The convection pattern is most easily characterized by one small cell and one large cell. The large cell has an almost circular perimeter and the small cell is crescent-shaped. The latter surrounds a portion of the circular cell. The placement of the small and large cells depends to first order on the sign of B_y.

When the IMF has a southward component the signature of a two-cell convection pattern is seen well beyond the dawn–dusk meridian into the nightside. Relationships between the nightside convection geometry and the IMF are only

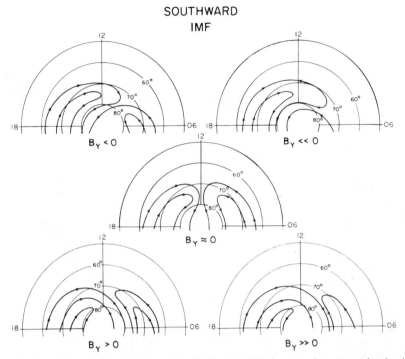

SOUTHWARD
IMF

Fig. 6.9. Schematic representation of the dayside high-latitude convection pattern showing its dependence on the y component of the IMF when B_z is south. [After Heelis (1984). Reproduced with permission of the American Geophysical Union.]

now being explored and have not been firmly established at this time. The nightside convection pattern does, however, have some distinctive geometric characteristics that are related to the field-aligned and horizontal currents, which are described later. Figure 6.10 illustrates the variety of convection geometries that are consistent with the currently available plasma flow data. It is found, for example, that the premidnight convection reversal occurs at slightly lower latitudes at later local times. This reversal can frequently take place over a narrow latitudinal region, leading to a crescent-shaped convection cell in the dusk sector. The dawnside convection reversal most frequently occupies a much larger latitudinal extent than the duskside reversal and thus the pattern in Fig. 6.10d may prevail. However, there is considerable variability in the pattern, as we have seen from the few examples shown in this section, and thus the alternative geometries in Fig. 6.10a–c can also exist.

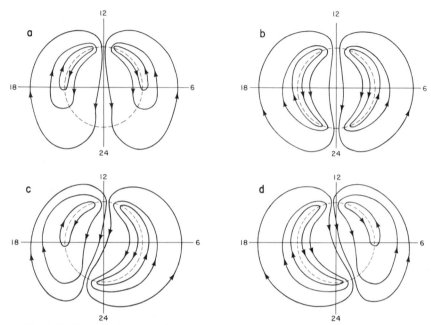

Fig. 6.10. The nightside convection pattern can have a variety of geometries. Patterns a and d are most frequently observed. [After Heelis and Hanson (1980). Reproduced with permission of the American Geophysical Union.]

6.2.2 Observations during Northward IMF

A significant departure from the previously described convection patterns is seen when the IMF has a northward component. In this case the convection velocity is usually much more structured and of smaller magnitude. When the identification of an organized convection pattern is possible it is, surprisingly, characterized by a sunward flow component (a dusk-to-dawn electric field) in the central polar cap. In addition, the entire region of significant plasma motion is confined to much higher latitudes than in the southward IMF case. Figure 6.11 shows several examples of the dawn-to-dusk electric field measured at high latitudes when the IMF had a northward component. The shaded regions show the existence of a dusk-to-dawn field and hence sunward flow deep in the polar cap. This is the major difference from what would otherwise be the signature of a weak two-cell convection pattern. Quite frequently only a two-cell convection pattern at lower latitudes is identifiable. At higher latitudes the flow is extremely structured at small scales, and observations can look more typically like those shown in Fig. 6.12. Despite the difficulties associated with the recognition of convection features during northward IMF, a qualitative description is possible as sum-

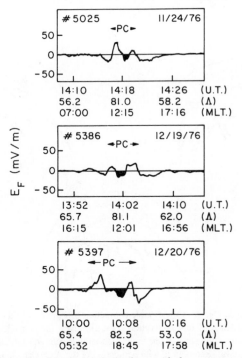

Fig. 6.11. Three examples of the high-latitude, dawn-to-dusk component of the electric field in the ionosphere for a northward IMF. The shaded regions can be interpreted as sunward plasma flow. [After Burke *et al.* (1979). Reproduced with permission of the American Geophysical Union.]

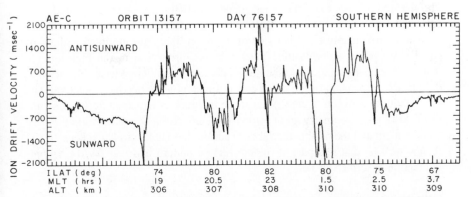

Fig. 6.12. High-latitude plasma drift velocity in the ionosphere shown for a northward IMF. The flow is extremely structured and does *not* indicate a simple two-cell convection pattern as is often the case during southward IMF. [After Heelis and Hanson (1980). Reproduced with permission of the American Geophysical Union.]

NORTHWARD
IMF

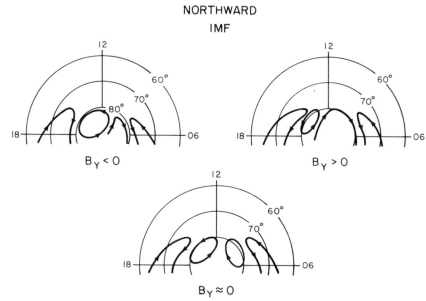

Fig. 6.13. The main feature of the dayside convection geometry when the IMF has a northward component is the existence of four convection cells. However, the dependence of the convection pattern on B_y leads to the dominance of one of the high-latitude cells and a three-celled pattern arises. [After Heelis *et al.* (1986). Reproduced with permission of the American Geophysical Union.]

marized in Fig. 6.13. When the IMF has a northward component, four convection cells can be identified in the dayside hemisphere. Two high-latitude cells circulate in a manner that produces sunward convection at the highest latitudes. The size and orientation of these cells depend to first order on the magnitude and sign of the y component of the IMF. When B_y is very small these two cells are located approximately on either side of the noon–midnight meridian and are of equal size. When B_y is positive the clockwise circulation in the high-latitude dawn cell tends to expand into the dusk side, and the anticlockwise cell virtually disappears. Expansion of the dawn cell is apparent as B_y becomes positive. The lower-latitude convection cells circulate in the manner expected during a southward IMF. Their geometry and position show no strong dependence on the IMF y component and the total potential drop across these cells rarely exceeds 10 kV. It may well be that this part of the convection pattern is driven by the viscous interaction described earlier.

6.3   Simple Models of Convection in the Magnetosphere

In our simple considerations of magnetosphere–solar wind coupling, the ionospheric plasma motion reflects the motion of the magnetospheric plasma to which it is magnetically connected. At the highest latitudes, antisunward convection is associated with similar flow in the magnetosheath, while sunward convection at lower latitudes is associated with flow in the plasma sheet. This relationship between plasma flows in different regions makes the concept of moving magnetic field lines or frozen-in flux extremely useful. However, if, as we suggested, the magnetic field lines associated with plasma flow in the polar cap are open and those associated with plasma flow in the auroral zones are closed, then this concept must break down wherever the plasma flows across the boundary between these two regions. When this breakdown occurs, the term "merging" or "reconnection" is frequently used to describe the phenomenon. A description of the plasma processes involved in a merging or reconnection region is beyond the scope of this book. It is convenient, however, to use these terms when describing moving magnetic field lines that change their identity either from an open to a closed topology or from one open topology to another.

6.3.1   Models for Southward IMF

We have seen that many ionospheric observations indicate that the IMF exercises significant control over the plasma convection pattern. From examination of the energetic electron precipitation patterns at high latitudes and direct detection of solar cosmic rays in the polar ionosphere, it is generally believed that the antisunward flow in the polar cap during a southward IMF exists on open magnetic field lines. The sunward flow in the auroral zone that completes the two-cell convection pattern is on closed field lines. When the sunward flow of the auroral zone crosses the polar cap and becomes antisunward, we use the term "merging" to describe the process by which closed field lines become open and connected to the interplanetary magnetic field (Dungey, 1961). The term "reconnection" describes the process by which open magnetic field lines are joined in the magnetotail to produce a closed field line. The reconnection region must correspond in the ionosphere to the region where antisunward flow crosses the polar cap boundary and becomes sunward. With these two processes in mind, Fig. 6.14a shows that a southward-pointing IMF field line (A–B) breaks and merges with the earth's field at point N1. Still attached to the solar wind, that field line moves across the polar cap and eventually rejoins a field line from the other hemisphere at point N2. During this time the interplanetary electric field penetrates down into the ionosphere, where it drives an antisunward flow. This picture is not to scale since the solar wind velocity is so large that by the time the foot point in the ionosphere traverses the polar cap, the other end of the field line is very far away from the earth. Hence, a long magnetic tail forms on the nightside. The time history of convecting plasma or flux tubes in the magnetosphere and ionosphere during the circulation of plasma in one cell of the two-cell convection pattern is shown in Fig. 6.14b. As noted above, on the dayside at point 1 a closed magnetic field line of the earth "breaks" and connects with the interplanetary

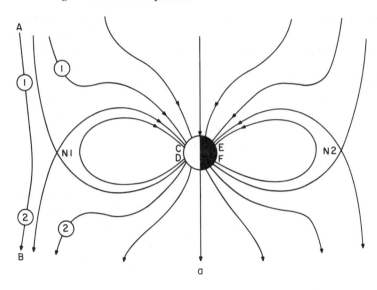

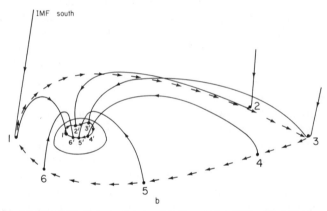

Fig. 6.14. (a) Schematic diagram showing dayside merging (N1) and nightside reconnection (N2) in the noon–midnight meridian. (Courtesy of D. P. Stern.) (b) Time history of convecting flux tubes that result from connection of a southward IMF to the earth's magnetic field. The convection along path 1, 2, 3 is on open field lines, while the path from 3, 4, 5, 6 to 1 is on closed field lines that connect to the southern hemisphere.

magnetic field to produce an open field line. Not shown is the magnetic field line from the south polar region which connects to the "other half" of the interplanetary field line. This point may alternatively be viewed as the location at which magnetospheric plasma moves from a region located on closed magnetic field lines to a region on open field lines. After merging takes place, the magnetospheric plasma moves antisunward along the path labeled 1, 2, and 3. Small arrows show the corresponding convective path of the plasma in the ionosphere as it moves from 1′ to 2′ to 3′. At point 3 open field lines from the northern and southern hemispheres reconnect in the equatorial plane and subsequently convect sunward in the equatorial plasma sheet as closed field lines. Again, one may alternatively view point 3 as a place where magnetospheric plasma makes a transition from being an open field line to being a closed field line. From either viewpoint, the subsequent convective paths of the plasma in the ionosphere and magnetosphere, respectively, are 3′, 4′, 5′, 6′, 1′ and 3, 4, 5, 6, 1. Note that in this figure the ionospheric flow is shown on a relatively enlarged sphere so that its features can be seen.

These basic elements of a two-cell convection pattern can also be attributed to the viscous interaction process mentioned earlier (Axford and Hines, 1961). In this process the momentum of the magnetosheath plasma is transmitted across the magnetopause by waves and diffusion, creating an effective viscosity. Plasma near the equatorial plane which is just inside the magnetopause, the so-called equatorial boundary layer, is set in motion antisunward. A return flow toward the sun occurs on lower latitude field lines and is driven by back pressure in the magnetosphere near midnight. The ionospheric convection signature of this process is qualitatively identical to that created by direct connection with the solar wind magnetic field or merging, except that even the antisunward convective paths in the magnetosphere are on closed field lines. Figure 6.15a shows the convection cycle in the magnetosphere for the viscous interaction mechanism along with an analogous process in fluid dynamics. Convection paths in the magnetosphere and ionosphere are shown in Fig. 6.15b for the viscous interaction driver. In this case there are no merging and reconnection regions but simply locations at which convection on closed field lines changes direction from sunward to antisunward at point 1 and from antisunward to sunward at point 3. The sunward convection from 4 to 5 to 6 is identical to that shown in Fig. 6.14, but the antisunward flow in the magnetosphere now exists on closed field lines in the equatorial plane as shown by the arrows connecting 1 to 2 to 3.

It can be seen from a comparison of the primed paths in Figs. 6.14 and 6.15 that observations of just the ionospheric convection signature cannot distinguish between these two convection mechanisms. However, simultaneous observations of the convection signature and the energetic particle environment indicate that the magnetic field topology is often open. Likewise, theoretical considerations of the physics involved in a viscous interaction indicate that the potential asso-

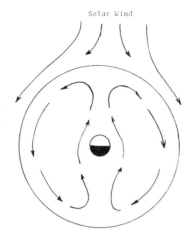

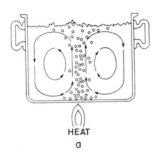

HEAT

a

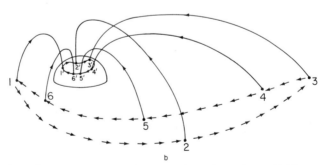

b

Fig. 6.15. (a) Schematic diagram showing how a viscous interaction could drive magnetospheric convection, along with a fluid analogy. (Courtesy of D. P. Stern.) (b) Time history of convecting flux tubes resulting from a viscous interaction. Here all the flux tubes are closed and also connect into the southern hemisphere, which is not shown.

ciated with this mechanism is usually less than 10 kV. Observed cross-polar cap potentials in excess of 60 kV and the dependence of the convection geometry on the y component of the solar wind magnetic field strongly suggest that the merging process dominates over viscous interaction when the IMF has a southward component.

Although a rigorous discussion of the way in which the solar wind and earth's magnetic fields interconnect is beyond the scope or requirements of this book, it is possible to understand the basic properties of the convection geometry from rather simple considerations. We start by assuming that field lines interconnect when they have antiparallel components in a plane perpendicular to the earth–sun line (the y–z plane). Figure 6.16 shows views from above the north pole of one of the earth's magnetic field lines that is connected to the IMF when the IMF y component is nonzero. In the situation for negative B_y, once merging has taken place the newly open field lines have extremely large curvature near the merging point. The resulting tension in the northern hemisphere open field lines will produce a duskward component in the plasma flow in the dayside magnetosphere. A corresponding duskward plasma flow will exist in the dayside polar cap of the northern hemisphere ionosphere. In the southern hemisphere the plasma flow at the magnetopause and in the ionosphere will be directed toward dawn when B_y is negative. Similar arguments can be applied to the situation when B_y is positive. In this case a dawnward flow component will be seen in the

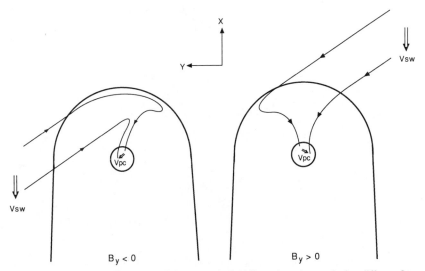

Fig. 6.16. Schematic illustration of the magnetic field line orientation producing different flow directions in the polar cap when B_y is positive versus the situation where B_y is negative.

northern ionospheric polar cap flow. This rather simple idea can account for the B_y dependence of the ionospheric flow direction seen near local noon in Fig. 6.9 and for the dependence of the flow speed at the dawn and dusk edges of the polar cap on B_y (Heppner, 1972). This dependence is such that in the northern hemisphere a larger flow speed is seen near the dawn side of the polar cap when B_y is positive and a larger flow speed is seen at the dusk side of the polar cap when B_y is negative. Also note that this simple model predicts that the B_y dependences seen in the northern hemisphere appear in the opposite sense for the southern hemisphere—a fact borne out by observation.

6.3.2 Models for Northward IMF

The pattern of the high-latitude circulation cells during northward IMF can again be understood by the rather simple concepts of merging and frozen-in flux used in the case of southward IMF. As we shall see, a number of different magnetic field topologies might exist during times of northward IMF. Experimental and theoretical investigations in this area are still being undertaken and we will describe here only some of the basic concepts involved.

One way in which the earth's magnetospheric field may connect to a northward IMF is shown in Fig. 6.17. Here magnetic field lines that extend into the tail of the magnetosphere have the right orientation for merging to occur with a northward IMF at point 1. Subsequent antisunward motion of open field lines to points 2 and 3 produces antisunward motion in the ionosphere from 1′ to 3′. At this point the open or closed state of the flux tube becomes an important issue. In Fig. 6.17 the field line remains open and convects down the tail and toward

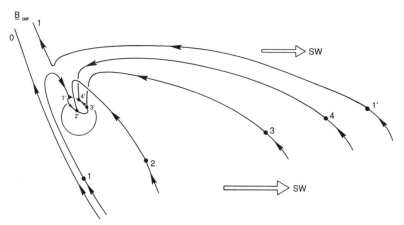

Fig. 6.17. Time history of convecting flux tubes (connecting to a northward IMF) that produce a dominant high-latitude convection cell. This must be combined with the pattern for viscous interaction shown in Fig. 6.15 to produce the observed three- or four-cell pattern.

the earth–sun line until it assumes the position of the original field tube (1). The associated convection is labeled 1, 2, 3, 4, 1 in which the antisunward flow toward the earth–sun line at the magnetopause corresponds to sunward flow 3′, 4′, 1′ in the ionosphere. Alternative or additional convection paths in which field tubes reconnect at point 3 and subsequently flow sunward in a plasma sheet that thickens along the earth–sun line have been proposed to explain the possible existence of closed magnetic field lines at the highest latitudes. It should be emphasized that the precise location of the transition from open to closed field lines in the ionosphere can be quite variable within the picture we have drawn.

The convective motion described above exists in addition to the viscous inter-action mechanism described earlier, and we must invoke both of them to explain the four-cell convection pattern that is most frequently observed. We note that the tendency of the flow of the inner two cells to be preferentially directed toward dawn or dusk depending on whether B_y is positive or negative can again be explained by the effective "tension" in connection to the IMF (see Fig. 6.16).

6.4 Empirical and Analytic Representations of High-Latitude Convection

As we shall see in Chapter 7, knowledge of the high-latitude ionospheric convection pattern is required to effectively calculate the distribution and composition of plasma in the F region and to determine the F-region neutral wind field and Joule heating rate. For such studies it is necessary to know the ionospheric flow velocity at all points at any given time. This requirement has led to the development of several empirical and semiempirical models for the convection pattern designed to mimic one or more of the features described in previous sections. These models are derived from a synthesis or statistical analysis of satellite and radar data. A global representation of the electrostatic potential distribution derived from satellite electric field measurements is shown in Fig. 6.18. This empirical model is valid for southward B_z and is shown for the IMF with B_y negative (a) and positive (b). It depicts several of the major properties of the convective flow that we have previously noted in individual cases. For example, the ion flow or electric field is larger on the dawn side than on the dusk side in the northern hemisphere when B_y is positive (i.e., the potential lines are closer together). Figure 6.18b also shows the existence of a crescent-shaped convection cell on the dusk side similar to that shown in Fig. 6.10d. A comparison of Figs. 6.18a and b shows the change in the flow configuration near noon when B_y changes sign, as depicted in Fig. 6.8c. Data such as this can be useful in computer modeling studies if numerical values of the potential are obtained over a latitude–local time grid.

More detailed distributions of the electrostatic potential in the sunward flow regions have been obtained from ground-based radars that probe this region of

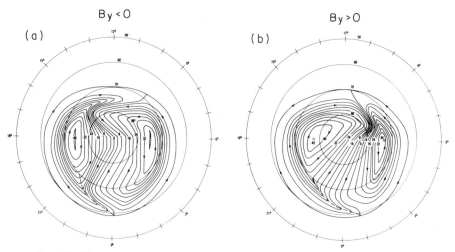

Fig. 6.18. Electrostatic potential contours derived from a synthesis of satellite measurements of the ionospheric electrostatic potential when B_y is negative (a) and when it is positive (b). [After Heppner and Maynard (1987). Reproduced with permission of the American Geophysical Union.]

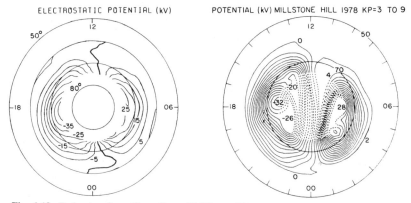

Fig. 6.19. Radar data from Chatanika and Millstone Hill used to derive the convection pattern in the auroral zone (sunward-flow regions). [After Foster (1983) and Oliver *et al.* (1983). Reproduced with permission of the American Geophysical Union.]

the ionosphere. Many years of data from these radars can be sorted by latitude and local time as well as by magnetic activity and solar wind conditions. Tabulated values of the coefficients of functional fits to these data are then available. Figure 6.19 shows examples of these average convection patterns derived from radars at Chatanika, Alaska and Millstone Hill. Although data such as these do not cover the entire high-latitude region, they do provide a means of assessing the effects of magnetic activity on the auroral zone flow, a task that is not easily accomplished using satellite data.

Analytic models use a collection of mathematical expressions to specify the electrostatic potential as a function of position in the high-latitude F region. The comparative ease with which they can be incorporated into computer code makes them most often used in F-region modeling studies, and here we provide illustrative examples. With this technique the polar cap is usually designated as a circle and outside this circle the potential ϕ decays via an inverse power law sine wave. That is,

$$\phi = \phi_0(\sin \theta/\sin \theta_0)^n \qquad (6.15a)$$

Here θ is magnetic colatitude, ϕ_0 is the potential at the polar cap boundary where $\theta = \theta_0$, and the index n typically takes values between -2 and -4. Inside the polar cap boundary, Volland (1975) showed that if

$$\phi = \phi_0(\sin \theta/\sin \theta_0) \sin(H) \qquad (6.15b)$$

where H is the local hour angle (MLT), then the lines of electrostatic equipotential depict convective flow that is antisunward inside the polar cap. This model can be modified to allow ϕ_0 to take different values on the dawn- and dusk-side polar cap boundaries and to allow the origin of the potential coordinate system to be displaced from the magnetic pole. It has been used extensively in F-region modeling studies, where account has also been taken of the offset between the geomagnetic and geographic poles.

To reproduce such observations as the crescent-shaped cell in the dusk sector and the predominance of eastward or westward flow at local noon, some modifications to the mathematical model can be made. These include the removal of any local time dependence of the potential in certain local time sectors and the removal of the latitudinal discontinuity in the potential gradient at the polar cap boundary. A variety of convection geometries are then available with realistic field-aligned current densities similar to those described in Section 6.5. The challenge of producing mathematical models that adequately describe the effects of changing B_y when the IMF is southward and the four-cell convection pattern when the IMF is northward still remains.

In our description in previous sections we have shown how the IMF exercises a large degree of control on the geometry or shape of the convection cells. This geometry is related to the distribution of electrostatic potential around the polar cap boundary and within the polar cap itself. These are the two fundamental properties addressed by empirical models. To date, our description has been only qualitative or empirical. There is, however, convincing evidence that merging or direct connection with the solar wind dynamo via the IMF is an important contributor to the flow in the polar cap. In this case we may reasonably ask what effect interplanetary conditions have on the magnitude of the drift velocities or, alternatively, on the size of the polar cap and the potential drop across it. In this area an analytic model can be extremely useful.

A model developed by Siscoe (1982) describes the region 1 and region 2 field-aligned currents introduced in Section 6.1 as two concentric rings. A potential ϕ is assumed to be distributed sinusoidally in local time around the region 1 ring. Then consideration of the conservation of magnetic flux in the ionosphere and the energy dissipated in the region 2 current loop can be used to show that in an equilibrium situation the radius of the region 1 circle, that is, the polar cap radius r, is related to the potential by the expression

$$r = r_0 \phi^{3/16} \tag{6.16}$$

Here r_0 depends on the ionospheric conductivity and the width of the auroral zone, but these quantities may be considered as constants to first order. The limited data which exist do not disagree with this relationship, but the exponent on the potential is extremely small and thus, for potential differences exceeding about 20 kV, the polar cap radius is almost independent of the magnitude of ϕ. Magnetic storms and substorms (discussed briefly in the next section) are known to cause expansions of the polar cap and the entire convection pattern, but even a qualitative description of this behavior has not yet been accomplished.

The polar cap potential drop is, however, found to be a strong function of the interplanetary magnetic field magnitude and orientation. From rather simple considerations one might expect that the potential difference depends on the area over which the IMF and geomagnetic field interconnect and the efficiency with which the electric field in the interplanetary medium is transferred across that area. Hill (1975) approached this problem by determining the dissipative component of the solar wind electric field at the magnetopause (i.e., the component parallel to the magnetopause current) that separates an internal earth's magnetic field $\mathbf{B}_2$ from an external field $\mathbf{B}_1$ (the IMF). He called this field the merging region field and denoted it by $\mathbf{E}_j$. He then showed that this field and the relative orientation of the magnetic fields on either side of the magnetopause would determine the effective potential transmitted across the boundary to the ionosphere.

The electric field in the merging region can be expressed as

$$E_j = E_0 \frac{\alpha^2(\alpha - \cos\theta)^2}{1 + \alpha^2 - 2\alpha\cos\theta} \tag{6.17}$$

where $\alpha = B_1/B_2$, θ is the angle between $\mathbf{B}_1$ and $\mathbf{B}_2$, and E_0 depends on the solar wind density and the magnetic field magnitudes. The total potential drop Φ across the polar cap due to the merging field is simply $\Phi = E_j d$, where d represents the length of the magnetopause merging region. One would therefore expect that when the IMF has no component perpendicular to the sheet current at the magnetopause (and hence $E_j = 0$), Φ should also be zero. Observations of the polar cap potential difference are plotted in Fig. 6.20 against values of E_j

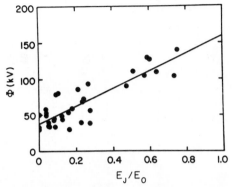

Fig. 6.20. Measured polar cap potential difference ϕ plotted as a function of the quantity E_j/E_0 defined in (6.18). The potential increases as the IMF becomes more southward but a potential still exists when the solar wind electric field parallel to the magnetopause current is zero. [After Rieff *et al.* (1981). Reproduced with permission of the American Geophysical Union.]

calculated for the measured solar wind and earth's magnetic fields at different times. The E_j is normalized to E_0, which depends on the plasma density in the solar wind and the earth's magnetic field. As the solar wind magnetic field changes, the observed potential does not go to zero when $E_j = 0$. This additional potential difference may be related to electric fields communicated by processes other than merging or to a time response of the magnetosphere–ionosphere system that does not allow it to change instantaneously when the IMF changes.

Figure 6.21 shows the results of studying the time response of the iono-sphere–magnetosphere system to a change in the north–south component of the IMF. It shows the polar cap potential difference measured as a function of time after the IMF turns northward. The data suggests that relatively large potentials, associated with a two-cell convection pattern, tend to persist for 1 or 2 h follow-ing a transition from southward to northward IMF. After this time quite small polar cap potential differences are observed. Bearing in mind the uncertainties in defining the polar cap these small values are consistent with a multicell con-vection pattern for a northward IMF. It is interesting to note that this time delay is consistent with the time required to reverse a two-cell convection pattern in the neutral gas that is established during a southward IMF (see Chapter 7). Thus a dynamo electric field might be produced in the ionosphere at high latitudes by a flywheel effect during transitions from previously stable IMF orientations. In such a case, $\mathbf{J} \cdot \mathbf{E} < 0$ in the ionosphere and the Poynting flux is upward.

The observations of the flow geometry when the IMF has a northward com-ponent are not sufficiently advanced to make quantitative analysis possible. As mentioned earlier, it is likely that processes other than merging (e.g., viscous interaction) may be important in producing the ionospheric motion in this case, and advances in this area are occurring at the time of this writing.

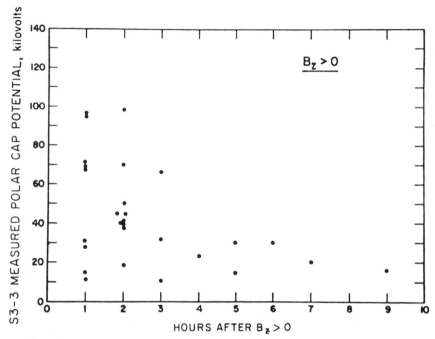

Fig. 6.21. The polar cap potential decreases slowly after the IMF turns northward. It finally reaches a value between 20 and 15 kV that is perhaps attributable to the viscous interaction process. [After Wygant *et al.* (1983). Reproduced with permission of the American Geophysical Union.]

6.5 Observations of Field-Aligned Currents

Field-aligned currents are essential to the linkage between the solar wind–magnetosphere system and the ionosphere. The ionosphere is not a passive element in the electric field mapping process between these regions, and in fact the electric field can be modified at the driver and throughout the system by the ionosphere. This is particularly true in the auroral zone, where the ionosphere and magnetosphere are linked and the hot plasma sheet particles can determine the ionospheric conductivity through particle precipitation. Study of these currents yields information comparable in importance to measurements of the convective motions.

Field-aligned currents, frequently called Birkeland currents after the man who first postulated their existence (see Section 6.1), can be detected by the magnetic perturbation they produce. A current flowing in a direction parallel to the earth's magnetic field will produce a magnetic perturbation δB in a direction perpendicular to the field. From the diagnostic equation (2.23c) in Chapter 2 the two quantities are related by

$$\mu_0 J_\parallel = (\nabla \times \delta \mathbf{B})_\parallel \tag{6.18}$$

The magnetic disturbances can be observed by satellite- and rocket-borne magnetometers that measure three mutually perpendicular components of the disturbance vector. A measurement along a single trajectory does not allow the determination of a curl, so in practice some approximations are necessary to derive $J_\parallel$.

Field-aligned currents frequently, but not always, occur as sheets whose dimensions in one horizontal direction greatly exceed those in the other (see Fig. 6.4). The magnetic disturbance is then predominantly parallel to the sheet, and Fig. 6.22 shows schematically the disturbance that would be measured on passing perpendicularly through the sheet. The current density is given by (6.18), which simplifies to $\mu_0 J_\parallel = (\partial/\partial x)(\delta B_y)$, and for the dimensions shown in Fig. 6.22 the disturbance would correspond to an outward current sheet of density 0.8 μA/m². Most observations are not, of course, made with the convenience of such perpendicular crossings of the current sheet, so care must be taken in the arbitrary application of (6.18). In general, the full perturbation vector is measured, so in many cases only a rotation of a coordinate system is required to determine $J_\parallel$. A statistical analysis of satellite magnetometer data of this kind has enabled a systematic picture of field-aligned current directions, magnitudes, and locations to be derived. As in the case of electric field or plasma flow observations, there exist some relatively stable characteristics in the patterns of field-aligned currents that persist at all times. In addition, the interplanetary magnetic field magnitude and orientation introduce changes in the location and configuration of the current systems. This should not be surprising considering the intimate relationships between currents and electric fields that were derived in Section 6.1. Since we have established the basic magnetic field topologies associated with the electric field configuration and given a qualitative description of the drivers that produce a dependence on the IMF orientation, we will confine ourselves here to a description of the field-aligned current morphology. In the final section we will describe the behavior of the horizontal currents.

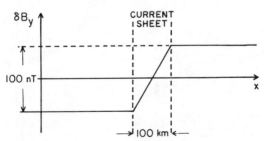

Fig. 6.22. A uniform 0.8 μA/m² current sheet in a plane perpendicular to the paper and occupying the 100-km extent shown here produces a perturbation change of 100 nT in the magnetic field.

6.5.1 Current Patterns for a Southward IMF

When the IMF has a southward component, Fig. 6.23 shows the most stable features of the field-aligned current system seen during times (a) when the internal state of the magnetosphere is quiet and (b) when it is more disturbed. The following features should be noted:

1. There exist essentially two concentric field-aligned current rings approximately the same as those predicted from the simple considerations of Section 6.1.

2. The inner ring, which may be considered to contain the driving currents for ionospheric convection, has current directed into the ionosphere in the morning hemisphere and away from the ionosphere in the evening hemisphere. These are the region 1 currents, identified in Section 6.1.

3. The outer ring, which results from the feedback between the ionosphere and the magnetospheric plasma sheet, has the opposite current direction in the morning and evening hemispheres from that of region 1. These are the region 2 currents identified in Section 6.1.

4. In the magnetic local time sector from 2200 to 2400 h, three regions of field-aligned current are formed by the overlap of the two rings seen at other local times.

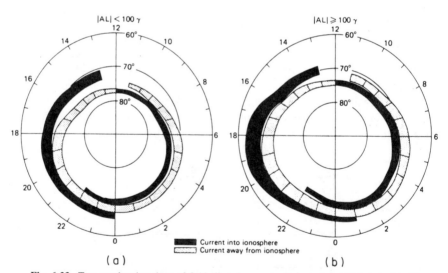

Fig. 6.23. Two overlapping rings of field-aligned currents always exist at high latitudes. They occur at higher latitudes and have smaller latitudinal extent during quiet times. The inner ring is termed region 1 current and the outer ring region 2. [After Iijima and Potemra (1978). Reproduced with permission of the American Geophysical Union.]

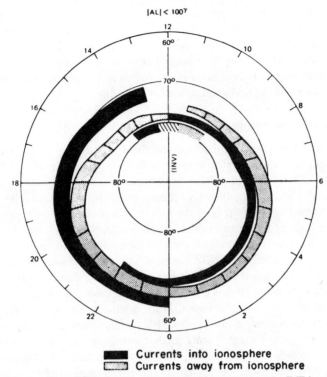

$|AL| < 100^{\gamma}$

■ Currents into ionosphere
▨ Currents away from ionosphere

Fig. 6.24. A more complete pattern of field-aligned currents when the IMF has a southward component and $B_y = 0.1$. An additional field-aligned current system resides near local noon. These "cusp" currents have a dominant polarity determined by the y component of the IMF. [After Iijima and Potemra (1976). Reproduced with permission of the American Geophysical Union.]

5.   During magnetically active periods there is a slight expansion of both current rings to lower latitudes. The region of overlap then increases to occupy the MLT region from 2200 to 0100 h. The main characteristics of the current system remain unchanged.

These properties of the field-aligned current system are common to both hemispheres; there is, however, an additional current system that shows a strong dependence on the IMF and has a dominant polarity that for strong B_y is opposite in the two hemispheres. This additional current system is called the cusp current and, as shown in Fig. 6.24, it lies just poleward of the region 1 currents near local noon. When the IMF y component is small, the cusp current system consists of two localized field-aligned current regions symmetrically located about local noon in both hemispheres. The current directions are opposite to the adjacent region 1 currents, being downward into the ionosphere in the afternoon and upward away from the ionosphere in the morning. Change in the sign of B_y significantly affects this distribution. Observations made in both the northern and southern hemispheres show that when $B_y > 0$ the cusp region current is extremely asymmetric and is predominantly upward in the northern hemisphere and downward in the southern hemisphere. When $B_y < 0$ the current is predominantly downward in the north and upward in the south.

6.5.2   *Current Patterns for a Northward IMF*

When the IMF has a northward component the distribution of field-aligned currents changes quite dramatically (as does the electric field distribution). The two-ring pattern of field-aligned currents is retained, but they are seen to move to higher latitudes and are significantly reduced in intensity. Perhaps more significant is a reorganization of the cusp currents in such a way that they expand over most of the high-latitude area enclosed by the region 1 currents and become as intense as the region 1 and region 2 currents themselves. This new cusp current system has been called the "NBZ" current system to denote the northward-directed B_z component that exists during its occurrence. The NBZ current system again shows a strong dependence on the sign of B_y, and Fig. 6.25 shows schematically its location and distribution in the southern hemisphere. Except for their dramatic redistribution in area, the NBZ currents show essentially the same behavior as the cusp currents, being predominantly downward in the southern hemisphere when B_y is positive and upward when B_y is negative. Conversely, we expect the NBZ current to be upward in the northern hemisphere when B_y is positive and downward when B_y is negative.

6.5.3   *Dependence on Magnetic Activity, IMF, and Season*

When the IMF has a southward component, some qualitative descriptions of the ionospheric field-aligned currents are available in terms of their dependence on the degree of magnetic activity, on the sign of B_y, and on season. In all these descriptions the field-aligned current density is regarded as uniform in each of the region 1 and region 2 rings so that a magnitude and a local time distribution can be ascribed to them. A comparison of the upper and lower panels of Fig. 6.26 shows that the local time distribution is only slightly affected by magnetic activity. The maximum region 1 current intensities occur between about 0800 and 1000 h on the morning side and between about 1400 and 1600 h on the afternoon side. At these times the region 1 currents are about a factor of 2 greater than the corresponding region 2 currents. During equinox and winter the region 1 and region 2 currents show relatively little dependence on local time except for the two region 1 current maxima. In the summer, however, it is found that the region 1 currents can exceed the region 2 currents at all local times,

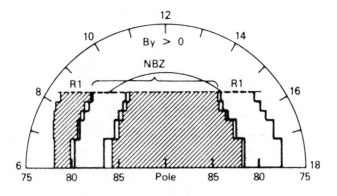

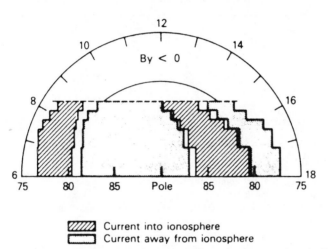

Fig. 6.25. Cusp currents in the southern hemisphere shown for the case when the IMF has a northward component. The cusp currents expand to fill the polar cusp and have been called NBZ currents. Their dominant polarity is determined by the y component of the IMF. [After Iijima *et al.* (1984). Reproduced with permission of the American Geophysical Union.]

although the region of maximum difference remains on the dayside. The region 2 currents show a much higher degree of variability with magnetic activity than do the region 1 currents. This can lead to situations on the nightside in which the region 2 currents exceed the region 1 currents during disturbed periods.

The magnitude of the region 1 currents shows a strong dependence on the interplanetary magnetic field, as might be expected from arguments similar to those applied for the polar cap potential difference. Figure 6.27 shows the de-

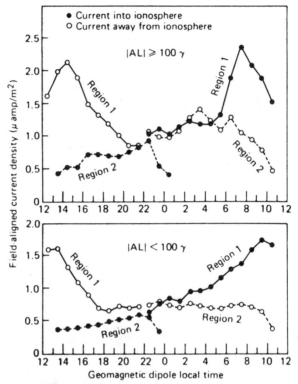

Fig. 6.26. Region 1 and region 2 currents plotted as a function of local time for quiet and disturbed geomagnetic conditions. The region 1 currents have maxima near 14.00 and 8.00 h where they exceed the region 2 currents. Elsewhere the region 1 and region 2 currents are about the same magnitude and show little variation with local time. [After Iijima and Potemra (1978). Reproduced with permission of the American Geophysical Union.]

pendence of the morningside and afternoonside region 1 currents on the magnitude of the solar wind electric field. These data show that the field-aligned currents increase as the electric field associated with the IMF increases from zero. Similar conclusions concerning the polar cap potential difference were drawn from Fig. 6.20, so it is not surprising that the driving currents should show the same behavior. As in the case of the polar cap potential difference, the region 1 current density does not reduce to zero when the solar wind electric field reduces to zero. This observation supports the existence of a mechanism in addition to merging—perhaps viscous interaction—that also provides driving currents for ionospheric convection.

The observation of a coherent pattern of field-aligned currents when the IMF

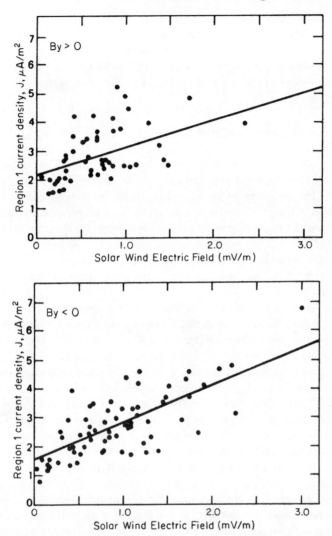

Fig. 6.27. Region 1 currents plotted against the solar wind electric field $[V_{SW}(B_y^2 + B_z^2)^{1/2}$ $\sin(\theta/2)]$ for $B_y > 0$ and $B_y < 0$. The region 1 currents show a dependence on the solar wind electric field similar to that of the polar cap potential drop. Notice that the region 1 currents do not reduce to zero when the solar wind electric field does. [After Iijima and Potemra (1982). Reproduced with permission of the American Geophysical Union.]

has a northward component is a relatively new discovery and no quantitative description of their behavior has been advanced. As in the case of the high-latitude convection pattern, there exists considerable variability in the observed data and it is likely that a rigorous treatment of the dependence of the currents on IMF orientation will be required to elucidate all their properties.

6.6   Horizontal Currents at High Latitudes

Equations (6.8) and (6.12) represent the fundamental relationships between horizontal and field-aligned currents and the electric field and conductivity in the ionosphere. The discussion of the dependence of the ionospheric conductivity on altitude given in Chapter 2 shows that the region over which substantial currents flow perpendicular to the magnetic field lines is restricted to the range from about 90 to about 130 km in the sunlit ionosphere and may extend up to 300 km when no appreciable local ionization source is present. This current flow is almost horizontal at high latitudes and produces a magnetic signature that can be observed on the ground.

Ground-based magnetometers function somewhat differently from those on satellites, but their output, three mutually perpendicular components of the magnetic perturbation from a normal steady baseline, is the same. These magnetic field perturbations are usually resolved along the geographic north, east, and vertically down directions and are denoted by H, D, and Z components, respectively. Sometimes a geomagnetic coordinate system is used, in which case the symbols X, Y, and Z denote the magnetic perturbations. It is impossible to derive the true horizontal ionospheric current distribution uniquely from ground magnetic perturbations, since they are a superposition of contributions from the horizontal ionospheric currents, field-aligned currents, distant currents in the magnetosphere, and currents induced in the earth's surface. For these reasons the ground magnetic perturbations are usually expressed in terms of "equivalent" ionospheric currents. The study of magnetic perturbations and their interpretations as current systems in the earth and in space is extremely complex and we will not discuss this topic in detail. However, magnetic perturbations are used to describe phenomena such as magnetic storms and substorms and to derive indices such as DST, K_p, and AE that describe the magnetic activity in the earth's environment. It is therefore necessary to discuss the meaning of these indices and the nature of the magnetic measurements. This discussion is located in Appendix B.

Essentially two techniques are utilized to derive the equivalent overhead horizontal current flowing in a thin shell near 100-km altitude. One method calculates the magnetic signature on the ground from a current flowing east–west in a small horizontal cell which in turn flows into field-aligned currents at the edges of the cell (Kisabeth and Rostoker, 1971). The field-aligned currents are assumed to flow along dipole magnetic field lines and subsequently close in the equatorial plane. By a "best-fit" process, this technique yields the total three-dimensional current system (horizontal and field aligned) that produces the measured ground magnetic perturbations.

The other, more widely used method expresses the overhead current $\mathbf{J}$ in terms of a divergence-free component sometimes called the equivalent current $\mathbf{J}_e$ and a curl-free component sometimes called the potential current $\mathbf{J}_p$. The potential current can be viewed as the closing current for the field-aligned currents. For vertical magnetic field lines it can be shown that the field-aligned and potential current circuit produces no magnetic perturbation at the ground. Further, if the ionospheric conductivity is uniform it can be shown that the potential current is the Pedersen current. In this case, since other magnetic effects are small, the equivalent current will largely represent the horizontal ionospheric Hall current. Since it is divergence-free, the equivalent current can be expressed in terms of a current function Γ such that

$$\mathbf{J}_e = \hat{r} \times \nabla\Gamma \qquad (6.19)$$

where $\hat{r}$ is the unit radial vector in a coordinate system with the origin at the center of the earth.

No current flows near the ground, so the magnetic perturbation there, $\delta\mathbf{B}$, can be expressed in terms of a magnetic potential ψ such that

$$\delta\mathbf{B} = -\nabla\psi \qquad (6.20)$$

There are straightforward mathematical relationships that uniquely relate Γ and ψ so that the horizontal equivalent current can be derived from the potential function ψ. The function ψ is determined by taking the divergence of (6.20) and solving the two-dimensional Poisson equation

$$\nabla^2\psi = -\nabla \cdot \delta\mathbf{B} \qquad (6.21)$$

The right-hand side of this equation is given by a global distribution of ground observations with suitable interpolation at each grid point.

These two quite different techniques yield very similar results for the horizontal equivalent ionospheric current. In principle, only the equivalent current can be derived from ground magnetic perturbations, and some model of ionospheric conductivity is required to proceed further. However, many features of the equivalent current system are useful in establishing self-consistency among the electrodynamic parameters and we will now describe these characteristics.

Our purpose here is to describe the dependence of the equivalent current distribution on magnetic activity and the interplanetary magnetic field orientation. In this way a self-consistent picture of the electrodynamic parameters $\mathbf{J}_\perp$, $\mathbf{J}_\parallel$, and $\mathbf{E}_\perp$ can be established. The effects of magnetic activity on the horizontal

EQUIVALENT CURRENT SYSTEM

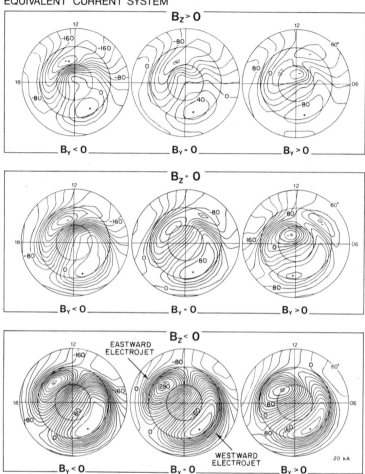

Fig. 6.28. Contours showing the horizontal equivalent currents for different orientations of B_z and B_y. The horizontal equivalent current shows many of the features of the high-latitude convection pattern. In fact, the features differ only where there exist substantial field-aligned currents and conductivity gradients. [After Friis-Christensen *et al.* (1985). Reproduced with permission of the American Geophysical Union.]

equivalent current manifest themselves principally in small-scale features that we will discuss later. The large-scale features of the horizontal current and its dependence on the IMF are shown in Fig. 6.28. A two-cell pattern of current flow exists for all orientations of the IMF. In each of the pictures the current flows in closed loops around a focus labeled " + ," located on the nightside near midnight, and around a focus labeled " $-$," located on the dayside near noon. The sense of circulation of the current is clockwise around the $+$ focus and anticlockwise around the $-$ focus. When the IMF has a southward component ($B_z < 0$) the current is much larger and the two-cell pattern is more well defined than when the IMF is northward. Examining the southward IMF case in more detail, we note that at the low-latitude extremes of the current system there exist bands of current flowing, respectively, eastward on the dusk side and westward on the dawn side. These current bands are colocated with the diffuse auroral zones and are called the eastward and westward electrojets. Note that the direction of the electrojet currents is opposite to that of the F-region plasma flow.

The fact that a two-cell current pattern is retained for all orientations of the IMF indicates that some element of the current pattern may be independent of the IMF. Statistical analysis of ground-based magnetograms shows this to be the case, and this portion of the current system is called the S_q^p current system, for "solar quiet polar." Detailed analysis shows that there may also exist semidiurnal contributions to this current that may be attributable to lunar variations, as is the case in the equatorial region.

For a southward IMF during times of relatively low magnetic activity (i.e., $K_p < 3$), the total current system is made up of the S_q^p system plus another system called the DP2 system; the DP2 current system is made up of the DPY and DPZ systems, which are strongly dependent on the y and z components of the IMF, respectively. The features of the IMF-dependent DPY current system dominate the horizontal current distribution and can be clearly seen if we subtract the IMF-independent contribution from equivalent current systems measured when B_y has a substantial positive component and when it has a negative component. Figure 6.29 shows the result of such an exercise, in which it can be seen that the IMF-dependent system is approximately zonal and flows across the local noon meridian. This "DPY current" flows from dawn to dusk when B_y is positive and from dusk to dawn when B_y is negative. The DPY current produces a rotation and an asymmetry in the current vortices so that the high-latitude sunward-flowing current is directed toward the prenoon sector for both polarities of the IMF. When the z component of the IMF is very small the two-cell horizontal current system is preserved, but the intensity of the current is much less than that existing when the IMF is strongly southward. As B_y changes from negative to positive the line of symmetry for the current vortices rotates clockwise as it does for a southward IMF. However, when B_z is very small the degree of rotation is much larger, leading to currents that flow almost parallel to the dawn–dusk meridian when B_y is positive.

When the IMF has a strong northward component, the horizontal currents associated with closure of the field-aligned NBZ current system can be observed from satellites but only in the summer polar cap on the ground. What is detect-

B_z South

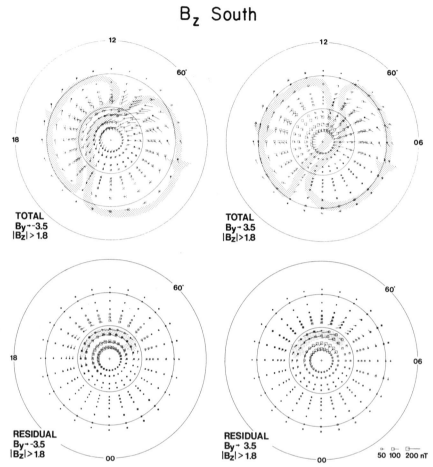

Fig. 6.29. The arrows indicate the direction of flow of horizontal currents in the high-latitude ionosphere. In the top two diagrams the total currents are indicated. Features of these currents that are present for any orientation of the IMF have been removed in the lower two diagarms, which therefore represent IMF-dependent horizontal currents. [After Friis-Christensen (1981). Reproduced with permission of Reidel, Hingham, Massachusetts.]

able on the ground is a low-intensity version of the two-cell current system existing when B_z is negative or very small. In addition there is evidence for an antisunward current in the center of the polar cap.

References

Axford, W. I., and Hines, C. O. (1961). A unifying theory of high-latitude geophysical phenomena and geomagnetic storms. *Can. J. Phys.* **39,** 1433.

Burke, W. J., Kelley, M. C., Sagalyn, R. C., Smiddy, M., and Lai, S. T. (1979). Polar cap electric field structures with a northward interplanetary magnetic field. *Geophys. Res. Lett.* **6,** 21.

Dungey, J. W. (1961). Interplanetary magnetic field and the auroral zones. *Phys. Rev. Lett.* **6,** 47.

Foster, J. C. (1983). An empirical electric field model derived from Chatanika radar data. *JGR, J. Geophys. Res.* **88,** 981.

Friis-Christensen, E. A. (1981). High latitude ionospheric currents. *In* "Exploration of the Polar Upper Atmosphere" (C. S. Deehr and J. A. Holtet, eds.). Reidel, Hingham, Massachusetts.

Friis-Christensen, E. A., Kamide, Y., Richmond, A. D., and Matsushita, S. (1985). Interplanetary magnetic field control of high latitude electric fields and currents determined from Greenland magnetometer data. *JGR, J. Geophys. Res.* **90,** 1325.

Heelis, R. A. (1984). The effects of interplanetary magnetic field orientation on dayside high latitude convection. *JGR, J. Geophys. Res.* **89,** 2873.

Heelis, R. A., and Hanson, W. B. (1980). High latitude ion convection in the nighttime F-region. *JGR, J. Geophys. Res.* **85,** 1995.

Heelis, R. A., Reiff, P. H., Winningham, J. D., and Hanson, W. B. (1986). Ionospheric convection signatures observed by DE-2 during northward interplanetary magnetic field. *JGR, J. Geophys. Res.* **91,** 5817.

Heppner, J. P. (1972). Polar cap electric field distributions related to the interplanetary magnetic field direction. *J. Geophys. Res.* **77,** 4877.

Heppner, J. P., and Maynard, N. C. (1987). Empirical high latitude electric field models. *JGR, J. Geophys. Res.* **92,** 4467.

Hill, T. W. (1975). Magnetic merging in a collisionless plasma, *JGR, J. Geophys. Res.* **80,** 4689.

Iijima, T., and Potemra, T. A. (1976). Field-aligned currents in the dayside cusp observed by Triad. *JGR, J. Geophys. Res.* **81,** 5971.

Iijima, T., and Potemra, T. A. (1978). Large-scale characteristics of field-aligned currents associated with substorms. *JGR, J. Geophys. Res.* **83,** 599.

Iijima, T., and Potemra, T. A. (1982). The relationship between interplanetary quantities and Birkeland current densities. *Geophys. Res. Lett.* **9,** 442.

Iijima, T., Potemra, T. A., Zanetti, L. J., and Bythrow, P. F. (1984). Large-scale Birkeland currents in the dayside polar region during strongly northward IMF: A new Birkeland current system. *JGR, J. Geophys. Res.* **89,** 7441.

Kisabeth, J. L., and Rostoker, G. (1971). Development of the polar electrojet during polar magnetic substorms. *J. Geophys. Res.* **76,** 6815.

Lysak, R. L. (1985). Auroral electrodynamics with current and voltage generators. *JGR, J. Geophys. Res.* **90,** 4178.

Mozer, F. S., and Lucht, P. (1974). The average auroral zone electric field. *JGR, J. Geophys. Res.* **79,** 1001.

Oliver, W. L., Holt, J. M., Wand, R. H., and Evans, J. V. (1983). Millstone Hill incoherent scatter observations of auroral convection over $60° < \Lambda < 75°$. 3. Average patterns versus K_p. *JGR, J. Geophys. Res.* **88,** 5505.

Reiff, P. H., Spiro, R. W., and Hill, T. W. (1981). Dependence of polar cap potential on interplanetary parameters. *JGR, J. Geophys. Res.* **86,** 7639.

Siscoe, G. L. (1982). Polar cap size and potential: A predicted relationship. *Geophys. Res. Lett.* **9,** 672.

Smiddy, M., Burke, W. J., Kelley, M. C., Saflekos, N. A., Gussenhoven, M. S., Hardy, D. A., and Rich, F. J. (1980). Effects of high-latitude conductivity on observed convection electric fields and Birkeland currents. *JGR, J. Geophys. Res.* **85,** 6811.

Sugiura, M., Maynard, N. C., Farthing, W. H., Heppner, J. P., Ledley, B. G., and Cahill, L. J. (1982). Initial results on the correlation between the electron and magnetic fields observed from the DE 2 satellite in the field-aligned current regions. *Geophys. Res. Lett.* **9,** 985.

Vickrey, J. F., Livingston, R. C., Walker, N. B., Potemra, T. A., Heelis, R. A., and Rich, F. J. (1985). On the current–voltage relationship of the magnetospheric generator at small spatial scales. *Geophys. Res. Lett.* **13,** 495.

Volland, H. (1975). Models of the global electric fields within the magnetosphere. *Ann. Geophys.* **31,** 159.

Wygant, J. R., Torbert, R. B., and Mozer, F. S. (1983). Comparison of S3-2 polar cap potential drops with the interplanetary magnetic field and models of magnetopause reconnection. *JGR, J. Geophys. Res.* **88,** 5727.

Chapter 7 | Effects of Plasma Flow at High Latitudes

In this chapter we describe some additional features of the high-latitude ionosphere that distinguish it from the lower-latitude regions which are in magnetic contact with the plasmasphere. We have defined the high-latitude ionosphere as that region of latitudes in which, at least some of the time, the plasma may flow in magnetic flux tubes that either have only one foot on the ground or are so extended that the effect on the local plasma of connection to the conjugate hemisphere is negligible.

From our knowledge of the variability of the high-latitude convection pattern we know that such a definition can include the ionosphere at invariant latitudes above about 50°. In this region the ionospheric plasma may at some time be subject to "boundary conditions" in the outer magnetosphere that allow rapid expansion along the magnetic field lines. We also know that within this region the plasma velocities perpendicular to the magnetic field can be of the order of 1 km/s. This gives the bulk thermal plasma some very different properties from those found at lower latitudes, and it is necessary to consider the details of ionospheric plasma motion both parallel and perpendicular to the magnetic field in order to understand them. The high plasma velocities which occur in this region also greatly affect neutral atmospheric dynamics.

7.1 Ionospheric Effects of Parallel Plasma Dynamics

7.1.1 Ionospheric Composition at High Latitudes

The effects of plasma motion along the magnetic field can be most dramatically seen in the ionospheric composition near an altitude of 1000 km as a function of

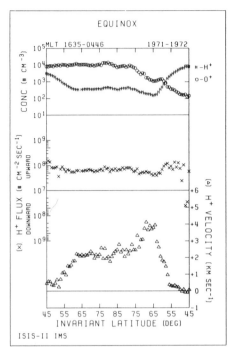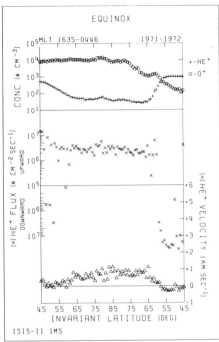

Fig. 7.1. The average equinox distribution of O^+, H^+, and He^+ at 1400 km plotted in the upper panels shows that the reduced concentration of light ions at high latitudes is accompanied by significant outward flow velocities. Quiet magnetic conditions pertained in the data sets used to construct these curves. [After Hoffman and Dodson (1980). Reproduced with permission of the American Geophysical Union.]

latitude. At this altitude the ionosphere has essentially three ion species: O^+, H^+, and He^+. Figure 7.1 shows the quiet-time average distribution of these ion species for satellite passes across the polar region from the dayside (~1635 magnetic local time) to the nightside (~0446 magnetic local time). The field-aligned flow velocities and number flux of the light ions are also given. The data were obtained near equinox by the ISIS II satellite, which was located at 1400-km altitude. These data show quite abrupt decreases in the H^+ and He^+ concentrations between 55° and 65° invariant latitudes to very low, almost constant concentrations near 100 cm^{-3} at higher latitudes. In the same latitude sector the O^+ concentration is nearly constant on the dayside and shows only a gradual change at night. This signature in the light ion species may be seen at all altitudes where the species H^+ or He^+ is detectable and is a persistent feature at all local times. It is called the light ion trough. The latitude gradient in the H^+ concentration defines the equatorward edge of the light ion trough and the magnitude of this gradient may change significantly as a function of local time and season. Its position is also strongly dependent on magnetic activity. As we shall see in Section 7.2, this magnetic dependence is a straightforward consequence of the variability of the convection paths of the plasma that, in fact, determine where the trough will be formed. This large area of depressed light ion concentrations relative to the O^+ concentration can be produced only by a field-aligned motion of the light ions relative to O^+. Satellite-borne ion mass spectrometers have observed this relative motion at altitudes below 2000 km as well as at much higher altitudes. The light ion velocities parallel to the magnetic field can be several kilometers per second and are described as the "polar wind." The velocity of the light ions can, in fact, be supersonic at times. The existence of the polar wind and the depressed light ion concentrations at high latitudes indicate that in this region the light ion population of the ionosphere is usually not in diffusive equilibrium along the magnetic field lines.

7.1.2 Hydrodynamic Theory of the Polar Wind

In Chapters 2 and 5 we discussed the forces that contribute to motion of the ionospheric plasma along the magnetic field. In these discussions we included the generation of an electric field with a component along the magnetic field lines which was created internally by the plasma. This "ambipolar" field does not drive a current parallel to the magnetic field but merely serves to equalize the forces on the ions and electrons so that they move together. In the inner magnetosphere the field lines are closed and relatively short in length. This allows the flux tubes to fill to a condition of diffusive equilibrium on a time scale of a day or so. Under normal conditions, then, the plasma is produced during the daytime and tends to flow slowly up into the inner magnetosphere. At night the flow is reversed and plasma tends to flow back into the ionosphere from above, replacing the ionospheric plasma against recombination losses. Charge exchange converts O^+ to H^+ and vice versa at the top of the ionosphere.

The crucial difference between the high- and low-latitude cases is that in the former the flux tubes are either open or so long that no semblance of diffusive equilibrium exists. In effect, whatever plasma is produced by sunlight or particle influx merely expands into a near vacuum.

Below about 2000 km we learned that collision frequencies and gyrofrequencies are sufficiently high that most problems allow neglect of the acceleration term in the equations of motion. In this region the plasma is said to be collision dominated. This means that the ions undergo several collisions in moving through one scale height. A plasma is collision dominated if

$$(V_i/H_i) \ll \nu_i$$

where V_i is the field-aligned ion velocity and H_i and ν_i are the ion scale height and appropriate collision frequency, respectively. However, since the upper boundary condition is crucial at high latitudes we cannot make this approximation or restrict ourselves to altitudes below 2000 km. Note that it is not our intent here to provide all the mathematical details involved in the theoretical formulation of parallel ion flow in such a situation; we only outline the initial steps and give some physical insight into the phenomenon based on experiments and simple model calculations.

In a collision-dominated plasma with multiple ion species, where one ion may move relative to another at a high speed which varies with altitude, the advective derivative in (2.5) must be retained. The velocity of the jth ion species can then be found from the hydrodynamic equation (2.22b)

$$\rho_j(\mathbf{V}_j \cdot \nabla)\mathbf{V}_j = -\nabla p_j + \rho_j \mathbf{g} + n_j q_j(\mathbf{E} + \mathbf{V}_j \times \mathbf{B}) - \sum_{j \neq k} \rho_j \nu_{jk}(\mathbf{V}_j - \mathbf{V}_k) \quad (7.1)$$

We drop the $\partial/\partial t$ term under an assumption of steady-state flow conditions in time. In the equation of motion for the electrons the terms containing the electron mass can be neglected, as can the acceleration and advective derivative terms. Then, as shown in Chapter 5, the ambipolar electric field component parallel to $\mathbf{B}$ is given by

$$E_\parallel = (1/n_e q_e)\nabla_\parallel p_e \quad (7.2)$$

Remember that since q_e is a negative number, $E_\parallel$ is opposite in direction to $\nabla_\parallel p_e$; that is, $E_\parallel$ points upward in the topside ionosphere. If we now let s be measured opposite to the magnetic field, the component of the equation of motion in the s direction for the jth ion species in the northern hemisphere is

$$V_{js}\frac{\partial V_{js}}{\partial s} + \frac{1}{n_e m_j}\frac{\partial p_e}{\partial s} + \frac{1}{n_j m_j}\frac{\partial p_j}{\partial s} - (\mathbf{g} \cdot \hat{B}) + \sum_{j \neq k} \nu_{jk}(V_{js} - V_{ks}) = 0 \quad (7.3)$$

Inspection of (7.3) reveals the forces acting on a minor ion. From right to left they arise from friction brought about by collisions with the other ion species, the component of gravity along the magnetic field, the partial pressure gradient in the minor ion species itself, and the ambipolar electric field generated between the electrons and the major ions. This upward electric force must ultimately be balanced by a positive (downward) partial pressure gradient in order to reach a state of diffusive equilibrium. While this situation can occur along the relatively short closed magnetic field lines that exist in the mid-latitude ionosphere, it may not be achievable along the open or highly distended field lines in the high-latitude ionosphere. A continuous outward flow of the minor light ions H^+ and He^+ can therefore occur at either subsonic or supersonic velocities in a manner consistent with the data such as those shown in Fig. 7.1.

A description of this ion motion can be obtained by considering the equations of continuity, motion, and energy for each species. These equations make up a closed system that is generally solved numerically (see Schunk, 1977, for a full description of these equations) given a boundary condition at the top of the ionosphere near 3000 km. If we apply some simplifying assumption, however, a description of the fundamental physics that is operating can be obtained. Consider the motion of a minor ion species, j, embedded in a major ion species, i, where the ith species is in diffusive equilibrium in an isothermal topside ionosphere. In such a case the electron and major ion gas is distributed along the magnetic field tube according to the expression

$$n_e = n_{e0} \exp[-(s - s_0)/H_p] \tag{7.4}$$

where n_{e0} is the electron density at some reference altitude s_0, H_p is the plasma scale height, and the variation of gravity with altitude has been neglected (Rishbeth and Garriott, 1969). From (2.8), the pressure for each species may be written $p_j = n_j k_B T_j$. If we assume also that the altitude is sufficiently high that neutral collisions may be neglected, the minor ions collide only with the major ions and (7.3) can be rewritten as

$$\nu_{ji} V_{js} + V_{js}\left(\frac{\partial V_{js}}{\partial s}\right) = (\mathbf{g} \cdot \hat{B}) + \left(\frac{k_B T_e}{m_j H_p}\right) - \left(\frac{k_B T_j}{m_j}\right)\left(\frac{1}{n_j}\right)\left(\frac{\partial n_j}{\partial s}\right) \tag{7.5}$$

Neglecting the perpendicular motion of the plasma, from (2.22a) the steady-state continuity equation for species j can be written in the form

$$A(P_j - L_j) = (\partial/\partial s)(A n_j V_{js}) \tag{7.6}$$

where A is the cross-sectional area of a magnetic flux tube, P_j the production term, and L_j the loss term. Integrating this equation along the magnetic flux tube from the reference altitude to any point s, we find

$$(1/Q_j)(\partial Q_j/\partial s) = (1/n_j)(\partial n_j/\partial s) + (1/V_{js})(\partial V_{js}/\partial s) + (1/A)(\partial A/\partial s) \tag{7.7}$$

where

$$Q_j = \int_{s_0}^{s} A(P_j - L_j)\,ds \tag{7.8}$$

Substituting for $(1/n_j)(\partial n_j/\partial s)$ in (7.5), we arrive finally at the equation

$$(V_{js}^2 - V_t^2)(1/V_{js})(\partial V_{js}/\partial s) = -\nu_{ji} V_{js} + \mathbf{g} \cdot \hat{B} - (k_B T_e/m_j H_p)$$
$$- V_t^2[(1/Q_j)(\partial Q_j/\partial s) - (1/A)(\partial A/\partial s)] \tag{7.9}$$

where the thermal speed of the jth species is given by

$$V_t = (k_B T_j/m_j)^{1/2} \tag{7.10}$$

This first-order differential equation for V_{js} has a singularity where $V_{js} = V_t$. This point is commonly called the supersonic transition in fluid theory. In a multiple ion species plasma the plasma sound speed is not given by V_t, but it is convenient here to refer to $V_{js} > V_t$ as representing supersonic flow of the jth

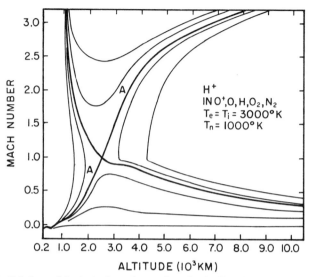

Fig. 7.2. Solutions of the hydrodynamic equations for the H^+ Mach number as a function of altitude in a gas containing O_2^+ ions as well as the neutral species O, H, O_2, and N_2. Curve A provides a transition from subsonic to supersonic flow. [After Banks and Holzer (1969). Reproduced with permission of the American Geophysical Union.]

species. Figure 7.2 shows the family of solutions of (7.9) using a Mach number $M_j = V_{js}/V_t$ plotted as a function of altitude. The physically meaningful solutions are those with $M_j < 1$ at all altitudes and one solution (marked as curve A) passing through the transitional point at $M_j = 1$ and for which the flow is subsonic at low altitudes and supersonic at higher altitudes. A similar analysis has been done for the solar wind in the pioneering work by Parker (1958).

Solutions of these so-called hydrodynamic equations have now reached a degree of sophistication that allows the details of heat transfer and energy balance to be considered in addition to the momentum and density distribution of various ion species. In such models the acceleration of minor ion species with mass less than the mean ion mass can occur where ion–ion collisions are sufficient to produce frictional heating and anisotropic ion temperatures. In the high-latitude ionosphere both H^+ and He^+ can be accelerated through the O^+ gas and be subjected to this heating process. The existence of different temperatures and temperature distributions in the ion species can induce motion in these species that, depending on the ion mass, either resists or assists their outward flow.

Figure 7.3 shows some typical results of such calculations performed assuming different outflow velocities for H^+ at 3000 km. Notice in Fig. 7.3b that the H^+ velocity increases rapidly in the region between 600 and 1400 km, where O^+ is the dominant ion (see Fig. 7.3a). This flow is induced by the O^+–electron polarization field. The flow of H^+ through O^+ produces frictional heating of the

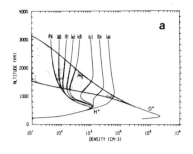
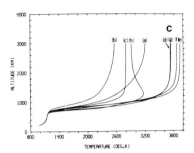
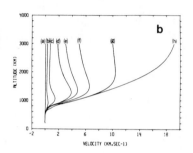
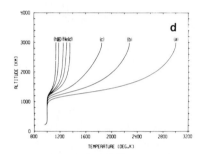

Fig. 7.3. Altitude distribution of (a) O^+ and H^+ densities, (b) H^+ velocity, (c) H^+ temperature, and (d) O^+ temperature from models of the polar wind. The curves labeled (a)–(h) correspond to assumed H^+ outflow velocities at 3000 km ranging from 0 to about 20 km/s. [After Raitt *et al.* (1975). Reproduced with permission of Pergamon Press.]

H^+ gas to temperatures above that of the O^+ ions in this region. At higher altitudes the distribution of the species is controlled by the specified H^+ outflow velocity, that is, by the boundary condition placed on the solution. It shows that at 3000 km the predicted H^+ polar wind can be supersonic since velocities in excess of 10 km/s are associated with temperatures less than 4000 K.

Two important considerations should be kept in mind when considering the usefulness of this hydrodynamic description of the polar wind. First, the polar wind does not simply result from an open field line geometry of field tubes with very large volumes. The minor ions H^+ and He^+, for example, receive very little acceleration if the electron population is very cold, as it might be in the winter polar region. Second, the theoretical descriptions of the polar wind are dependent on some boundary condition that is usually applied at the top of the modeled region near 3000 km, where the ionosphere becomes essentially collisionless. The effects of the acceleration processes and the distribution of the ion species below this altitude are dependent on these boundary conditions. Ultimately, boundary conditions must be consistent with conditions in the outer magnetosphere, where the field tubes are in contact with a much lower-density plasma that has a variety of possible energy distributions quite different from those existing at lower altitudes.

7.1.3   High-Altitude Observations: What Are the Real Boundary Conditions?

In the previous section we dealt with observations and theoretical treatment related to parallel ion flow in a region where ion–ion collisions were important. With such a treatment it is possible to understand the appearance of ions flowing supersonically from the ionosphere to the magnetosphere. These ions would, however, be expected to have isotropic temperature distributions. With this in mind, it is worthwhile to examine the variety of ion populations that have been observed at high altitudes in the magnetosphere. Satellite measurements have revealed a number of ion populations, apparently of ionospheric origin, that have quite different energy distributions than expected from the theoretical treatment outlined in the previous section.

Figure 7.4 shows a representative sample of the energy distribution and flow direction of ionospheric ion species observed at mid to high latitudes and at altitudes in excess of 7000 km by instruments on the Dynamics Explorer 1 satellite (Chappell *et al.*, 1982). In Fig. 7.4A the top panel shows the flux of H^+ at energies between 0 and 500 eV indicated on the right. The lower panel shows the direction with respect to the spacecraft velocity (labeled RAM) from which the ions are seen. The most intense fluxes (on the right-hand side of the plot) were observed inside the plasmasphere (L values < 4.6 as listed below the plot). The lower panel shows a peak flux in the RAM direction, indicating the dominance of relatively low-energy plasma (temperature about 3000 K) with a flow velocity much less than the spacecraft velocity. Just outside the plasmasphere the ionospheric ions are seen to have a variety of arrival directions. At even larger L values the ions were grouped at a flow angle of $-90°$, which corresponds to flow outward along the magnetic field lines at high velocities. This result is similar to the theoretical discussion above, but the ions have much higher temperatures than have been considered. The remaining panels emphasize the flow direction data. In Fig. 7.4B the fluxes of H^+ and He^+ ions both maximize near $-90°$, which again corresponds to an outward flow velocity in excess of the satellite velocity of 3 km/s. In Fig. 7.4C the same feature is seen, except it is somewhat more erratic. Notice also that in the lower panel fluxes of H^+ are occasionally seen at all angles with respect to the satellite velocity. This indicates that the thermal speed of these ions must be in excess of the satellite velocity. From these observations it is clear that alternative formulations to the hydrodynamic one we have described so far are required to understand adequately how the variety of energies and motions of ionospheric ions might arise. Finally, we show that sometimes a flow can be detected even when low temperatures of H^+ and He^+ pertain. These are the conditions as modeled by the hydrodynamic theories. These unique measurement conditions arise when the spacecraft po-

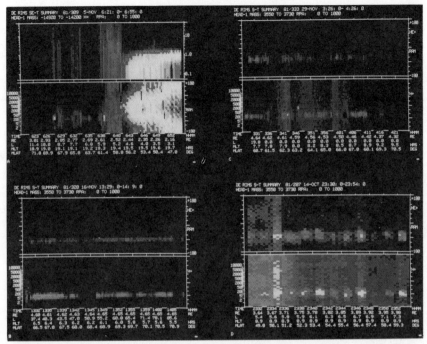

Fig. 7.4. Data from Dynamics Explorer 1 showing (A) the energy and pitch angle distribution of H$^+$ for a pass crossing the plasmasphere, (B and C) pitch angle distributions of H$^+$ and He$^+$ showing outflowing ions outside the plasmasphere with energies of several electron volts, and (D) pitch angle distributions of H$^+$ and He$^+$ illustrating the detection of outward-flowing ionospheric ions of energy less than 1 eV similar to those predicted in the theoretical models. [After Chappell *et al.* (1982). Reproduced with permission of the American Geophysical Union.]

tential is near zero volts and hence is ideal for measuring low temperatures. These data are shown in Fig. 7.4D by the enhanced ion fluxes seen at 90° from the RAM direction when the instrumentation is capable of detection of low-energy ions.

7.1.4   More Advanced Theories

In the collisionless regime, a number of different formulations describing the motion of plasma along **B** are possible. The additional physics that must be contained within such a formulation is the possibility for anisotropic pressure in the flowing species. That is, the parallel and perpendicular temperatures of the ion species must be treated separately. Such a treatment can be included in the approaches described next.

The kinetic approach derives all the plasma parameters from the collisionless Boltzmann equation, which is sometimes called Vlasov's equation. In this formulation the velocity distribution of the ions is specified at a lower boundary called the exobase or baropause, below which the plasma is assumed to be collision dominated. Then the velocity distribution is given at any point above this boundary in terms of the boundary condition, the total energy in the system, and the magnetic moment of the particles (Lemaire and Scherer, 1970). This is one of the adiabatic invariants discussed briefly in Chapter 2 which must remain conserved if spatial and temporal changes occur slowly. In brief, the quantity $\mu = K_\perp/B$ must remain the same, where μ is the magnetic moment, $K_\perp$ is the perpendicular energy, $\frac{1}{2}mV_\perp^2$, of the particle, and B is the magnitude of the magnetic field. As a particle moves up along the magnetic field line, then, the fact that B decreases requires $K_\perp$ to decrease as well. Ignoring gravity for the present, if μ and the particle energy are both conserved, the parallel velocity must increase corresponding to the decrease in $K_\perp$. This is the origin of the mirror force in a converging magnetic field geometry (see Chapter 2). Rocket experiments employing barium ion jets have shown quite excellent agreement with the theory of low-altitude ion motion along **B**, including the gravitational force as well as the mirror force in the equation of motion. The same experiments have also revealed that well above the ionosphere, around 5000-km altitude, a region of high parallel electric field is sometimes encountered which accelerates the barium ions upward and electrons downward. We have chosen not to include a detailed study of this region in the present text, although it certainly is a crucial factor in the production of discrete auroral electron precipitation and upward ion acceleration.

An alternative approach to the collisionless flow description is to use the classical hydrodynamic equation, (7.3). However, to take account of possible anisotropic temperatures, the energy equation must also be included in the calculations, and the effects of stress or viscosity must be included. This formulation was first employed by Holzer *et al.* (1971).

Finally, it is possible to use generalized transport equations to describe the flow of plasma in both the collision-dominated and collisionless regimes. These equations allow a complete description of the plasma and are generally used for computer modeling of the polar wind. They can be shown to reduce to the simpler hydrodynamic and kinetic equations in the collision-dominated and collisionless limits. Comparing the collisionless hydrodynamic equations to the kinetic equations, Holzer *et al.* (1971) have shown that they yield essentially the same solutions for density, flow velocity, and temperature in the case where the electrons are assumed to have an isotropic temperature distribution and to be in equilibrium with the generated electric fields.

With reasonable choices of boundary conditions it is possible to explain the observed fluxes of relatively cold (<1 eV) ions that have been observed from satellites in terms of any of these theories. The appearance of ionospheric ions with energies in the range 10 to 50 eV, however, requires some further theoretical treatment. In this regard we note that the properties of the plasma described so far have been derived assuming steady-state conditions. The convective motion of the plasma at high latitudes, which is impressed on the ionosphere from very high altitudes, will produce quite rapid changes in the upper boundary so that conditions approaching vacuum expansion might be expected. Time-dependent modeling of this expansion using a simple kinetic approach provides some useful insight into what might occur. In this approach the electrons are assumed to be in equilibrium with any parallel ambipolar electric fields that they create. At time $t = 0$ a constant-density plasma in the region $s < 0$ is allowed to expand into a vacuum in the region $s > 0$. Then the electron distribution will be given by

$$n_e = n_{e0} \exp[e\phi(s)/k_B T_e] \qquad (7.11)$$

where n_{e0} is the electron concentration at the location where the potential $\phi(s)$ is zero, and T_e is the electron temperature. If we assume that the electric field resulting from gradients in the potential $\phi(s)$ is the only force affecting the ions, then their equations of momentum and continuity are simply

$$\partial V_{is}/\partial t + V_{is}(\partial V_{is}/\partial s) = -(e/m_i)\ \partial\phi(s)/\partial s \qquad (7.12)$$

$$\partial n_i/\partial t = -\partial/\partial s(n_i V_i) \qquad (7.13)$$

Solutions to these equations can be expressed in terms of a dimensionless "self-similar variable"

$$\varepsilon = (s/t)(m_i/k_B T_e)^{1/2} \qquad (7.14)$$

such that

$$n_i = n_e = n_{e0} \exp[-(\varepsilon + 1)] \qquad (7.15)$$

$$V_{is} = (k_B T_e/m_i)^{1/2}(\varepsilon + 1) \qquad (7.16)$$

$$\phi = -(k_B T_e/e)(\varepsilon + 1) \qquad (7.17)$$

The properties of these solutions are shown in Fig. 7.5 (Singh and Schunk, 1983), where the electron concentration, electric field, and field-aligned ion velocity are shown as a function of s. For $t > 0$ a rarefaction wave moves into the plasma region $s < 0$ as the plasma expands into the region $s > 0$. The electric field is uniform in the region $s > -t(k_B T_e/m_i)^{1/2}$ but differentiation of (7.17) shows that it decreases linearly as time increases. This electric field can accelerate the ions as shown in the bottom plot of Fig. 7.5 to produce some very energetic ions far from the expansion region, a result which may explain some of the energetic ions seen in the DE-1 data.

While far from complete, developments of this nature lend confidence in our ability to understand eventually the various energetic ionospheric ion populations existing at high latitudes and high altitudes in the terrestrial magnetosphere. A

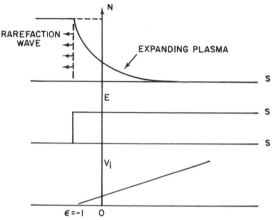

Fig. 7.5. Schematic representation of the self-similar solutions for number density, electric field, and velocity in an expanding plasma. [After Singh and Schunk (1982). Reproduced with permission of the American Geophysical Union.]

potentially important factor barely taken into consideration to date is the role of oxygen and other ion cyclotron waves in heating and accelerating ionospheric plasma. We touch on this briefly in Chapter 8, but the topic is likely to be of great importance in understanding the role of the ionosphere in supplying magnetospheric plasma. Similarly, the parallel electric fields mentioned above which are responsible for accelerating electrons downward in the aurora will also accelerate ions outward into the magnetosphere.

7.2 Ionospheric Effects of Perpendicular Plasma Dynamics

7.2.1 The Role of Horizontal Transport

In Chapter 6 we described in some detail the convective motion of the ionospheric plasma that results from an electric field of solar wind origin. The large-scale convection pattern at high latitudes is fixed with respect to the earth–sun line and the magnetic dipole axis. In this coordinate system the convection pattern is dependent on latitude and longitude (also referred to as magnetic local time). The plasma moves both toward and away from the sun and into and out of the auroral zone. Since the solar radiation and auroral particle precipitation are the most important ionization sources, the plasma convection velocity is thus one of the most important factors affecting the ion distribution and temperature at all altitudes.

In addition to the ionospheric motion imposed from the magnetosphere, the earth's atmosphere and portions of its plasma environment corotate about the geographic axis. We thus have a rather complicated situation in which one source of motion and ionization is most easily specified in a geographic reference frame and another is most easily specified in a geomagnetic reference frame. During the course of a day the geomagnetic pole rotates around the geographic pole, and in a coordinate system fixed with respect to the sun there is therefore a universal time dependence (in addition to latitude and local time dependence) in both the magnetospheric convection velocity and the auroral zone ionization rate. At high latitudes the relatively high ion velocities produce frictional heating due to collisions between the ions and the neutral gas. These collisions also affect the rate at which ions recombine with the neutral gas. We will deal with these collisional effects later. We will not deal in detail with the production and loss processes for plasma in the ionosphere but will describe the convection effects in sufficient detail that their effects on the plasma composition, distribution, and temperature can be recognized.

To understand the net plasma distribution at high latitudes we will consider a frame of reference in which the earth–sun line and the geographic pole are fixed. This frame is essentially fixed in inertial space since its motion, consisting of rotation around the sun once per year, is negligible compared to the convective motion of the plasma during a 1-day cycle. In this reference frame the plasma has a corotation component that is independent of local time and a magnetospheric component that is dependent on local time and universal time as the magnetic pole rotates around the geographic pole. Figure 7.6 shows several different plasma convection paths that result from this complex addition of corotation and magnetospheric convection. Each path is traced for a 24-h period to illustrate that the paths do not necessarily close and certainly are quite different from the convection paths illustrated in Chapter 6, in which the motion of the coordinate system was removed. The starting points for these three trajectories may be roughly summarized as follows: a, dusk convection cell; b, polar cap; c, dawn convection cell. Also shown in each portion of the figure is the position of the solar terminator in the northern hemisphere during winter, summer, and equinox. Notice that in this inertial frame the plasma can move in and out of sunlight as well as in and out of the auroral zone. A complex distribution of plasma production, transport, and loss which has local time and universal time dependences results from this motion.

With the goal of understanding some of the consequences of this plasma motion, we will consider some large-scale features observed in the high-latitude F region. Figure 7.7 shows the total ion concentration and the perpendicular components of the plasma drift velocity observed by a satellite traversing the high-latitude ionosphere. The plot is made in a rotating frame such that a corotating plasma would register zero velocity. In the time interval labeled B, the zonal flow velocity indicated by the upper curve is weak and to the west. At the point where the curve crosses the dashed line the plasma is flowing toward the west

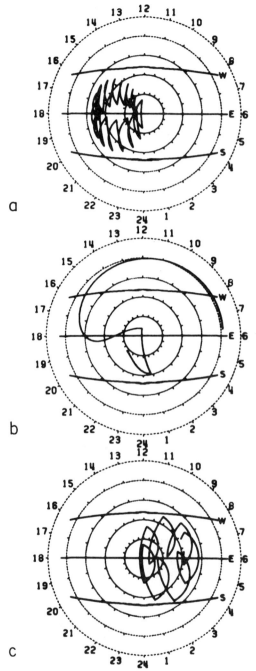

Fig. 7.6. Several complex trajectories that plasma may undergo in a period of 24 h due to the displacement of the geomagnetic and geographic poles. [After Sojka *et al.* (1979). Reproduced with permission of the American Geophysical Union.]

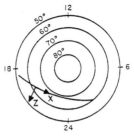

ATMOSPHERE EXPLORER-C
R PA DRIFT METER
ORBIT 5595   DAY 75048
NORTHERN HEMISPHERE

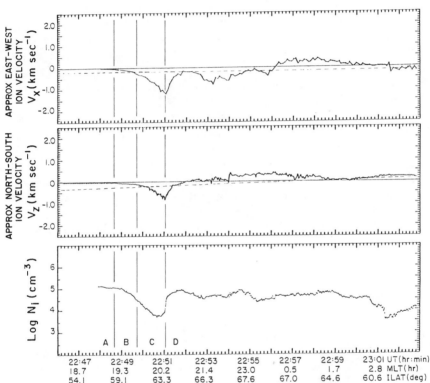

Fig. 7.7. Plasma velocity and density data from AE-C which show that the mid-latitude trough begins to form in a region where the plasma moves eastward but at a speed less than corotation. This region is labeled B. [After Spiro *et al.* (1978). Reproduced with permission of the American Geophysical Union.]

with the velocity at which the earth is rotating toward the east at that latitude. This means that the dusk terminator is moving *westward* at the same velocity as the plasma and therefore that the plasma sees solar conditions which are independent of time. As in the case here, the equatorward edge of the high-latitude convection region is frequently colocated with a region of depleted density, which usually has a sharp poleward edge and a more gradual decline at its equatorward side. The total plasma concentration depletion can exceed an order of magnitude in the F region and is called the mid-latitude trough. Examination of the plasma drift characteristics within the premidnight trough in Fig. 7.7 shows that the plasma depletion lies within a region in which there is a transition from either eastward or corotating plasma drift equatorward of the auroral zone to westward drift in the auroral zone. At such a transition zone the plasma flows very slowly and chemical recombination will have more time to reduce the ion concentration than in regions where the plasma moves more quickly or where auroral zone ionization exists. Figure 7.8 shows a schematic diagram of a modeled set of convection paths in the evening trough region which are similar to the data shown in Fig. 7.7. On the right-hand side the results of a simple chemical decay model are given for the plasma content taking into account its motion along these paths. The results confirm that plasma stagnation in the nighttime region just equatorward of the auroral zone is a likely candidate for the quiet-time trough formation. We discuss the storm-time trough in Chapter 8.

The convection paths that give rise to the trough in the premidnight sector result quite naturally from the addition of the essentially eastward corotation that dominates the flow characteristics at low latitudes and the essentially westward flow from the magnetospheric field that dominates the flow characteristics at high latitudes. Figure 7.9 shows the results of adding these two convection sources in a magnetic coordinate frame. In this figure the premidnight trough would result from the very slow flow in the nighttime region of the trajectory labeled II. The illustrated flow pattern of course has considerable variations due to the universal time effects mentioned earlier and to the temporal changes in the magnetospheric

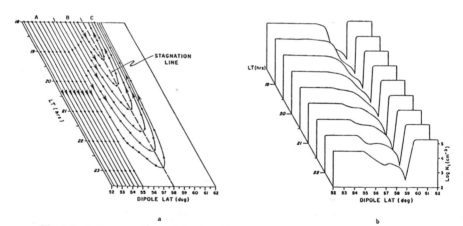

Fig. 7.8. Model convection paths having the same properties as those observed in Fig. 7.7 are shown in (a). The resulting number density profile due to the almost stagnant flow is shown in (b). [After Spiro *et al.* (1978). Reproduced with permission of the American Geophysical Union.]

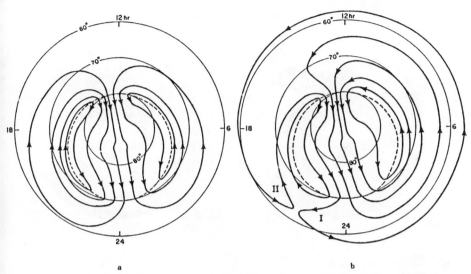

a   b

Fig. 7.9. The model convection shown in Fig. 7.7 can be created by the addition of a simple two-cell convection pattern shown in (a) with a corotation velocity. The result shown in (b) produces typical trough flow characteristics along the contour labeled II. [After Spiro *et al.* (1978). Reproduced with permission of the American Geophysical Union.]

electric field. These variations can explain the variety of latitudinal profiles of the mid-latitude trough that are observed. It should also be remembered that a mid-latitude trough is observed in regions other than the premidnight sector, so it is unlikely that the slow flow convection trajectories described here are always responsible for formation of the trough. In fact, local plasma stagnation can readily account for the mid-latitude trough only in the premidnight sector, where the trough signature and associated flow paths are easily recognizable. At other local times the mid-latitude trough may be due to low plasma densities produced elsewhere and transported to the observation region in the complex flow pattern produced by the offset between the dipole axis and the earth's rotation axis.

The rather simple concept of opposing flows from corotation and magnetospheric sources that can account for the premidnight trough also pertains inside the polar cap in the postmidnight sector. In the winter season the antisunward flow in the postmidnight region of the polar cap takes place in darkness and is directed approximately to the west. This flow is opposed by the eastward corotation velocity and can therefore, at certain longitudes and universal times, produce plasma stagnation. In the dark polar cap there are very few sources of ionization, and thus we might expect a plasma depletion to occur in a similar manner to the mid-latitude trough. Figure 7.10 shows the maximum and minimum O^+ concentrations in the southern winter hemisphere. These observations,

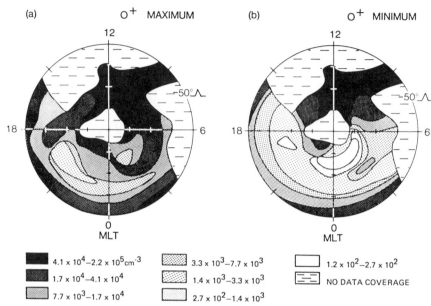

Fig. 7.10. Examination of the O⁺ concentration of the high-latitude winter ionosphere near 300 km shows two regions of plasma depletions that may result from plasma convection—the mid-latitude trough seen near 60° invariant latitude and between 1800 and 2400 h, and the high-latitude hole seen near 80° invariant latitude and 0700 hours. [From Brinton *et al.* (1978). Reproduced with permission of the American Geophysical Union.]

made on an Atmospheric Explorer satellite near 300-km altitude, clearly show the mid-latitude trough in the O⁺ concentration and, as discussed above, a large plasma depletion at latitudes above the auroral zone in the postmidnight polar cap. This polar cap depletion is called the high-latitude or polar hole and is the site of the lowest plasma concentration in the winter high-latitude ionosphere. It is a persistent feature in the winter, appears sporadically at equinox, and almost never appears in the summer. This occurrence pattern can easily be attributed to the movement of the solar terminator, which places this location always in darkness in the winter, partially in darkness at equinox, and completely in sunlight during the summer. Figure 7.11 summarizes some characteristics of the total ion concentration and its relationships to a typical high-latitude convection pattern.

7.2.2   Ion Heating Due to Collisions

When ionic and neutral particles collide, the amount of energy exchanged depends on the relative velocity between the two particles. At high latitudes this relative velocity can become quite high due to the magnetospheric convection

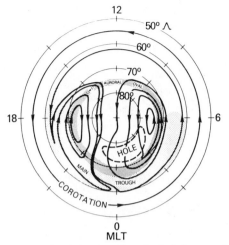

Fig. 7.11. The plasma stagnation that may produce the depletions seen in Fig. 7.10 occur when the magnetospheric convection pattern and the corotation velocity are combined as shown here. [After Brinton *et al.* (1978). Reproduced with permission of the American Geophysical Union.]

electric field. Then this energy exchange can significantly affect the ion temperature, the ion composition, and even the neutral wind.

The ion temperature at high latitudes is determined principally by frictional heating, which occurs whenever there exists a relative velocity between the ion and neutral gases, and by heat exchange with the neutral gas and electron gas. The former can be described equally well as Joule heating due to the Pedersen current $J_p = \sigma_p E$. When all these effects are considered, the ion temperature can be expressed as (St.-Maurice and Hanson, 1982),

$$T_i = T_{eq} + (m_n \phi_{in}/3k_B \psi_{in})|\mathbf{V}_i - \mathbf{V}_n|^2 \tag{7.18}$$

where

$$T_{eq} = T_n + [(m_i + m_n)\nu_{ie}/m_i \nu_{in}\psi_{in}](T_e - T_i) \tag{7.19}$$

Here the dimensionless parameters ϕ_{in} and ψ_{in} depend on the nature of the collisional interactions between the ions and the neutral gas, but above about 200 km, where O^+ collides principally with atomic oxygen, both these parameters are approximately unity.

These expressions show that the ion temperature will increase from its equilibrium value whenever there is a relative velocity between the ion and neutral gases. This relationship between the ion drift velocity and the ion temperature can easily be seen at high latitudes, where the magnetospheric electric field can rapidly produce ion velocities that greatly exceed the neutral gas velocity. One such example is shown in Fig. 7.12, where an extremely good correlation be-

AE-C   ORBIT   20193   DATE 77236

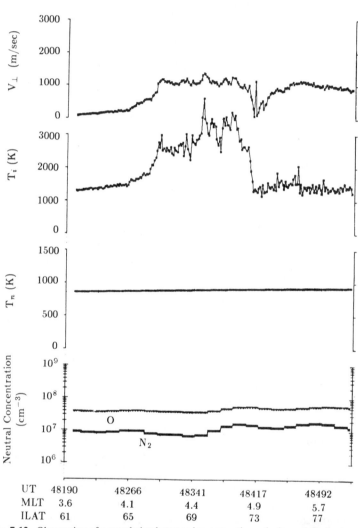

Fig. 7.12. Observation of a correlation between ion convection velocity and the ion temperature is extremely good evidence for Joule heating in the region of sunward flow. [After St.-Maurice and Hanson (1982). Reproduced with permission of the American Geophysical Union.]

tween ion temperature and ion velocity can be seen. One should keep in mind, however, that such a correlation can occur only when the neutral gas velocity is much smaller than the ion velocity. At high latitudes the same ion–neutral collisions that produce frictional heating of the ion gas also tend to set the neutral gas in motion, leading to an equalization of the ion and neutral velocities in a steady state. When this occurs the ion gas can be moving quite rapidly with almost no frictional heating. Figure 7.13 shows such an event where the ion temperature maintains its equilibrium value throughout a region of antisunward flow where the ion velocity is in excess of 1 km/s. In this example, simultaneous measurement of the neutral gas velocity shows that it also is moving at high speed and in the same direction as the ions. The momentum transfer between the plasma and the neutral gas is described in more detail in Section 7.3. As we shall see, if the high-latitude ion convection pattern remains stable for several hours, then the ion–neutral collisions can in fact set up a similar convection pattern in the neutral gas. However, if we see the evidence of frictional heating in the high-latitude polar cap region as shown in Fig. 7.12, then it is reasonable to conclude

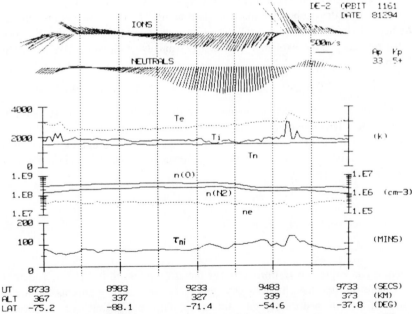

Fig. 7.13. Under stable conditions the ion and neutral velocities can become comparable, as observed in the polar cap here by DE-2. Then little Joule heating occurs even when the ion velocity is large. [After Killeen *et al.* (1984). Reproduced with permission of the American Geophysical Union.]

that the observed ion convection pattern has been established within the last hour or two and the neutrals are not yet moving with the same velocity as the ions.

7.2.3 Velocity-Dependent Recombination

In addition to frictional heating, the relative energy with which the ion and neutral particles collide can affect the chemical reaction rates of the ions. This is particularly true in the ionospheric F region, where the rates for the charge exchange reactions

$$O^+ + N_2 \rightarrow NO^+ + N \qquad \text{(with rate } k_1)$$

$$O^+ + O_2 \rightarrow O_2^+ + O \qquad \text{(with rate } k_2)$$

are extremely sensitive to the relative energy of the reacting particles. If these rates are enhanced, then the plasma density will decrease since NO^+ and O_2^+ recombine quite rapidly due to dissociative recombination. The plasma ion composition will also change if O^+ is converted to molecular species via charge exchange.

Laboratory studies of these reactions express their rates in terms of an effective temperature. The experiments are performed by passing one gas through the other in a drift tube in a manner quite similar to the drift of ions through the neutrals in the ionosphere. The most important reaction rate in the F region is the rate for charge exchange between O^+ and N_2, which can be written in the form

$$k_1 = 8.0 \times 10^{-14}(T_{eff}/300)^2 \quad cm^3/s \qquad \text{for } T_{eff} > 750 \quad K$$

To determine the effective temperature for the ionosphere, we must take into account the relative velocity of the neutral gas with respect to the ions. In the F region above about 200 km, the most crucial reactants are N_2 and O^+. If we then express the relative ion–neutral velocity in terms of the electric field in the neutral frame such that

$$\left| \mathbf{V}_{i\perp} - \mathbf{U}_\perp \right| = E'_\perp/B$$

then the effective temperature is given by

$$T_{eff} = T_n + 0.33E'^2_\perp$$

where $E'_\perp$ is expressed in millivolts per meter (Schunk et al., 1975). It can be seen therefore that the charge exchange rate k_1 increases roughly as the fourth power of the ion–neutral velocity difference. This highly sensitive variation in the charge exchange rate is reflected quite dramatically both in the absolute ion concentration and in the relative ion composition at high latitudes. Figure 7.14 shows an example of this effect in the high-latitude F region. Near 250 km O^+ is the dominant ion by at least an order of magnitude over NO^+ for almost

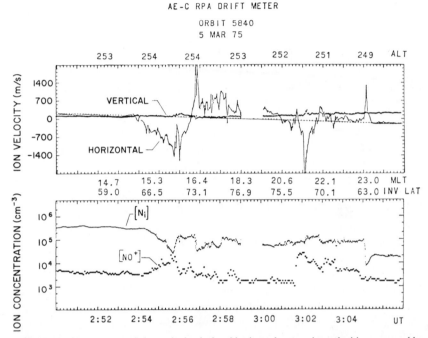

Fig. 7.14. The upper panel shows the vertical and horizontal convection velocities measured by the AE-C satellite. The lower panel shows the total ion concentration comprising O^+ and the molecular ions as well as the NO^+ concentration itself. Notice that when the horizontal ion drift is large the relative abundance of NO^+ increases due to the enhancement of the O^+ recombination rate. (Courtesy of R. Heelis and W. Hanson.)

the entire pass. In regions of large drift, however, we observe an increase in the relative concentration of NO^+ and a decrease in the total ion concentration. The decrease in total ion concentration is due to an increase in the recombination of NO^+.

7.3   Electrodynamic Forcing of the Neutral Atmosphere

7.3.1 $J \times B$ *Forcing*

The fact that the plasma is often in motion with high velocities in the polar region has important consequences for thermospheric dynamics. We cannot hope to treat this fascinating topic in the detail it deserves but will only outline some of its most important aspects.

In Chapters 3 and 5 we pointed out that the plasma acts as a drag on the thermosphere at low and middle latitudes. We showed that this drag force was equivalent to a $\mathbf{J} \times \mathbf{B}$ force on the neutral gas where

$$\mathbf{J} = \boldsymbol{\sigma} \cdot (\mathbf{U} \times \mathbf{B})$$

If this current is not divergence free, we found that electric fields build up and usually reduce the total current with the result that the $\mathbf{J} \times \mathbf{B}$ force, $\mathbf{F}_j$, decreases. Neutral winds in such a case therefore create electric fields via a *dynamo* process.

At high latitudes, the electric fields are impressed from outside the iono-sphere/thermosphere system and the possibility for a *motor* arises. In fact, the origin of the force on the neutrals is again the $\mathbf{J} \times \mathbf{B}$ force, only in the high-latitude case the expression

$$\mathbf{J} = \boldsymbol{\sigma} \cdot (\mathbf{E} + \mathbf{U} \times \mathbf{B})$$

is usually dominated by the impressed electric field $\mathbf{E}$. In the F region, where the conductivity tensor $\boldsymbol{\sigma}$ is diagonal, the force becomes

$$\mathbf{F} = \sigma_P \mathbf{E}' \times \mathbf{B}$$

where $\mathbf{E}'$ is the electric field in the neutral frame of reference. A data set taken in the evening auroral oval which shows the importance of electrodynamic forc-ing is presented in Figs. 7.15a and b. The former is an all-sky camera picture showing the location of several chemical releases made from the same rocket. The rocket was launched from Poker Flat, Alaska at 1810 local time on Febru-ary 28, 1978. The latter is a schematic diagram which describes the various features in the photograph. The two visible barium (Ba) ion clouds have been driven by the $\mathbf{E} \times \mathbf{B}$ drift at high velocity toward the west, since $\mathbf{E}$ was the usual northward electric field at that time period. The two circular strontium (Sr) neu-tral clouds have hardly moved since their release, which was at the same time and altitude (≥ 200 km) as the ion clouds. However, the trimethyl aluminum (TMA) trail shows a high-velocity region streaming in the same direction as the ion cloud at about one-third the plasma cloud velocity. More quantitative data from the same experiment are given in Fig. 7.16. The three solid tracks in the figure show sequential locations of the barium ion clouds and are labeled I205, I268, and I214, corresponding to the three ion (I) cloud altitudes in kilometers. Minute markers show the temporal progression of the three clouds, which moved parallel to the auroral oval at about 1 km/s. The two dashed lines marked N131 and N140 show the neutral (N) trail locations at the peak in the two neutral velocity profiles, which occurred at 131 and 140 km, respectively. Notice that the southern trail moved very nearly parallel to the $\mathbf{E} \times \mathbf{B}$ direction with a ve-locity equal to about 27% of the velocity of the ion cloud nearest in horizontal location. The northern trail had a speed equal to 44% of the nearest ion cloud velocity and a direction rotated slightly poleward of the $\mathbf{E} \times \mathbf{B}$ velocity. The high-

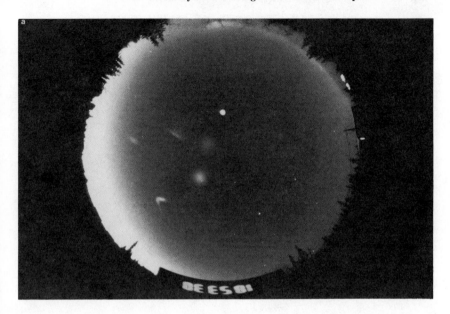

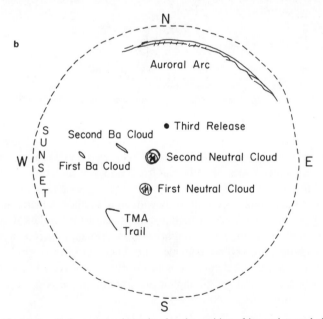

Fig. 7.15. (a) An all-sky camera photo showing the position of ion and neutral clouds in the BaTMAn experiment several minutes after the second barium release. The bright dot [see (b)] is the third barium release at the detonation time. (b) Key for identifying features in part (a).

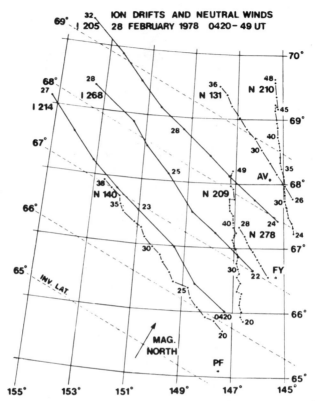

Fig. 7.16. Ground tracks of the various releases during the BaTMAn experiment shown in Fig. 7.15. [After Mikkelsen *et al.* (1981a). Reproduced with permission of the American Geophysical Union.]

altitude neutral clouds (N210, N278, and N209) were much slower and had a direction of motion considerably different from the $\mathbf{E} \times \mathbf{B}$ direction.

Some of these results may be explained in a straightforward manner. First note that since the $\mathbf{E}$ field maps uniformly with altitude, the main altitude dependence in $\mathbf{F}$ is in the σ term. The simultaneous Chatanika radar data showed that this term (primarily due to σ_P) peaked at 140-km altitude, which is in excellent agreement with the observed peaks in the neutral wind profiles. A crude estimate of the time required to accelerate the neutrals may be derived as follows. The acceleration of the neutral gas by the electromagnetic force is given by an expression of the form

$$\rho(\partial u/\partial t) \simeq |\mathbf{J} \times \mathbf{B}| = \sigma_P E B$$

where u is the zonal velocity and E is the meridional electric field component.

Since from (2.40b) the F-region Pedersen conductivity is given approximately by $\sigma_P = nm_i\nu_{in}/B^2$, this relationship may be written

$$\delta u/\delta t = (n\nu_{in}/n_n)(E/B)$$

where we have assumed that the ion mass and the neutral mass are the same. Now since ν_{in} is proportional to the neutral density n_n, the latter cancels out and we have interesting result that the neutral acceleration depends only on the plasma density n. Using $\nu_{in} \simeq 5 \times 10^{-10}n_n$ (see Appendix B), we can estimate the time for acceleration to a velocity $\delta u = \frac{1}{3}(E/B)$ to be $\delta t_{1/3} \simeq 10^{10}/15n$. For a peak density $n = 3 \times 10^5$ cm^{-3}, as was the case in the Alaskan experiment, $\delta t_{1/3} \simeq 2000$ s, which is quite short.

This estimate ignores many other terms in the equation of motion, but it does show that $\mathbf{J} \times \mathbf{B}$ forcing is significant. Viscosity is particularly important in the dynamics since it spreads the strong height-dependent $\mathbf{J} \times \mathbf{B}$ forcing to other altitudes. Mikkelsen *et al.* (1981a,b) have studied this event with a two-dimensional numerical model using the local forcing deduced from the radar and rocket data. They found good agreement with the calculated and observed zonal wind (see Fig. 7.17). Their model included the pressure gradient, a curvature effect due to the shape of the auroral oval and advection of momentum in the vertical and meridional directions. The agreement was not quite as good between

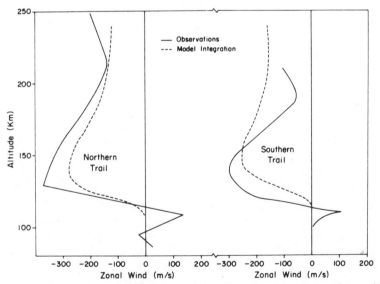

Fig. 7.17. Comparison of the observed zonal wind profiles at the northern and southern positions on February 28 with the results of the modeling. [After Mikkelsen *et al.* (1981b). Reproduced with permission of the American Geophysical Union.]

the observations and the model wind in the meridional component. Furthermore, when applied to a second data set which was obtained 3 days later on March 2, 1978, the agreement was poor in both components. The primary difference in the two events was the *location* of the forcing in longitude. In the February 28 data shown in Figs. 7.15–7.17, a substorm occurred near College, Alaska and local forcing was an adequate approximation. On March 2, however, the substorm was far to the east of College. Since the observed winds were very similar, Mikkelsen *et al.* concluded that advection of momentum in the zonal direction was crucial and that a three-dimensional model was essential for progress in understanding electrodynamic forcing of the upper atmosphere.

7.3.2   Global Observations and Simulations

Heppner and Miller (1982) collected a number of barium neutral cloud measurements in the thermosphere and compared them to several convection models. One such comparison is shown in Fig. 7.18. The wind vectors are superimposed on a representative convection pattern in Fig. 7.18a at the observed location. In Fig. 7.18b the wind locations were arbitrarily shifted by 2 h to take into account the sluggish response of the neutrals (e.g., see the calculation of δt above). The vectors show a much clearer relationship to the ion flow in Fig. 7.18b. One obvious difference between the plasma flow and the measured winds, even in Fig. 7.18b, is that the strong neutral flow seems to continue at latitudes well

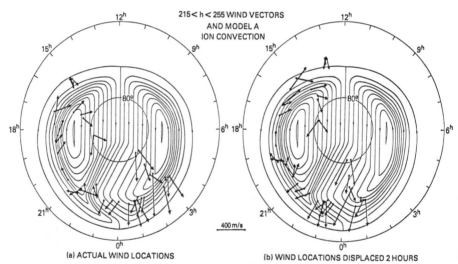

Fig. 7.18. (a) Compilation of barium neutral cloud velocities for a number of local times superimposed on a plasma convection model. In (b) the same data are plotted 2 h earlier in local time. [After Heppner and Miller (1982). Reproduced with permission of the American Geophysical Union.]

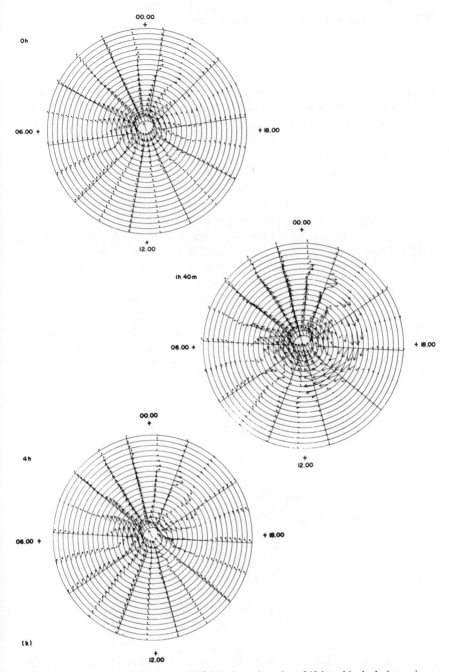

Fig. 7.19. Plot of wind vectors from 50° latitude to the pole at 240-km altitude during a simulated substorm. Plots are shown at intervals of 0, 1 h 40 min, and 4 h. Scale: an arrow the same length as 2° in latitude corresponds to 160 m/s. [After Fuller-Rowell and Rees (1981). Reproduced with permission of Pergamon Press.]

below the auroral oval. This occurs because there is no comparable force to "turn" the neutral wind when the flow leaves the sunward convection part of the two-celled plasma flow pattern. Notice that something of a flywheel effect will occur in this case since the wind in the subauroral region can create electric fields via the F-region dynamo process discussed in Chapters 3 and 5. This effect has been suggested by Gonzales *et al.* (1978) to explain some of the anomalous electric field observations in the evening sector (see Chapter 5).

Although the primary electrodynamic effect seems to be momentum transfer via the $\mathbf{J} \times \mathbf{B}$ force, changes in pressure due to Joule heating of the neutral gas discussed in Section 7.2 also affect the neutral atmospheric dynamics. This effect may be more important in producing vertical upwelling and neutral composition changes than in the generation of strong neutral winds.

Three-dimensional models of neutral wind forcing due to realistic electric field patterns are now available which include all the important physical processes. An example of such a calculation is given in the series of plots in Fig. 7.19, which are polar diagrams of the wind velocity at 240-km altitude for various time intervals after a model magnetic substorm which starts at $t = 0$ and lasts for 2 h. The initial plot at $t + 20$ min is representative of the solar forced winds. By $t + 100$ min a clear vortex has formed in the dusk sector where wind speeds exceed 300 m/s. Very low velocities are found in the morning hours where the substorm effects are counteracting the solar forcing. Two hours after the storm ended, the pattern returned virtually to the initial state.

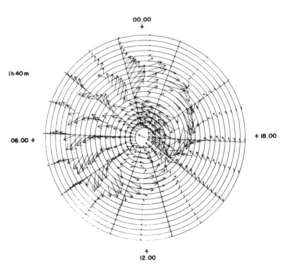

Fig. 7.20. Plot of wind vectors from 50° latitude to the pole at 120-km altitude during a simulated substorm. The plot shown is at a time of 1 h 40 min after substorm onset. Scale: an arrow the same length as 2° in latitude corresponds to 40 m/s. [After Fuller-Rowell and Rees (1981). Reproduced with permission of Pergamon Press.]

Winds in the E region for the same simulation are shown in Fig. 7.20 at $t +$ 100 min. Note that the scale differs by a factor of four relative to the F-region calculation. The wind vectors show a clear double-cell pattern with a line of symmetry running from 0300 to 1500. The symmetry of the two vortices is due to the lower value of pure solar forcing at E-region heights. The rotation of the pattern away from a noon–midnight symmetry is due to the importance of the Hall conductivity in the $\mathbf{J} \times \mathbf{B}$ force.

With computing power increasing quickly and becoming more readily available, a number of the three-dimensional models now exist. Models such as those discussed here will prove quite valuable in developing further insights into the electrodynamic forcing of the neutral atmosphere.

7.4 Summary

In this chapter we have discussed a variety of observed phenomena in the high-latitude ionosphere and a number of physical processes that are important in this region. It should be clear that some knowledge of the prior convective history of the plasma and the magnetic field topology it has encountered is essential to understanding or explaining many phenomena.

When considering events below the F peak, the local solar zenith angle, the time at which plasma may encounter the auroral ionization source, the magnitude of the drifts along its convective path, and the time such drifts would allow the plasma to reside in a production region must all be taken into account. The magnetic field topology is a minor consideration. The mid-latitude trough, for example, can be understood in this manner. Above the F peak the magnetic field topology is an important consideration, as is the plasma temperature. They will both affect the field-aligned motion of the plasma and its topside distribution. The light ion trough can be understood with these considerations. Progress toward a more complete understanding of the high-latitude ionospheric plasma will undoubtedly require multipoint measurements that provide some information about the time history of the plasma motion and the ionization sources it encounters. Many data sets of this nature are available from simultaneous measurements made by rockets, satellites, and ground-based instrumentation, but they must be integrated into a study that includes a versatile computer model in order to sort out the most important processes.

The effect of plasma dynamics on the neutral atmosphere can also be quite dramatic. Numerous measurements have shown extremely strong winds in the high-latitude thermosphere. Simultaneous plasma velocity measurements show clearly that these winds are driven by electric fields which are impressed from above through interactions between the earth's magnetosphere and the solar wind. This control of the earth's upper atmosphere by interplanetary processes is a fascinating example of the interaction between a flowing plasma and a planetary atmosphere.

References

Banks, P. M., and Holzer, T. E. (1969). Features of plasma transport in the upper atmosphere. *J. Geophys. Res.* **74**, 6304.

Brinton, H. C., Grebowsky, J. M., and Brace, L. H. (1978). The high-latitude winter F region at 300 km: Thermal plasma observations from AE-C. *JGR, J. Geophys. Res.* **83**, 4767.

Chappell, C. R., Green, J. L., Johnson, J. F. E., and Waite, J. H., Jr. (1982). Pitch angle variations in magnetospheric thermal plasma—initial observations from Dynamics Explorer-1. *Geophys. Res. Lett.* **9**, 937.

Fuller-Rowell, T. J., and Rees, D. (1981). A three-dimensional simulation of the global dynamical response of the thermosphere to a geomagnetic substorm. *J. Atmos. Terr. Phys.* **43**, 701.

Gonzales, C. A., Kelley, M. C., Carpenter, L. A., and Holzworth, R. H. (1978). Evidence for a magnetospheric effect on mid-latitude electric fields. *JGR, J. Geophys. Res.* **83**, 4397.

Heppner, J. P., and Miller, M. L. (1982). Thermospheric winds at high latitudes from chemical release observations. *JGR, J. Geophys. Res.* **87**, 1633.

Hoffman, J. H., and Dodson, W. H. (1980). Light ion concentrations and fluxes in the polar regions during magnetically quiet times. *JGR, J. Geophys. Res.* **85**, 626.

Holzer, T. E., Fedder, J. A., and Banks, P. M. (1971). A comparison of kinetic and hydrodynamic models of an expanding ion exosphere. *J. Geophys. Res.* **76**, 2453.

Killeen, T. L., Hays, P. B., Carignan, G. R., Heelis, R. A., Hanson, W. B., Spencer, N. W., and Brace, L. H. (1984). Ion–neutral coupling in the high latitude F-region: Evaluation of ion heating terms from Dynamics Explorer-2. *JGR, J. Geophys. Res.* **89**, 7495.

Lemaire, J., and Scherer, M. (1970). Simple model for an ion-exosphere. *Planet. Space Sci.* **18**, 103.

Mikkelsen, J. S., Jorgensen, T. S., Kelley, M. C., Larsen, M. F., Pereira, E., and Vickrey, J. F. (1981a). Neutral winds and electric fields in the dusk auroral oval. 1. Measurements. *JGR, J. Geophys. Res.* **86**, 1513.

Mikkelsen, J. S., Jorgensen, T. S., Kelley, M. C., Larsen, M. F., and Pereira, E. (1981b). Neutral winds and electric fields in the dusk auroral oval. 2. Theory and model. *JGR, J. Geophys. Res.* **86**, 1525.

Parker, R. N. (1958). Dynamics of the interplanetary gas and magnetic fields. *Astrophys. J.* **128**, 664.

Raitt, W. J., Schunk, R. W., and Banks, P. M. (1975). A comparison of the temperature and density structure in high and low speed thermal proton flows. *Planet. Space Sci.* **23**, 1103.

Rishbeth, H., and Garriott, O. K. (1969). "Introduction to Ionospheric Physics," Int. Geophys. Ser., Vol. 14. Academic Press, New York.

St.-Maurice, J. P., and Hanson, W. B. (1982). Ion frictional heating at high latitudes and its possible use for an in-situ determination of neutral thermospheric winds and temperatures. *JGR, J. Geophys. Res.* **87**, 7580.

Schunk, R. W. (1977). Mathematical structure of transport equations for multispecies flows. *Rev. Geophys. Space Phys.* **15**, 429.

Schunk, R. W., Raitt, W. J., and Banks, P. M. (1975). Effect of electric fields on the daytime high-latitude E and F regions. *JGR, J. Geophys. Res.* **80**, 3121.

Singh, N., and Schunk, R. W. (1982). Numerical calculations relevant to the initial expansion of the polar wind. *JGR, J. Geophys. Res.* **87**, 9154.

Sojka, J. J., Raitt, W. J., and Schunk, R. W. (1979). Effect of displaced geomagnetic and geographic poles on high-latitude plasma convection and ionospheric depletions. *JGR, J. Geophys. Res.* **84**, 5943.

Spiro, R. W., Heelis, R. A., and Hanson, W. B. (1978). Ion convection and the formation of the mid-latitude F-region ionization trough. *JGR, J. Geophys. Res.* **83**, 4255.

Chapter 8 | Instabilities and Structure in the High-Latitude Ionosphere

The high-latitude sector is extremely rich in plasma instabilities and other processes which act to create structure in the ionosphere. Our primary interest lies in the horizontal variation of plasma density and electric fields. Since the magnetic field is nearly vertical, the horizontal structure is equivalent to variations perpendicular to the magnetic field. We use the generic term "structure" since terms such as "waves" or plasma density "irregularities" conjure up specific sources. In fact, a very long list of processes contribute to the generation of horizontal structures in the plasma density and velocity field of the high-latitude ionosphere. It is essentially impossible to treat the topic in its entirety since our understanding is developing very quickly. Our hope instead is to give a reasonable hint at the breadth of the phenomena involved and, within the various subtopics, to treat a few important processes in some detail. As was the case in Chapter 4, we start with the F region and follow with the E region in the second portion of the chapter.

8.1 Planetary and Large-Scale Structures in the High-Latitude F Region

In this section we review the physical processes which create horizontal variations in the high-latitude plasma density at the largest scales. We somewhat arbitrarily define the planetary scale to be larger than 1000 km and large-scale processes to have perpendicular wavelengths in the range $30\,\mathrm{km} \leq \lambda \leq 1000\,\mathrm{km}$. In these regimes production, loss, and transport dominate the ionospheric processes.

Ionospheric plasma instabilities are not very important in this scale size regime (although some local ionospheric processes do contribute at the lower end). Of course, this is not to say that *magnetospheric* plasma instabilities are unimportant, since they control some of the precipitation and turbulence in the flow patterns which are impressed on the ionosphere. Detailed study of the associated magnetospheric physics is beyond the scope of the present text, although we do refer in passing to some of the most important processes which occur.

8.1.1   Planetary Scale Structure in the High-Latitude Ionosphere

We have already discussed variations of plasma flow and plasma density variations at planetary scales in Chapters 6 and 7. We briefly summarize the material here for completeness. The velocity field applied to the ionosphere from the solar wind–magnetosphere interaction has a number of discernible patterns at planetary scales. However, which of these patterns applies at a given time is highly dependent upon conditions in the interplanetary medium. The most crucial parameter is the sign of the north–south component (B_z) of the interplanetary magnetic field (IMF). When the IMF is southward for any extended time (e.g., tens of minutes) the classic two-celled convection pattern is imposed on the ionosphere. The magnetic field lines which thread the high-latitude ionosphere spread out enormously as they recede from the earth. This means that the pattern in the ionosphere represents a focused version of the electrodynamical processes which create the electric field pattern. Unfortunately, the topological mapping that occurs is quite complicated since some field lines are open and some are closed. The ionospheric pattern actually is somewhat less complicated than the source which produces it.

For steady B_z south, the two-cell ionospheric flow pattern is more or less fixed with respect to the sun–earth line. The earth and its neutral atmosphere rotate under this plasma flow pattern once a day (ignoring acceleration of the neutrals by the plasma). This creates a planetary scale, diurnally varying plasma flow field in the ionosphere. Now even with B_z held south, the classic symmetric two-celled flow shifts with respect to the sun–earth line as the other components of the IMF vary (particularly the component parallel to the earth's orbit, B_y). This shifting of the flow field can occur within one or two Alfven travel times from ionospheric altitudes to the generator. Likewise as B_z and the velocity of the solar wind change, the rate of energy transfer to the magnetosphere varies from minute to minute and a flow field which is highly variable in space and time results.

When B_z changes sign, the major source of energy transfer ceases but other effects take over. A viscous interaction seems to create a small two-cell pattern and the connection of field lines to the IMF in regions of the magnetosphere far from the ecliptic plane also creates multiple cells with planetary scales (Burke *et al.*, 1979).

To gauge the effect of these complex flow fields on plasma content we first need to discuss the processes which create and destroy plasma on planetary scales. The most important source is photoionization by sunlight. In a nonrotating frame this region is bounded by the terminator, which moves across the polar region on a seasonal basis. On the dark side of this line, recombination rapidly destroys plasma below 200 km but only very slowly erodes the F-peak region. The time constant is roughly 1 hr at 300 km.

Now when the planetary scale flow is imposed on this source and loss pattern, it is clear that solar plasma can be transported for vast distances into and clear through regions of total darkness before recombination can play much of a role. In this way, planetary scale structuring of the plasma occurs which is much more complex and interesting than the simple terminator effect would have been.

A more subtle process can create planetary scale deletions of plasma as well. Since the dipole magnetic axis of the earth is offset by $11°$ from the rotation axis and the plasma flow is organized by the magnetic geometry, in the winter time some convection patterns have flux tubes which are never illuminated by sunlight. Then, very deep plasma depletions can occur due to recombination, yielding peak plasma densities as low as 10^2 cm^{-3} with He$^+$ the dominant ion (see Fig. 7.11).

The two-cell convection pattern is associated with another planetary scale plasma source, impact ionization by particle precipitation in the auroral oval. From a visual perspective, this band of light around the polar region expands and thickens with increasing B_z south and shrinks when B_z is northward. Much of the plasma in this oval is created so low in the atmosphere (≤ 200 km) that it is short-lived. Nonetheless, the lowest energy precipitating particles produce plasma high enough in altitude to create an important F-layer plasma source—particularly in winter.

8.1.2 Some Effects of Plasma Transport and Loss on the Large-Scale Horizontal Structure of the Ionosphere

One of the most fascinating aspects of the high-latitude ionosphere is its interaction with the various magnetospheric regions to which it is connected by magnetic field lines. In portions of the ionosphere which are not sunlit, the influx of precipitating particles is one of the dominant sources of the ionospheric plasma; the other major source is transport from either a sunlit region or a region where particles are precipitating. One might at first suspect that solar-produced plasma would display very little structure in the F region since the ionization is long lived and the source is smoothly varying. However, since the flow field varies drastically in time and space, even solar-produced plasma may become horizontally structured. It is, in fact, difficult to separate particle precipitation zones from electric field patterns since they are intimately related. We concentrate on the latter in this section and then follow with some comment on precipitation in the next section.

As discussed in Chapters 6 and 7, perpendicular electric fields occur throughout the auroral zone and polar cap. Their existence results in transport of F-region plasma from production zones to areas where one might not expect to find much plasma at all. A good example is found in the winter polar cap. F-region plasma is not produced by sunlight at all in this region and there is only a very weak particle input, called "polar rain," when B_z is southward. However, observations show considerable structured plasma in the polar cap. The red light emission at 6300 Å due to recombination is sufficiently large to make this plasma visible to image-intensified camera systems. In Chapter 5 we showed that localized low-density regions could be observed as an absence of these emissions and pointed out that these airglow depletions were due to equatorial spread F wedges or bubbles. At high latitudes it is the enhancements of plasma which are detected and which can be tracked by optical techniques. The all-sky camera photographs in Fig. 8.1 show such a plasma patch marching across the field of view over Thule, Greenland. These structures are of the order of 1000 km across. In Section 8.2 we study the plasma instabilities which may develop on the edges of these regions of enhanced plasma density, but here we are interested in their origin and lifetime. This is determined by both recombination and diffusion. As noted by Rishbeth and Garriott (1969), a Chapman-like F layer subject to diffusion, gravity, and loss will preserve its shape and decay at a rate given by $e^{-\beta(z_0)t}$ where $\beta(z_0)$ is the recombination rate at the altitude z_0 of the peak in the plasma density. The decay time constant is about 1 h. Numerical calculations verify this behavior (Schunk *et al.*, 1976; Schunk and Sojka, 1987). The typical dawn–dusk electric field across the polar cap with B_z south is 25 mV/m, which corresponds to 500 m/s velocity. In 1 h then a convecting patch can move 1800 km, just over 16° of latitude. Clearly, in such a rapid flow pattern both the dayside auroral oval and the sunlit dayside of the high-latitude zone can supply plasma to the dark polar cap regions with only modest loss due to recombination, gravity, and parallel diffusion.

Diffusion of plasma perpendicular to **B** is much more complex and interesting and we reserve Section 8.2.3 to its study. A quick estimate, however, shows that large structures will easily survive transport over long distances without perpendicular diffusive loss. The time scale for perpendicular diffusion, τ_D, is given by

$$\tau_D = (k^2 D_\perp)^{-1}$$

For the F region the classical diffusion coefficient has an upper limit given by the ion diffusion coefficient $D_{i\perp}$ (see Section 8.2.3), which has typical values in the range 10–50 m^2/s. Even for $D_\perp = 50$ m^2/s, a 10-km wavelength structure has a time constant of 14 h. The evolution of large-scale features is therefore

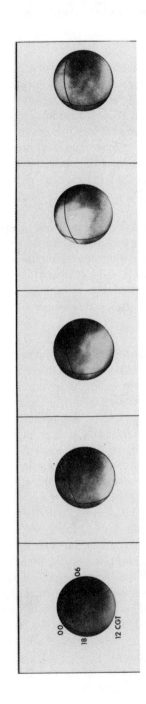

DRIFT OF POLAR CAP IONIZATION PATCH

THULE, 22 JANUARY 1982

6300 Å ALLSKY PHOTOMETER IMAGES

10:06 10:11 10:21 10:36 10:41 UT

Fig. 8.1. All-sky (155° field of view) 6300-Å images at 5-min intervals which illustrate large-scale patch structure and drift in the polar cap. The dawn–dusk and noon–midnight meridians are projected into the images at a height of 250 km. [After Weber *et al.* (1986). Reproduced with permission of the American Geophysical Union.]

controlled by gravity, production, loss, and transport rather than perpendicular diffusion.

We still need to discuss how the patches shown in Fig. 8.1 are formed. We postpone for the moment structured sources of ionization to consider only electric field effects. For example, large-scale ($\lambda \le 1000$ km) organization of the F-region plasma can occur when the flow field has spatial variations in that same scale. Such plasma structure can even evolve out of a uniform sunlit region if different portions of the plasma move into dark areas at different speeds. As shown in Fig. 2.2, this advection effect can cause the local plasma density to vary if

$$\partial n/\partial t = -(\mathbf{V} \cdot \nabla_\perp n) \ne 0 \tag{8.1}$$

even when the production and loss terms vanish. Equation (8.1) holds for an incompressible fluid, a valid approximation for the F region. For sunlit conditions $\nabla_\perp n$ is determined by the solar depression angle, and the typical perpendicular gradient scale length $L_\perp = [(1/n)(dn/dx)]^{-1}$ is several hundred kilometers. In (8.1) $\mathbf{V} = \mathbf{E} \times \mathbf{B}/B^2$ where $\mathbf{E}$ is the applied electric field. The "growth rate" $\gamma_s(k)$ for structuring at a given wave number $k > (L_\perp)^{-1}$ is then given by

$$\gamma_s(k) = E(k)/L_\perp B \tag{8.2}$$

where $E(k)$ is the value of the electric field component orthogonal to $\nabla_\perp n$. This expression shows that density structure will always arise at some k value when the flow field has structure at that value of k, since a weak horizontal gradient in the plasma density almost always exists due to variations in solar illumination, neutral winds, etc. on a planetary scale. A graphic illustration of plasma structuring by a flow field is shown in Fig. 8.2. Here an initially cylindrical plasma blob is placed in a spatially varying flow characteristic of the dusk-side auroral oval. With time, the structure becomes very elongated in the magnetic east–west direction. Since the flow is incompressible, the elongation results in a very short scale size in the meridional direction. If the flow field is turbulent, as it usually is, patches may be formed and be carried off by the mean flow into the polar cap.

Plasma loss through recombination can also create structure when coupled to a spatially varying flow field. This process was discussed in detail in Chapter 7, where the mid-latitude plasma trough and the polar hole phenomena were shown to occur when the plasma remains for a long time on flux tubes with no sunlight and no particle precipitation. This can occur on flux tubes which flow west at the same speed as the earth rotates (the trough) or which circulate around a vortex that is entirely in darkness (the polar hole). Then even a slow decay rate is sufficient to deplete the plasma density.

The mid-latitude nightside trough can become particularly deep during magnetic storms, when very large electric fields have been observed to build up at

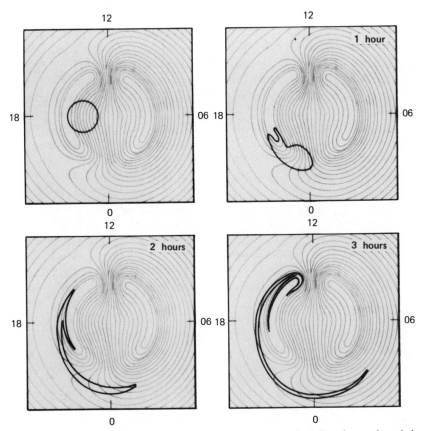

Fig. 8.2. Distortion of a circular blob of ionization as it convects from the polar cap through the auroral zone. The first panel shows the initial conditions and the assumed convection model. [After Robinson *et al.* (1985). Reproduced with permission of the American Geophysical Union.]

the equatorward edge of the plasma sheet–ring current system in the magneto-sphere (Smiddy *et al.*, 1977). This electric field points radially outward from the earth in the equatorial plane. Mapped to the ionosphere, this localized electric field is poleward and causes intense ionospheric flow toward the dusk terminator. In fact, the largest ionospheric electric field ever reported, 350 mV/m, occurred in such an event (Rich *et al.*, 1980). Satellite data taken during such an event are reproduced in Fig. 8.3a. The top panels show the intense localized electric field (250 mV/m) near $L = 4$ ($60°$ invariant latitude) with magnetic field, den-sity, and energetic particle influx data plotted below. The large pulse corresponds to a potential drop of about 25 keV across the boundary. Notice that the iono-spheric plasma composition changes from H^+ at lower latitudes to O^+ at higher

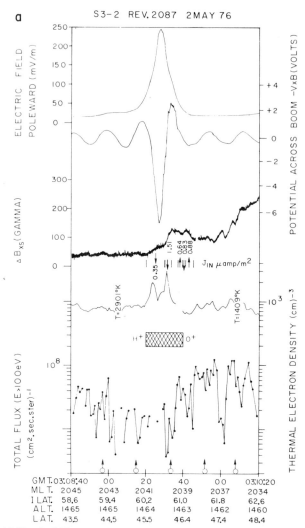

Fig. 8.3. (a) Electric field, magnetic field, thermal electron, and energetic electron data obtained on the S3-2 satellite near the intense poleward electric field for revolution 2087 on May 2, 1976. The transition from H^+ to O^+ (shaded region) is coincident with the strong electric field. The calculated field-aligned current (J_{IN}) is displayed under the magnetometer trace with arrows indicating current into (downward arrow) or out of (upward arrow) the ionosphere. [After Rich *et al.* (1980). Reproduced with permission of the American Geophysical Union.] (b) Millstone Hill radar-measured electric field and electron density profiles at an altitude of 309 km (top) and corresponding electron and ion temperature profiles (bottom). Both parts use a coordinate system centered at the maximum in the electric field. [After Providakes *et al.* (1989). Reproduced with permission of the American Geophysical Union.]

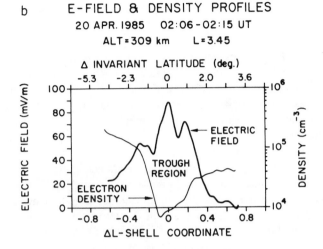

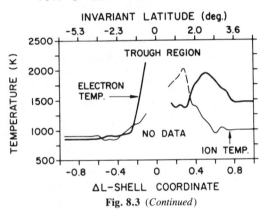

Fig. 8.3 (*Continued*)

latitudes at the same location. At ionospheric altitudes the width of this high-flow region is between 50 and 200 km. Although small field-aligned currents were detected during this pass, the current is usually much less (Providakes *et al.*, 1989; Pfaff, personal communication, 1989).

In such a region the electric field-enhanced recombination effect discussed in Chapter 7 then acts to help create a deep decrease in the electron density in this sector. An example of such an event is shown in the upper panel of Fig. 8.3b, where the plasma density and the meridional electric field are plotted from data taken during an azimuth scan of the Millstone Hill radar (Providakes *et al.*,

1989). This event occurred during a large solar minimum magnetic storm in which auroras were seen in the Washington, D.C. area of the United States, which is well south of the radar location. There is a clear anticorrelation between the observed plasma density and the flow velocity, which is at least in part due to the velocity-dependent recombination rate. The ion and electron temperatures plotted in the lower panel are also elevated in the trough region. This deep trough is in some sense the ionospheric image of the interface region which bounds two quite different plasma populations in the magnetosphere: the cool, dense, co-rotating plasmasphere and the hot, tenuous, rapidly moving plasma sheet. It is not surprising then that a number of interesting phenomena occur at this location. The extremely large localized electric fields mentioned above are one such phenomenon, and it is interesting to note that the potential difference across the region is the order of the temperature difference between the plasmasphere and the plasma sheet. A complete discussion of this interesting region is beyond the scope of this text but we can touch on a few topics of particular interest.

The ionospheric region in contact with the plasma sheet is rendered visible by the widespread particle precipitation which causes the atmospheric emissions referred to as the diffuse aurora. At times the equatorward edge of the diffuse aurora is structured in very interesting patterns (Lui *et al.*, 1982; Kelley, 1986). Two photographs taken in visible light from the DMSP satellite (800-km altitude) which illustrate this effect are shown in Fig. 8.4. Note that the equatorial edge of the diffuse aurora is scalloped in a highly nonlinear fashion. Typical observed wavelengths of this feature range from 200 to 800 km. This seems to be a clear example of ionospheric F-region structuring due to a magnetospheric process. The argument is as follows. Since the plasma sheet is the source of diffuse auroral precipitation, if the inner edge of the plasma sheet is distorted the light emissions and the production of plasma (see Section 8.1.3) in the ionosphere will mirror that distortion. Consequently, the F-region plasma will take on a horizontal structure with the same scale size. These relationships have been verified by simultaneous observations of the undulations by radar and optical means (Providakes *et al.*, 1989). This longitudinal structure will be superimposed on the latitudinal structure of the trough itself.

Indeed, ionospheric observations of plasma at the poleward edge of the ionospheric trough also show that it is often structured. One example of the latitudinal structure of this region is given in Fig. 8.5, which shows the development and meridional motion of enhancements in the F-region plasma density. Initially there is a dense E-region layer beneath the F-region structure, indicating that it is colocated with precipitating electrons. As the plasma blob drifted equatorward the plasma density became quite low in the E region below it. A plasma trough separates this high-latitude plasma from the solar-produced plasma in the equatorward region. The similar boundary structure shown in Fig. 8.6a was also probed with an east–west scan of the Chatanika radar, data from which are

DMSP DATA: 1227
May 2, 1976

DMSP DATA: 1227
May 2, 1976

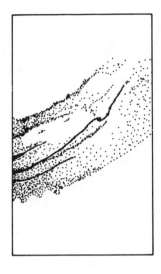

DMSP DATA: 0203
May 3, 1976

DMSP DATA: 0203
May 3, 1976

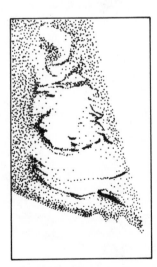

Fig. 8.4. DMSP photos showing well-developed undulations of the equatorward edge of the diffuse aurora. An artist's conception is shown on the left-hand side of each panel. [After Kelley (1986). Reproduced with permission of the American Geophysical Union.]

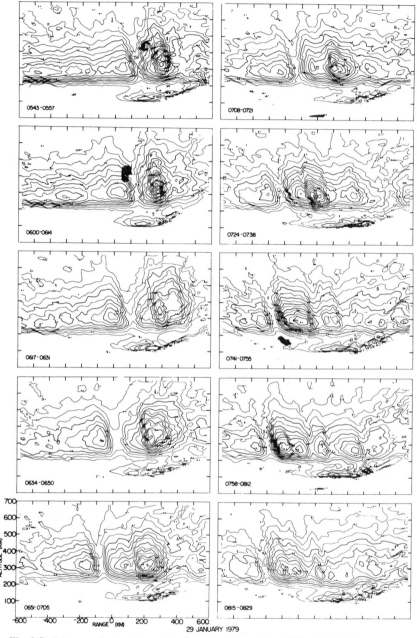

Fig. 8.5. Sequence of radar electron density contour maps from 0543 to 0829 UT, January 29, 1979. [After Weber *et al.* (1985). Reproduced with permission of the American Geophysical Union.]

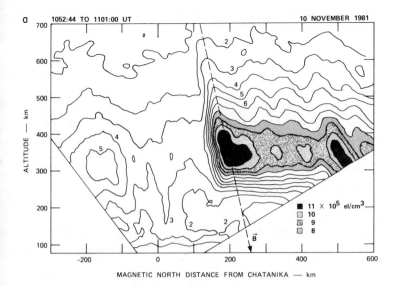

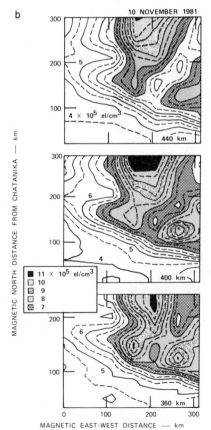

MAGNETIC EAST-WEST DISTANCE — km

Fig. 8.6. (a) Latitudinal distribution of plasma density at the edge of the auroral zone on November 10, 1981. (b) Contour maps of plasma density distribution (at 360-, 400-, and 440-km altitudes) that reveal east–west plasma structure at large scale sizes during the event presented in (a). (Figures courtesy of R. Tsunoda.)

shown in Fig. 8.6b. The plasma density contours are clearly modulated in the zonal direction with a wavelength of about 200 km. A detailed study of this event by Tsunoda (1989) shows that ionospheric instabilities are not the origin of the east–west density modulation. Such boundary blobs are also observed on the poleward edge of the auroral oval (Robinson et al., 1984). We have discussed these boundary effects in this section since variations in the magnetospheric flow field most likely organize the hot plasma sheet plasma, which then imposes a mirrorlike structure in the ionosphere via impact ionization. There is evidence that the undulations on the equatorward edge of the auroral oval seen in high-altitude photographs (e.g., Fig. 8.4) occur at the time when the large localized electric fields exist and that the latter drive some magnetospheric instability (e.g., Kelvin–Helmholtz) which causes undulations (Kelley, 1986).

The localized temperature enhancements in the trough shown in Fig. 8.3b are also very interesting and are not completely understood at this time. One theory (Cole, 1965) is that the ionospheric electrons are in good thermal contact with the magnetosphere along the magnetic field lines. In the trough, where the electron density is low, this input of heat raises the electron temperature more than, say, in the region poleward of the trough, where the heat input is also large but the electron density is high. The enhanced ion temperatures may be explained by a combination of thermal exchange with electrons and Joule heating due to the large electric field in the region of interest. Care must be exercised in interpreting ion temperatures from radar data, since the ion composition changes, shear flow on the sampled volume, and anisotropic ion distribution functions can all lead to incorrect interpretation of the incoherent scatter spectrum (Providakes et al., 1989).

8.1.3   Horizontal Plasma Variations Due to Localized Plasma Production and Heating

When energetic particles precipitate into the atmosphere, energy loss occurs primarily via ionizing collisions. To a crude first approximation, for every 35 eV lost, one ion–electron pair is produced. A single kilo-electron-volt electron would thus yield about 30 pairs. The type of ion produced depends on the altitude where the collision occurs, since the atmospheric constituents vary with height. The initial particle energy determines how deeply it can penetrate into the atmosphere. For F-region physics we are most concerned with "soft" particles, that is, electrons with energy less than about 500 eV, since they deposit their energy at the highest altitudes. Vertical transport by diffusion is crucial, however, since even soft electrons produce plasma at relatively low altitudes. Electron heat conduction is also very important since secondary electrons are much more energetic than the ionospheric electrons and heat the latter quickly. This can create a redistribution of plasma at high altitudes which mirrors the horizontal precipitation pattern.

As discussed briefly in Chapter 1, soft electrons most copiously precipitate in the dayside cusp region, where magnetosheath plasma is in direct contact with the ionosphere. This occurs in the dayside auroral oval at invariant latitudes in the range 70–74°. Our interest here is not so much in the ionization process, which is more the topic of classical aeronomy (see, for example, Banks and Kockarts, 1973), as in the resulting horizontal variation in plasma content and structure. Strong evidence that the equatorward boundary of structured plasma in the dayside cusp region is colocated with the equatorward boundary of 300-eV electron flux was given by Dyson and Winningham (1974). Rocket data from the cusp region are even more striking. In Fig. 8.7 the plasma density and the total electron energy flux below 1 keV are plotted along a rocket trajectory (Kelley *et al.*, 1982a). The two waveforms are clearly related and display an outer scale in the range $\lambda = 75-100$ km. Identifying a break point in the spectrum at $k_0 = 2\pi/30$ km, we see that the power spectral density of the plasma

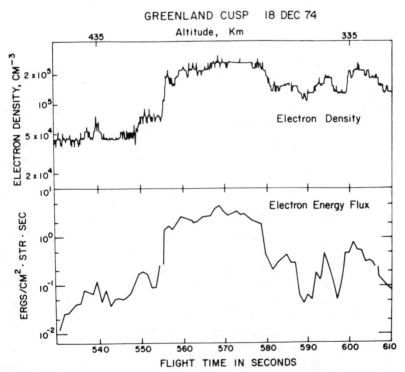

Fig. 8.7. Simultaneous *in situ* measurements of electron density and precipitating auroral energy flux in the polar cusp. [After Kelley *et al.* (1982a). Reproduced with permission of the American Geophysical Union.]

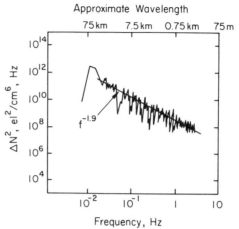

Fig. 8.8. Spectrum of irregularities obtained from analysis of the electron density measurements shown in Fig. 8.7. [After Kelley *et al.* (1982a). Reproduced with permission of the American Geophysical Union.]

density plotted in Fig. 8.8 has a maximum at $k < k_0$ and displays a power law dependence of the form -1.9 ± 0.2 for $k > k_0$.

This good correlation between plasma density and electron flux is at first glance somewhat surprising, since the altitude of the observation is about a hundred kilometers higher than the altitude where most of the ionization occurs. This correlation, however, has been confirmed in another dayside auroral oval rocket experiment using the results of several independent plasma density instruments, so there is no question of contamination in the probe response. The particle flux and plasma density data from this latter flight are shown in Fig. 8.9 and clearly show a colocation of structured F-region plasma and particle precipitation zones. Roble and Rees (1977) have made a time-dependent model for the soft electron flux case in which the typical energy is the order of 100 eV. We may use this calculation to investigate the time scale required to build up a density increase of 1.5×10^5 cm^{-3} at about 375 km as required by the data in Fig. 8.7. Roble and Rees found a local production rate of 300 cm^{-3} s^{-1} for a total energy input of 1 erg/cm·s. For 5 ergs, then, as indicated by the particle detector data, the rate will be about 1.5×10^3 cm^{-3} s^{-1}. Roughly 100 s is therefore required. As another example, the enhancements in the data of Fig. 8.9 have been quantitatively modeled by LaBelle *et al.* (1989) assuming a steady electron flux for about 2 min. Auroral arcs are very dynamic but this time scale is not at all unreasonable, and we conclude that structured soft electron input can create a very irregular F-layer plasma in the dayside auroral oval. Conversely, the hard spectrum studied by Roble and Rees (1977), which is typical of the heart of the

SONDRESTROM GREENLAND   23 JANUARY 1985

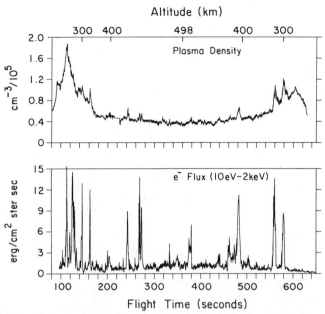

Fig. 8.9. Local plasma density along a rocket trajectory in the dayside auroral oval along with precipitating electron fluxes. (Figure courtesy of G. Earle.)

nighttime auroral oval, had a production rate of only 100 cm^{-3} s^{-1} at 400 km even for a total energy input of 8 ergs/cm$^2 \cdot$s.

The poleward edge of the nighttime auroral oval, on the other hand, is also characterized by intense soft electron fluxes (Tanskanen *et al.*, 1981) which are capable of producing F-region plasma enhancements. This source may be related to the numerous examples of plasma blobs in the nighttime oval detected by the Chatanika radar which are similar to the one shown in Fig. 8.10a. A spectral analysis of this particular radar map was performed by sampling the data with a horizontal cut through the data set at 350-km altitude. The power spectrum is given in Fig. 8.10b and is nearly identical to the rocket power spectrum shown in Fig. 8.8, which was obtained in the cusp region. These plasma blobs convect into and through the radar beam over a wide range of local times. Such enhancements could have been created at the poleward edge of the auroral oval and then convected into the field of view. They are much more common in the Alaskan sector at solar maximum, which may be due to the larger scale height of the neutral atmosphere and the corresponding increase in F-region plasma production when soft particles precipitate.

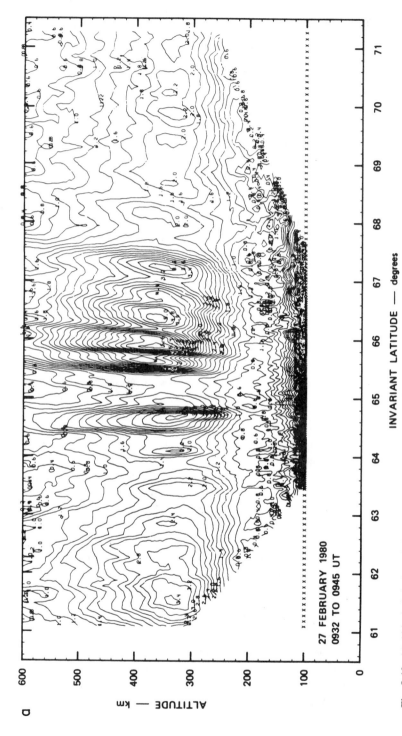

Fig. 8.10. (a) Altitude/latitude variation of electron density in the midnight sector auroral zone measured by the Chatanika radar. (b) Spectrum of electron density irregularities obtained by analyzing the radar measurements of the latitudinal variations of electron density at 350-km altitude shown in (a). [After Kelley *et al.* (1982a). Reproduced with permission of the American Geophysical Union.]

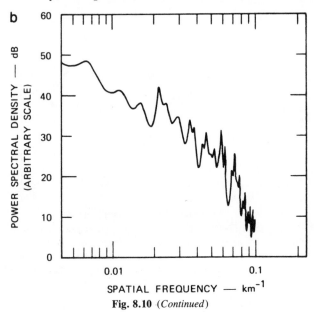

Fig. 8.10 (*Continued*)

Although plasma production by impact ionization is necessary to produce nightside plasma content enhancements on a given flux tube, a localized upper F-region enhancement could be created merely by heating. Consider a gas in diffusive equilibrium above some rigid boundary at height $h = 0$ and characterized by a constant temperature T. The concentration is then given by $n = n_0 e^{-h/H}$, where $H = kT/Mg$. If this gas is now heated to the temperature T' but no new material is added, the distribution will be of the form

$$n' = n_0' e^{-h/H'}$$

It is easy to show by integrating that the total column content N measured in particles per square meter is given by

$$N = n_0 H$$

In the case that N is conserved we have

$$n_0 H = n_0' H'$$

and the heated gas therefore has the distribution

$$n' = (n_0 T/T') e^{-h/H'}$$

This expression shows that at the surface ($h = 0$) the hotter gas is less dense than the colder gas. However, at heights greater than $h = H$, the hot-gas density

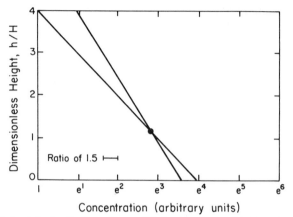

Fig. 8.11. Hydrostatic distributions of gases differing by a factor of 1.5 in an isothermal atmosphere at two different temperatures. The total quantities of the gases are the same at both temperatures. The concentrations are plotted on a natural logarithmic scale, with an arbitrary zero. The height scale is labeled in terms of reduced height for the colder gas. [Adapted from Rishbeth and Garriott (1969).]

actually increases relative to that of the cold gas. A plot showing this effect is given in Fig. 8.11. The units on the left are measured in scale heights of the colder gas. The two gas concentrations are equal one scale height above the surface.

Although a greatly simplified case, the example given above shows that localized horizontal electron temperature increases may be mirrored by a horizontally localized density enhancement in the F region above. How quickly this occurs depends on the altitude of the ionizing collisions, since the electron thermal transfer depends on whether the neutral gas or the ionospheric electrons absorb the excess heat. Since soft electrons deposit their energy at higher altitudes, they will be more efficient in creating F-region enhancements by the heat transfer effect described here.

We turn now to a discussion of polar cap aurora. When B_z is northward, the auroral oval contracts and particle precipitation is much reduced in the auroral oval. However, auroral activity actually *increases* in the polar cap region during B_z north conditions. There seem to be two basic states of the polar cap when B_z is north, an ordered multicell flow pattern and a chaotic flow (see Chapter 6 for details). In the ordered case a very long sun-aligned auroral arc can form in the polar cap. An example is shown in Fig. 8.12, which is a photograph of the polar region taken on the Dynamics Explorer 1 satellite. The circular auroral oval is linked from day to night by a sun-aligned auroral emission. The entire feature is referred to as a theta aurora. The particle energies associated with these auroras

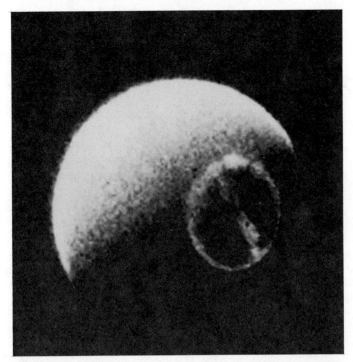

Fig. 8.12. Image of the auroral oval and a transpolar arc in the southern hemisphere at 0022 UT on May 11, 1983. The combination of continuous luminosities around the auroral oval and a transpolar arc can produce a pattern of luminosities which, when viewed from polar latitudes, is reminiscent of the Greek letter θ. Such auroral distributions are generally referred to as "theta" auroras. The dayglow and auroral emissions observed in this image are predominantly from neutral atomic oxygen at 130.4 and 135.6 nm. (Figure courtesy of L. A. Frank, J. D. Craven, and R. L. Rairden, University of Iowa.)

are in the "soft" category (≤ 100 eV) and both the plasma production and auroral airglow emissions therefore occur at lower F-region or upper E-region altitudes. A cut across one of these polar cap features showing the plasma density structure with a gray-scale representation is presented in Fig. 8.13. The arc was located at a ground range of -200 km. The plasma density peaked near 130 km in this portion of the scan, while almost no E-region plasma was observed outside the arc. Conversely, there was considerable plasma above 280 km on flux tubes adjacent to the arc but very little F-region plasma above the aurora. The horizontal structure of the ionosphere is thus very complex near polar cap auroras and is determined by a combination of production and transport. For example, the abrupt boundary between the two regions in Fig. 8.13 suggests that the over-

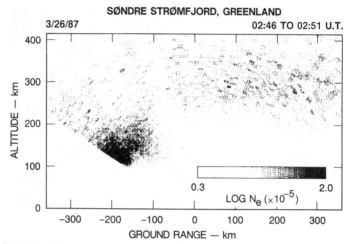

SØNDRE STRØMFJORD, GREENLAND
3/26/87   02:46 TO 02:51 U.T.

Fig. 8.13. Real-time gray-scale radar map made with a polar cap aurora in the field of view. The strong E region at a ground range of −200 km is the arc. (Figure courtesy of C. Heinselman and M. McCready.)

lying F-region plasma originated at quite different locations. There is also tantalizing evidence for a hole in the F region near the interface between these two structures.

Other experiments also show that sun-aligned arcs seem to occur in a region of shear in the plasma flow. In fact, the data suggest that electrons precipitate in a region where the flow vorticity has a positive sign (Burke *et al.*, 1982). This may be understood as follows. The flow vorticity is defined as the curl of the velocity field

$$\mathbf{W} = \nabla \times \mathbf{V} \tag{8.3}$$

As usual let the z direction be downward parallel to the magnetic field and the $\hat{a}_x$ direction along the arc. For $E_z = 0$, the plasma flow velocity $\mathbf{E} \times \mathbf{B}/B^2$ is independent of z. If we consider a flow field parallel everywhere to the arc and in the $\hat{a}_x$ direction, then the vorticity has only a z component,

$$\mathbf{W} = -(\partial V_x/\partial y)\hat{a}_z = W_z \hat{a}_z \tag{8.4a}$$

Suppose for simplicity we take the height-integrated Pedersen conductivity to be uniform in the horizontal plane. This would occur for a weak arc or for an arc in the sunlit hemisphere. Then since $\mathbf{V} = \mathbf{E} \times \mathbf{B}/B^2$ we have

$$J_z = \Sigma_P(\partial E_y/\partial y) = \Sigma_P B(\partial V_x/\partial y)$$

and hence

$$J_z = -\Sigma_P B W_z \tag{8.4b}$$

This equation shows that J_z is out of the northern hemisphere ionosphere when W_z is positive and into the ionosphere when W_z is negative. Since precipitating electrons carry current away from the ionosphere, this relationship suggests that an arc should occur when the vorticity is positive. This dependence has indeed been reported for polar cap arcs by Burke *et al.* (1982). It is of interest also to note that, from Poisson's equation, this result is equivalent to the statement that electron precipitation occurs in regions of net negative charge density (see Section 2.4). One possible explanation for this relationship is that the vorticity in the magnetospheric flow creates charge separation which leads to field-aligned currents. These currents then become unstable at some altitude, which results in particle acceleration via parallel electric fields. The experimental situation is complicated by the conductivity gradients which are created by the particle precipitation. In fact, detailed observations near a winter polar cap arc show that both conductivity gradients and structured electric fields are important in the horizontal current divergence (Weber *et al.*, 1988).

So far we have discussed the production of F-region plasma in the cusp (dayside oval), in the polar cap, and at the edge of the nightside auroral oval. In the heart of the nightside oval the situation is very chaotic, due in part to the role of substorm activity in the midnight sector. The series of Dynamics Explorer 1 images of the auroral oval in Fig. 8.14 shows some of the dynamical features in a typical substorm. As a crude first approximation there are three classes of oval precipitation, which may be described by their optical and plasma signatures. The diffuse aurora is characterized by a widespread, nearly uniform particle influx from the plasma sheet. Since the diffuse aurora mirrors the hot plasma sheet, it is also characterized by the same flow pattern, which is roughly zonally westward before midnight and zonally eastward after midnight. Imbedded in this plasma are regions of discrete auroral arcs which are usually aligned east–west and are often associated with potential drops along the magnetic field which accelerate electrons into the atmosphere. This acceleration zone has been located by satellite and rocket techniques at altitudes ranging from 2000 to 8000 km. This process occurs throughout the auroral oval. The auroral arcs in the dayside oval are associated with acceleration zones having several hundred to a thousand volts of potential drop. In the midnight sector the potential is higher and most of the plasma in this sector is produced by accelerated electrons striking neutrals in the E region. The intense E-region plasma density increase shown in Fig. 1.5 is of this type. Quiet auroral arcs are usually east–west aligned, that is, much smaller in latitudinal extent than in longitude. A third rough class of auroras are active auroral forms associated with substorms. Other than the obvious comment that more plasma is produced in a bright aurora, the active auroral forms are often extremely contorted. The examples in Fig. 1.4 and Fig. 8.14 show this very clearly. The plasma density produced by particle precipitation in the E and F regions will then also be quite structured. So many factors contribute to the

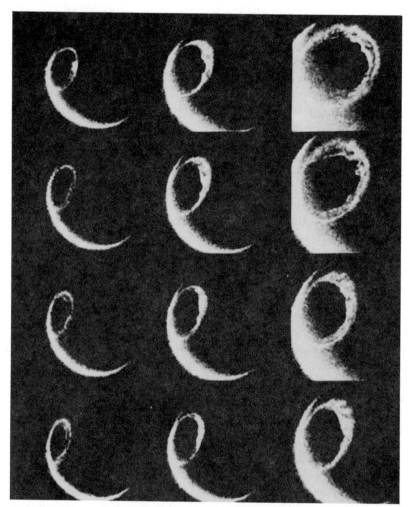

Fig. 8.14. Twelve consecutive images at ultraviolet wavelengths 123–160 nm record the development of an auroral substorm. The sequence begins at 0529 UT on April 2, 1982 (upper left image) as the NASA/GSFC spacecraft Dynamics Explorer 1 first views the auroral oval from the late evening side of the dark hemisphere at low northern latitudes near apogee (3.65 earth radii altitude) and then from progressively greater latitudes as the spacecraft proceeds inbound over the auroral oval toward perigee. The poleward bulge at onset of the auroral substorm is observed beginning at 0605 UT (fourth frame). In successive 12-min images the substorm is observed to expand rapidly in latitude and longitude. (Figure courtesy of L. A. Frank, J. D. Craven, and R. L. Rairden, University of Iowa.)

horizontal structuring in the nightside oval that it is often difficult to separate out cause and effect in that sector. In some cases the character of the density profile itself gives a clue to the origin or at least the time history of the plasma. For example, if the E-region density is high, production must be occurring simultaneously since recombination is very fast at low altitudes. Likewise, if electron temperature data are available, for example, from an incoherent scatter radar, they may be used to determine the history of F-region plasma. If precipitation is occurring, T_e will be well elevated over the ion and neutral temperatures. If the long-lived F-region plasma has been produced elsewhere and has been advected into the field of view, however, the electron temperature will have had time to equilibrate with the other gases and a low T_e is expected.

We conclude that production and heating of plasma by long-lived soft particle precipitation creates much of the F-region horizontal structure in the auroral oval at scales from 30 to 100 km. Larger-scale features are more likely due to a combination of solar production, chemical losses, and transport by large-scale electric field patterns. These concepts have been verified in a series of computer model calculations by Schunk and Sojka (1987).

8.2   Intermediate-Scale Structure in the High-Latitude F Region

The ordering of high-latitude plasma at scales above 30 km has been described in detail in the previous part of this chapter. Here we look into the processes which create structure in the range 0.1–30 km. There is a close analogy here to equatorial spread F phenomena in the following sense. At the largest scales in the equatorial case aeronomical processes such as production, recombination, and electrodynamic transport act in concert with neutral wind phenomena such as gravity waves to create the large-scale patterns. The intermediate-scale structuring then proceeds primarily via the generalized Rayleigh–Taylor instability. At high latitudes, we have seen that aeronomical processes are further supplemented by structured particle precipitation and structured large-scale flow patterns. It should not be surprising then that intermediate-scale structuring in the auroral zone and polar cap is also more complicated than at the magnetic equator. In a turbulent neutral fluid, viscosity eventually limits the scale sizes where structure occurs to values larger than some wavelength corresponding to the so-called viscous cutoff. In a plasma there are many more degrees of freedom in the system and new sources of structure may arise at various wavelength scales in the medium. In Sections 8.2.1 and 8.2.2 we concentrate on mechanisms which generate structure at intermediate scales, scales which lie between the size of the auroral oval and the size of auroral arcs. Since the linear theory for plasma instabilities in this range appeals to diffusive damping as a limiting mechanism, we spend some time discussing cross-field plasma diffusion and images in Section 8.2.3.

8.2.1   The Generalized $E \times B$ Instability at High Latitudes

Before delving into the differences, some similarities to equatorial spread F are to be noted. Figure 8.15 shows three sets of plasma density patterns detected along the flight path of the AUREOL-3 satellite (Cerisier *et al.*, 1985). Notice that the structure seems to occur preferentially on only one sense of the gradients. In Fig. 8.16 we have turned one of the vertical equatorial spread F rocket profiles on its side for comparison with the horizontal satellite data in Fig. 8.15. The similarity is striking in that both profiles show a strong tendency for irregularity development on a preferred direction of the density gradient.

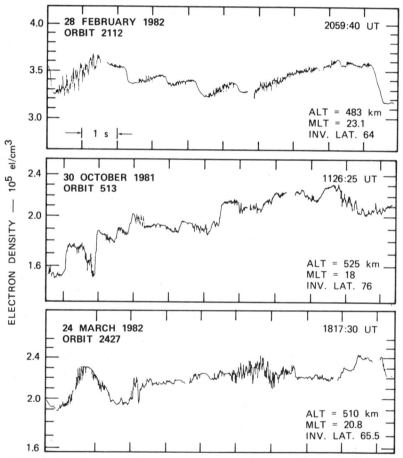

Fig. 8.15. Examples of the preferential structuring of high-latitude plasma density enhancements on gradients of a particular sign. [Adapted from Cerisier *et al.* (1985). Reproduced with permission of the American Geophysical Union.]

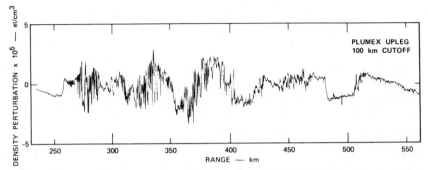

Fig. 8.16. A vertical plasma density profile during equatorial spread F "tipped on its side" to compare with Fig. 8.15. [After Kelley *et al.* (1982b). Reproduced with permission of the American Geophysical Union.]

Gravity plays very little role in high-latitude phenomena since **g** is essentially parallel to **B.** We therefore concentrate our analysis on the electric field-driven aspects of the instability in this chapter. We refer to the process as a generalized **E × B** instability and include neutral winds, electric fields, and field-aligned currents. As noted earlier, when (**E′ × B**) has a component parallel to ∇n, that is, when

$$(\mathbf{E'} \times \mathbf{B}) \cdot \nabla n > 0 \qquad (8.5)$$

the system is unstable. Referring back to Fig. 4.8a, we see that in this case perturbation electric fields develop when a Pedersen current flows perpendicular to the zero-order density gradient in the presence of a small disturbance. In effect, a high-density region polarizes in such a way that it has a slower drift velocity than the background plasma. A high-density region therefore lags behind and seems to grow with respect to the background density as it drifts "down" the gradient. Low-density regions move in the opposite direction and seem to grow with respect to the background. Indeed, based on the AUREOL-3 satellite measurements of the electric field, the event in the upper panel of Fig. 8.15 was such that the instability condition in (8.5) was satisfied since a 12-mV/m eastward zonal electric field component was measured and the density gradient was almost certainly poleward. We say almost certainly since only one component of ∇n is measurable from a polar-orbiting satellite. Cerisier *et al.* (1985) have studied six events of this type and found four of them to have (8.5) satisfied, to the best of their knowledge, given the ambiguity in ∇n. It seems clear that the **E × B** process can and does occur in the high-latitude sector and most of the discussion of the generalized Rayleigh–Taylor instability in Chapter 4 is directly applicable at high latitudes. However, there is one important caveat to raise. The satellite measures only **E,** not **E′** = **E** + **U × B,** and the role of the neutral wind is just not known in most experiments. Usually one argues that since $|\mathbf{E} \times \mathbf{B}/B^2| > |\mathbf{U}|$ at high latitudes, the neutral wind plays only a minor role. However, at the boundaries of the auroral oval, where the plasma flow changes direction, electrodynamically driven neutral winds will not change direction as quickly as the plasma flow and could act as a source for instability. In the midnight sector, for example, we expect a southward neutral wind driven by ion drag in the polar cap. At the convection boundary $\mathbf{E}$ is in the meridian plane and hence plays no role in the stability of the poleward plasma density gradient which often exists at that boundary. However, if as argued above $\mathbf{U}$ is in the southward direction, $\mathbf{E}' \times \mathbf{B}$ is poleward and hence the poleward wall of the trough could be unstable to the generalized $\mathbf{E} \times \mathbf{B}$ process.

Concerning details of the $\mathbf{E} \times \mathbf{B}$ instability, there are some aspects of this process which differ from its development at equatorial and middle latitudes. One feature involves the relatively high E-region Pedersen conductivity which exists at high latitudes. This affects both the growth rate and the loss rate of plasma structure. For example, a simple expression (Vickrey and Kelley, 1982) for the linear growth rate of the one-dimensional $\mathbf{E} \times \mathbf{B}$ instability for waves perpendicular to the gradient is of the form

$$\gamma = \frac{E_0'}{BL}\left(\frac{M-1}{M}\right) - k^2 D_\perp \qquad (8.6)$$

where E_0'/B is the component of $\mathbf{E}' \times \mathbf{B}/B^2$ parallel to ∇n, L the inverse gradient scale length, $D_\perp$ the height-averaged perpendicular diffusion coefficient, and $M = (\Sigma_P^E + \Sigma_P^F)/\Sigma_P^F = (1 + \Sigma_P^E/\Sigma_P^F)$. In the definition of M, Σ_P^F is the field-line-integrated Pedersen conductivity in the F region and Σ_P^E is the field-line-integrated Pedersen conductivity in the E region. If $\Sigma_P^E > \Sigma_P^F$, which is almost always the case in the auroral oval due to particle precipitation, the E region tends to short out the perturbation electric field $\delta\mathbf{E}$, produced in an $\mathbf{E} \times \mathbf{B}$ instability, and the growth rate becomes small. This is easy to understand since "growth" in the case of the $\mathbf{E} \times \mathbf{B}$ instability is just due to advection of high-density plasma down a gradient and low-density plasma up a gradient. The rate of change of the density is given by

$$\partial n/\partial t = -\delta\mathbf{V} \cdot \nabla n \simeq \gamma\,\delta n \qquad (8.7)$$

where $\delta\mathbf{V} = \delta\mathbf{E} \times \mathbf{B}/B^2$. If the perturbation charges which set up $\delta\mathbf{E}$ are shorted out by current flow to the E region, the growth rate becomes small. The diffusive damping term also depends on the E region, but in a subtle fashion which we postpone until Section 8.2.3. Tsunoda (1989) has compared M values for the auroral case with those for equatorial spread F and for large mid-latitude barium releases. For equatorial spread F he quotes values in the range 10 to 10^4, while for low-altitude barium releases M ranges from 60 for a 1-kg release to 470 for a 48-kg release. Auroral F-region plasma enhancements, on the other hand, have M values less than 1.2 for reasonable E-region densities ($\Sigma_P^E = 5$ mho). Even for nighttime conditions with no E layer $M \leq 10$ for F-region densities as high as 10^6 cm^{-3}. On this basis alone it is clear that local plasma instabilities may be less dominant in the auroral zone than at other latitudes.

The shorting process discussed above is accomplished by field-aligned currents due to the E region which short out both the driving fields, δE, and the ambipolar electric fields (see Section 8.2.3) which limit diffusion across **B.** In addition to these *structure*-related currents there are also large-scale field-aligned currents which link the magnetospheric and solar wind generators to the ionospheric load. These were discussed in some detail in Chapter 6. Here we investigate the role which such applied currents play with regard to the **E × B** instability. When field-aligned currents are included, the process is referred to in the literature as the current convective instability (Ossakow and Chaturvedi, 1979).

In brief, when the ion and electron species have different drift velocities along the magnetic field line (and hence a net parallel current is carried by the thermal plasma), the **E × B** instability is modified in such a way that even plasma gradients stable to the **E × B** instability may in principle be unstable to the current convective process. In the case of the pure **E × B** process the most unstable wave has $k_{\parallel} = 0$. For the current convective instability to operate, a finite $k_{\parallel}$ is required.

To understand why this instability occurs we isolate the effect of an upward (northern hemisphere) field-aligned current acting alone as shown in Fig. 8.17.

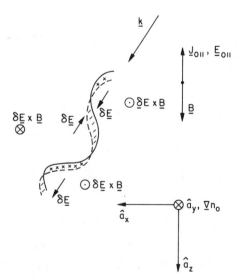

Fig. 8.17. Electrostatic wave with finite (exaggerated) $k_{\parallel}$ propagating in the presence of a field-aligned current. If a zero-order density gradient exists pointed into the page the wave is unstable.

The finite $k_\parallel$ is exaggerated to make the effect easier to illustrate. We assume a density perturbation of the form

$$\delta n(s,\ t) = n'e^{i(ks - \omega t)}$$

where s is a distance measured in the direction parallel to $\mathbf{k}$. The perturbation vector $\mathbf{k}$ is confined to the x–z plane while ∇n is in the y direction. Because both elements σ_0 and σ_P of the tensor conductivity are proportional to the density, it follows that they will depend on s in a similar manner,

$$\delta\sigma_0(s,\ t) = \sigma_0 e^{i(ks - \omega t)}$$

$$\delta\sigma_P(s,\ t) = \sigma_P e^{i(ks - \omega t)}$$

Our goal is to determine the perturbation electric field $\delta\mathbf{E}_k$, which is parallel to $\mathbf{k}$ since we are dealing with an electrostatic wave,

$$\delta\mathbf{E}_k(s,\ t) = \delta E_k\ e^{i(ks - \omega t)}$$

We now solve for δE_k in terms of the initial perturbation in δn by setting $\nabla \cdot \mathbf{J} = 0$. In the x–z plane,

$$\mathbf{J} = \sigma_0 E_z \hat{a}_z + \sigma_P E_x \hat{a}_x$$

Since the zero-order parallel electric field is upward in the case illustrated by Fig. 8.17, we have

$$\mathbf{J} = (\sigma_0 + \delta\sigma_0\ e^{i(k_x x + k_z z - \omega t)})(-E_\parallel + \delta E_k \sin\theta e^{i(k_x x + k_z z - \omega t)})\hat{a}_z$$
$$+ (\sigma_P + \delta\sigma_P\ e^{i(k_x x + k_z z - \omega t)})(\delta E_k \cos\theta e^{i(k_x x + k_z z - \omega t)})\hat{a}_x$$

Evaluating $\nabla \cdot \mathbf{J}$, dropping second-order terms, and setting the result equal to zero yields

$$(-ik_z\ \delta\sigma_0\ E_\parallel + ik_z\sigma_0\ \delta E_k \sin\theta + ik_x\sigma_P\ \delta E_k \cos\theta)e^{i(k_x x + k_z z - \omega t)} = 0$$

Substituting $k_z = k \sin\theta$ and $k_x = k \cos\theta$ and solving for δE_k,

$$\delta E_k = E_\parallel\ \delta\sigma_0 \sin\theta / (\sigma_0 \sin^2\theta + \sigma_P \cos^2\theta)$$

which may be written

$$\delta E_k = E_\parallel \left(\frac{\delta\sigma_0}{\sigma_0}\right)\frac{\sin\theta}{\sin^2\theta + (\sigma_P/\sigma_0)\cos^2\theta} \tag{8.8}$$

Our goal is not to carry out a full algebraic analysis of the process, which is quite messy, but rather to gain some physical insight. To proceed, however, we need one result of the detailed linear analysis made by Ossakow and Chaturvedi (1979). They linearized the continuity equation, the momentum equation, and $\nabla \cdot \mathbf{J} = 0$ and showed

$$\theta = \frac{k_{\parallel}}{k_{\perp}} = \left[\frac{\nu_i}{\Omega_i} \left(\frac{\Omega_i}{\nu_i} + \frac{\Omega_e}{\nu_e} \right)^{-1} \right]^{1/2}$$

Substituting $\sigma_P = (ne^2\nu_i/M\Omega_i^2)$ and $\sigma_0 = (ne^2/m\nu_e)$ in this expression yields

$$\theta = \left[\frac{\sigma_P}{\sigma_0} \left(1 + \frac{\Omega_i\nu_e}{\Omega_e\nu_i} \right) \right]^{-1/2} \simeq \left(\frac{\sigma_P}{2\sigma_0} \right)^{1/2}$$

where in the last step we use the fact that for typical F-region conditions $\Omega_i\nu_e/\Omega_e\nu_i \approx 1$. Substituting this result into (8.8) and using the fact that $\theta \simeq \sin\theta$ and $\cos\theta \simeq 1$,

$$\delta E_k = (\delta\sigma_0/\sigma_0)[\sqrt{2}E_{\parallel}/3(\sigma_P/\sigma_0)^{1/2}]$$

and finally since $\sigma_0 \propto n$ we have for maximum growth

$$\delta E_k = (0.47)(\delta n/n)E_{\parallel}/(\sigma_P/\sigma_0)^{1/2} \tag{8.9}$$

This electric field is very nearly perpendicular to $\mathbf{B}$ since $k_{\perp} \gg k_{\parallel}$ and, referring to Fig. 8.17, we see that it is such that δE_k is in the $\hat{a}_x$ direction when $\delta n/n > 0$. The perturbation $\mathbf{E} \times \mathbf{B}$ drift in this phase of the wave is thus equal to $+(\delta E_k/B)\hat{a}_y$, which means that a high-density region moves *down* the gradient to lower-density regions and therefore grows in relative amplitude. Therefore the plasma is unstable.

In this same configuration, suppose a background perpendicular zero-order electric field $\mathbf{E}_{0\perp} = E_{0\perp}\hat{a}_x$ existed in addition to the zero-order parallel electric field $E_{0\parallel}$. A review of the $\mathbf{E} \times \mathbf{B}$ instability theory presented in Chapters 4 and 5 shows that the growth rate is given by

$$\gamma_{EB} = -E_{0\perp}/BL = +V_{0y}/L$$

That is, if the perpendicular electric field is in the $\hat{a}_x$ direction the system is stable, while if it is in the $-\hat{a}_x$ direction instability occurs.

We now may compare the growth rate of the pure current convective instability to the classical $\mathbf{E} \times \mathbf{B}$ process. The latter we write in the form

$$\gamma_{EB} = E_{0\perp}/BL = V_{0\perp}/L$$

where $V_{0\perp}$ is the magnitude of the zero-order drift parallel to ∇n. Using (8.7), which can be written $\gamma = (\delta E_k/BL)(\delta n/n)^{-1}$, and substituting δE_k from (8.9) yields

$$\gamma_{cc} = \frac{\sqrt{2}}{3} \frac{E_{\parallel}}{BL(\sigma_P/\sigma_0)^{1/2}}$$

The usual practice is to compare these two growth rates in terms of $V_{0\perp}$ and the zero-order parallel drift velocity difference of the ions and electrons, which may be written

$$V_\parallel = \sigma_0 E_\parallel / ne$$

Then

$$\frac{|\gamma_{EB}|}{|\gamma_{cc}|} = \frac{3}{\sqrt{2}} \frac{V_{0\perp} B (\sigma_P \sigma_0)^{1/2}}{V_\parallel ne} \tag{8.10}$$

This result makes it easy to compare the two effects. First, we note that if both processes are unstable (8.10) will show which one is more important. Second, if $E_{0\perp}$ is in the stable configuration while $J_\parallel$ is destabilizing, instability is still predicted if (8.10) is less than one. To estimate the magnitude of $J_\parallel$ required to overcome, say, a stabilizing 10-mV/m perpendicular electric field ($V_{0\perp} = 400$ m/s) we use $V_\parallel = j_\parallel / ne$ and set $n = 5 \times 10^4$ cm^{-3} to find $J_\parallel \geq 7$ μA/m^2. This is a sizable current but is not out of the question for the auroral zone.

Using the same expressions for σ_P and σ_0 as used above, $(\sigma_P \sigma_0)^{1/2} = (\Omega_e \nu_i / \Omega_i \nu_e)^{1/2}$ and Eq. (8.10) becomes,

$$\frac{|\gamma_{EB}|}{|\gamma_{cc}|} = \frac{3}{\sqrt{2}} \frac{V_{0\perp}}{V_\parallel} \left(\frac{\Omega_e \nu_i}{\Omega_i \nu_e} \right)^{1/2}$$

Finally, once again we note that the quantity in parentheses is approximately equal to unity in the F region and we have

$$\gamma_{EB} / \gamma_{cc} \simeq 2 V_{0\perp} / V_\parallel$$

which is identical to the result usually quoted. A complete analysis (Ossakow and Chaturvedi, 1979; Vickrey et al., 1980) yields the following expression for the local growth rate of the current convective instability including the possible existence of $\mathbf{E}_{0\perp}$:

$$\gamma_{cc} = \frac{(-1/L)[(-E_{0\perp}/B)(\nu_{in}/\Omega_i) + V_\parallel \theta_{max}]}{(\Omega_i/\nu_{in} + \Omega_e/\nu_{ei}) \theta_{max}^2 + \nu_{in} \Omega_i}$$

$$- \left(\frac{k_\perp^2 \nu_{ei}}{\Omega_e \Omega_i} \right) C_s^2 - \frac{k_\parallel^2 C_s^2}{\nu_{in}} \{ 1 + (\nu_{in}^2/\Omega_i^2)/[(\nu_{ei}\nu_{in}/\Omega_e\Omega_i) + \theta_{max}^2] \} \tag{8.11}$$

where $L^{-1} = 1/n(dn/dy)$, C_s is the ion acoustic speed, ν_{in} and ν_{ei} are the ion–neutral and electron–ion collision frequencies, Ω_e and Ω_i are the electron and ion gyrofrequencies, $E_{0\perp}$ is the component of the perpendicular electric field in the $\nabla n \times \mathbf{B}$ direction, and the ratio of parallel to perpendicular wave numbers for maximum growth, θ_{max}, is given by

$$\theta_{max} = -\left(\frac{E_{0\perp}}{BV_\parallel} \right) \left(\frac{\nu_{in}}{\Omega_i} \right) \pm \left[\left(\frac{E_{0\perp}}{BV_\parallel} \right)^2 \left(\frac{\nu_{in}}{\Omega_i} \right)^2 + \frac{\nu_{in}/\Omega_i}{\Omega_i/\nu_{in} + \Omega_e/\nu_{en}} \right]^{1/2} \tag{8.12}$$

If we let $V_{\parallel}$ go to zero in (8.12) and choose the negative sign, then $\theta_{max} = 0$. This result yields the flute mode ($k_{\parallel} = 0$) $\mathbf{E} \times \mathbf{B}$ instability. Indeed, if we set $\theta_{max} = 0$ in (8.11) it reduces to the growth rate of the gradient drift instability. Thus, as already noted, a field-aligned current can serve to destabilize a plasma configuration that is otherwise stable to the $\mathbf{E} \times \mathbf{B}$ instability, or to enhance the growth rate of an already unstable situation. Choice of the negative sign in (8.12) corresponds to a damped acoustic mode.

The linear local theory described here has a preferential tendency to create unstable conditions at low plasma densities, since for a fixed $J_{\parallel}$, $V_{\parallel}$ is inversely proportional to n. The region above the F peak is therefore a region of high growth rate. However, the generalized process could very easily be *stable* on the same magnetic field line at or below the F peak. It is very clear then that a local theory is not very suitable. Such nonlocal effects in general reduce the growth rate. Note also that E-region shorting is every bit as important to the current convective instability as it is to the $\mathbf{E} \times \mathbf{B}$ instability and is not included in (8.11). Also, the growth rate given in (8.11) uses a perpendicular diffusion coefficient corresponding to a nonconducting E region, which may also underestimate the damping (see Section 8.2.3). Taken together, these negative aspects have led to the conclusion that although it is an interesting physical process, the current convective instability can only rarely overcome the stabilizing effects of an unfavorable $\mathbf{E} \times \mathbf{B}$ geometry. Finally, it should be noted that the instability only occurs when the current is carried by thermal plasma and does not apply when energetic electrons carry the bulk of an upward current.

8.2.2 Turbulent Mixing as an Alternative to Plasma Instabilities

It is well known that the magnetospheric flow pattern is seldom laminar but rather is usually somewhere between turbulent and extremely turbulent! The effect of this turbulence is to mix any existing density gradient due to solar production or particle impact ionization regardless of the sign of the gradient. We can estimate the turbulence level necessary to accomplish a higher level of density structure than that provided by the $\mathbf{E} \times \mathbf{B}$ process as follows. We assume that the latter has the appropriate sign for instability but, of course, this will happen only 50% of the time! The simple form of the linear growth rate for the $\mathbf{E} \times \mathbf{B}$ instability is γ_{EB} given above. If that same gradient is structured by a spatial electric field pattern characterized by some wave number spectrum $E(k)$, then we may use the continuity equation

$$\partial n / \partial t = -\mathbf{V} \cdot \nabla n$$

to define a mixing growth rate $\gamma_m(k)$ which is a function of k:

$$\gamma_m(k) = E(k)/BL$$

To compare the mixing growth rate to the linear $\mathbf{E} \times \mathbf{B}$ growth rate which is independent of k we must integrate $\gamma_m(k)$ over some portion of the turbulent spectrum. Then

$$\Gamma_m = \left[\int_{\Delta k} E^2(k) \, dk \right]^{1/2} \bigg/ BL \qquad (8.13)$$

is the appropriate growth rate provided Δk is chosen properly. Since the important wavelength range is for $k > k_L = 2\pi/L$, we integrate from k_L to ∞. The ratio of turbulent mixing to the plasma instability source is then

$$R = \Gamma_m/\gamma_{EB} = (1/E_{0\perp}) \left(\int_{k_L}^{\infty} E^2(k) \, dk \right)^{1/2} = \frac{E_{rms}}{E_{0\perp}} \qquad (8.14)$$

The bracketed term is the rms electric field between k_L and $\infty (E_{rms})$. Turbulent mixing is important when $R \geq 1$, that is, when the mixing growth rate is comparable to the linear growth rate of the plasma instability. The mixing process will even dominate the physics on an unstable gradient if the rms electric field in the range Δk exceeds $E_{0\perp}$, the quasi-dc component perpendicular to ∇n. For a stable gradient the mixing effect will, of course, be the dominant source and R need not be greater than 1. Furthermore, since turbulent fields are supplied by a low impedance magnetosphere generator, they will not be shorted out by the E region.

Surprisingly, very few electric field spectra exist in the literature. One example from the dawn auroral oval is presented in Fig. 8.18. The upper plot comes from periods of the flight near auroral arcs (note that the scale is broken). The spectra show evidence for two turbulent processes, one operating at long wavelengths yielding a $k^{-5/3}$ power law and one which injects turbulence in the ionosphere at the scale of individual auroral arcs. The latter process seems to result in a k^{-3} power law form for large $k > k_b \approx 290$ km^{-1}. For k values smaller than k_b, the spectrum is nearly flat until it "runs into" the $k^{-5/3}$ power law. These two power laws are in agreement with predictions of two-dimensional turbulence theory. The k-space integrals corresponding to these two regimes yield rms electric fields of 10 and 9 mV/m, respectively. Such values for E_{rms} indicate a significant role for turbulent mixing, since $E_{rms}/E_{0\perp} \approx 0.5$ for this event.

The fact of the matter is that auroral arcs are inextricably intertwined with structured electric fields. At the altitude of the auroral acceleration zone (≈ 3000–5000 km) in fact, the electric field is turbulent in a much larger region than that of the electron beams themselves. The acceleration zone is imbedded in this turbulent plasma which partially extends to ionospheric heights. It seems clear that any ionospheric plasma gradient *must* be mixed by those applied turbulent electric fields.

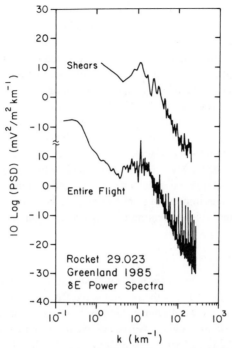

Fig. 8.18. Wave number spectra of the electric field from a rocket flight in the dayside auroral oval. The lower plot represents a large spatial region, while the upper plot shows smaller-scale spectra near auroral arcs. The latter are imbedded in the region sampled to produce the lower plot. Note that the scales are shifted. (Figure courtesy of G. Earle.)

The ionosphere is not passive in this context, however. Detailed study has shown that the magnetosphere behaves as a current generator at these scales (Vickrey *et al.*, 1986). This in turn implies that the turbulent electric field and flow velocity are inversely proportional to the ionospheric conductivity. This is evident when one compares winter and summer hemispheres. The winter ionosphere has a much more structured velocity field than does the summer iono-sphere. One hypothesis is that the solar wind driven flow in the magnetosphere has its own turbulent characteristics which are moderated by the gross E-region conductivity due to solar lighting conditions, diffuse auroral precipitation, and polar rain. Superposed on this are localized regions of turbulence associated with auroral arcs which create plasma and strongly mix it at the same time. At this scale the E region itself becomes dominated by the precipitation and the conduc-tivity and the flow field and plasma content are intertwined in a fundamental way. A review (Kintner and Seyler, 1985) discussing the interrelationship be-tween plasma fluid turbulence and plasma density variations in the ionosphere has been published. Although no direct conclusions were reached, a framework was provided for this study and progress should be steady in this area.

Further evidence for a turbulent mixing process at scales on the border between "large" and "intermediate" has been presented by Tsunoda et al. (1985). They studied a highly structured F region which was characterized by a stable $\mathbf{E} \times \mathbf{B}$ situation. They concluded that a driven mixing process was occurring but one which was *not* explained by the $\mathbf{E} \times \mathbf{B}$ instability. Similarly, referring to the data in Fig. 8.1, the high scintillation levels which were reported in association with plasma patches in the polar cap were found throughout the patches. That is, kilometer-scale irregularities were located on gradients with either sign, even though the mean patch convection is associated with a particular $\mathbf{E} \times \mathbf{B}$ flow direction. Regions of plasma shear are also irregular (Basu et al., 1986). We therefore conclude that impressed turbulent electric fields of magnetospheric origin play an important role in creating ionospheric structure at high latitudes.

8.2.3   Diffusion and Image Formation

As noted above, the lifetime of a horizontal structure in the high-latitude ionosphere depends on how quickly the plasma can diffuse across the magnetic field. Also, whether a particular plasma instability is stable or unstable at a given k value depends on whether some positive contribution to the growth rate exceeds diffusive damping due to terms of the form $-k^2 D_\perp$. In Chapter 5 we pointed out that parallel diffusion which operates with a time scale $(k_\parallel^2 D_\parallel)^{-1}$ is determined by the ion diffusion coefficient and that

$$D_\parallel = 2D_i$$

The factor of 2 comes from the fact that an ambipolar electric field builds up when electrons attempt to diffuse away quickly parallel to $\mathbf{B}$ due to their small mass. In effect, each electron has to drag a heavy ion with it along $\mathbf{B}$ and the *plasma* diffusion coefficient is then determined by the ions.

Perpendicular to $\mathbf{B}$, an analogous process occurs. The equations of conservation of momentum (2.22b) for each species in the moving reference frame of the neutral wind are

$$0 = -k_B T_j \, \nabla n + nM_j \mathbf{g} + nq_j \mathbf{E}' + nq_j(\mathbf{V}'_j \times \mathbf{B}) - nM_j \nu_{jn} \mathbf{V}'_j$$

where we have taken $dV'_j/dt = 0$. Let $\mathbf{B}$ be in the z direction and neglect the effect of gravity and neutral winds. Then in the directions perpendicular to $\mathbf{B}$,

$$0 = -k_B T_j (\partial n/\partial x) + nq_j E_x + nq_j V_{jy} B - nM_j \nu_{jn} V_{jx}$$

$$0 = -k_B T_j (\partial n/\partial y) + nq_j E_y - nq_j V_{jx} B - nM_j \nu_{jn} V_{jy}$$

Substituting for V_{jy} in the second equation and substituting into the first yields an expression involving only V_{jx}:

$$V_{jx}\left(1 + \frac{q_j^2 B^2}{M_j^2 \nu_{jn}^2}\right) = \frac{-k_B T_j}{M_j \nu_{jn}} \frac{1}{n} \frac{\partial n}{\partial x} + \frac{q_j}{M_j \nu_{jn}} E_x$$

$$- \frac{q_j B k_B T_j}{M_j^2 \nu_{jn}^2} \frac{1}{n} \frac{\partial n}{\partial y} + \frac{q_j^2 B^2}{M_j^2 \nu_{jn}^2} E_y$$

Similarly, we can derive

$$V_{jy}\left(1 + \frac{q_j^2 B^2}{M_j^2 \nu_{jn}^2}\right) = \frac{-k_B T_j}{M_j \nu_{jn}} \frac{1}{n} \frac{\partial n}{\partial y} + \frac{q_j}{M_j \nu_{jn}} E_y$$

$$+ \frac{q_j B k_B T_j}{M_j^2 \nu_{jn}^2} \frac{1}{n} \frac{\partial n}{\partial x} - \frac{q_j^2 B^2}{M_j^2 \nu_{jn}^2} E_x$$

Remembering that the $\mathbf{E} \times \mathbf{B}$ and the diamagnetic drift velocities can be written

$$\mathbf{V}_{Ej} = \mathbf{E} \times \mathbf{B}/B^2, \qquad \mathbf{V}_{Dj} = -\frac{k_B T_j}{q_j B^2 n} \nabla n \times \mathbf{B}$$

the equation for the perpendicular velocity can be recombined, giving

$$\mathbf{V}_{j\perp}\left(1 + \frac{q_j^2 B^2}{M_j^2 \nu_{jn}^2}\right) = -\frac{k_B T_j}{M_j \nu_{jn} n} \nabla_\perp n + \frac{q_j}{M_j \nu_{jn}} \mathbf{E}_\perp$$

$$+ \frac{q_j^2 B^2}{M_j^2 \nu_{jn}^2} \mathbf{V}_D + \frac{q_j^2 B^2}{M_j^2 \nu_{jn}^2} \mathbf{V}_E$$

Defining the perpendicular diffusion coefficient for motion antiparallel to a density gradient by an expression of the form $\mathbf{V} = -D(\nabla n/n)$ yields,

$$D_{j\perp} = k_B T_j/M_j \nu_{jn}(1 + \Omega_j^2/\nu_{jn}^2) \qquad (8.15a)$$

Now for F-region altitudes $\Omega_j \gg \nu_{jn}$ and this expression becomes

$$D_{j\perp} = (k_B T_j/M_j)(\nu_{jn}/\Omega_j^2)$$

Finally, since the species gyroradius is given by the expression

$$r_{gj} = (k_B T_j/M_j)^{1/2} \Omega_j^{-1} \qquad (8.15b)$$

the perpendicular diffusion coefficient of some species j may be written

$$D_{j\perp} = r_{gj}^2 \nu_j \qquad (8.15c)$$

where r_{gj} is the gyroradius and ν_j is the collision frequency. Intuitively this makes sense, since a particle will random walk one gyroradius between each collision. Since r_{gj} is much larger for ions than for electrons, the ions tend to diffuse more rapidly down a density gradient which exists perpendicular to the magnetic field than do the electrons. However, when this occurs an electric field builds up parallel to ∇n, so the ion motion across the field lines is retarded and the electron motion is enhanced. The result (derived below) is that an isolated *plasma* structure diffuses across a magnetic field with a diffusion coefficient equal to the low electron diffusion coefficient.

The discussion above assumes that the F-region plasma is the only plasma in the system. However, if an E region exists due to production of plasma by sunlight or production by auroral particle precipitation, the ambipolar electric field discussed above will be shorted out and the ions may then diffuse rapidly across the magnetic field. The electrons still cannot move across **B** but they *can* move along the magnetic field to complete the circuit through the E region and therefore neutralize the ambipolar electric field. The entire process is illustrated schematically in Fig. 8.19a. The converging arrows in the E region represent the ambipolar electric field which maps down to the E region due to the high conductivity parallel to **B**. Those arrows also show the direction of the E-region ion current which completes the circuit. The inner and outer sets of arrows show the electron flow direction. The line through the F-region structure represents the vertical magnetic field line.

A quantitative description of this diffusion process must allow for altitude variations in all the relevant quantities. Vickrey and Kelley (1982) have performed such calculations for a model ionosphere and computed the typical effective diffusion coefficient plotted in Fig. 8.19b as a function of the ratio of E-region to F-region conductivities. The effective (field-line-integrated) diffusion coefficient increases from roughly 1 m²/s to more than 10 m²/s as the E-region conductivity increases. Also plotted is the *e*-folding time, $\tau_\perp = (k_\perp^2 D_\perp)^{-1}$, for a horizontal structure with a characteristic perpendicular wavelength of 1 km. Even for high E-region conductivity the diffusion time scale is several hours for such a structure, and it could survive considerable transport around the high-latitude zone. In this particular calculation, the F layer had a Chapman distribution characterized by the parameters listed in the plot. Kelley *et al.* (1982a) used this semiquantitative approach to calculate the relative amplitude of kilometer-scale structures produced at various places in the auroral oval which were then allowed to convect through regions of varying E-region conductivity. The resulting amplitude distribution depended on season and local time because of the peculiar trajectories of some flux tubes (see Chapter 6).

Further advances in this area required a self-consistent calculation of the ambipolar electric field since, unlike the simplification used by Vickrey and Kelley (1982), which is valid only for large structures, the electric field mapping process is scale size dependent (see Chapter 2). In addition, the E region is not a passive medium but rather one in which structures can form that mirror the F-region irregularities. These "images" were first pointed out in the context of barium cloud striation physics by Goldman *et al.* (1976).

In the remainder of this section we follow a unified description of diffusion which includes F- and E-region coupling (Heelis *et al.*, 1985). First it should be

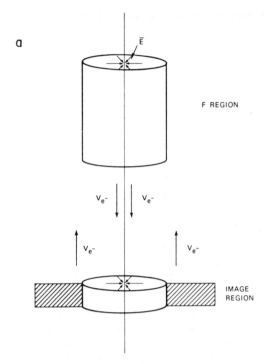

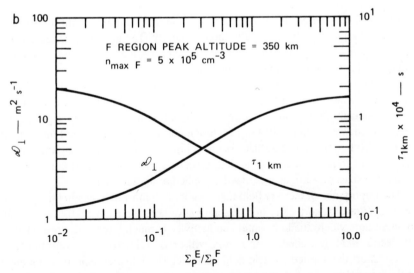

Fig. 8.19. (a) Schematic diagram showing the redistribution of electrons along the magnetic field during the image formation process as well as the ambipolar electric field. The arrows in the image region also show the direction of the ion Pedersen velocity. (b) Dependence of cross-field diffusion rate $D_\perp$ and lifetime of 1-km scale irregularities τ_{1km} on the ratio of E- to F-region Pedersen conductivities. The F-layer altitude and peak density are fixed. [After Vickrey and Kelley (1982). Reproduced with permission of the American Geophysical Union.]

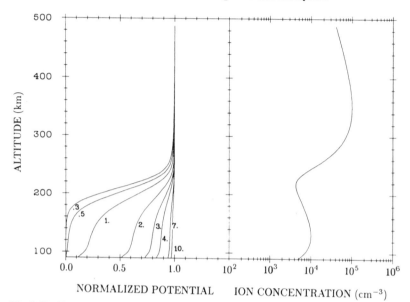

Fig. 8.20. The variation in electrostatic potential from different scale size electric fields is shown in the left panel. The right panel shows the ionospheric ion concentration profile. The potentials were applied at 500-km altitude. (Figure courtesy of R. Heelis.)

emphasized that the electrical conductivity along the magnetic field lines is not infinite but rather is a finite, usually large, quantity. The mapping characteristics of electric fields can easily be deduced by considering the current continuity equation and employing analysis methods such as those discussed by Farley (1959) and presented in Chapter 2. Figure 8.20 shows some solutions for different scale size electric field structures applied at 500-km altitude. The ionospheric plasma concentration profile used in the calculation is shown in the right panel. As expected, we see that a 10-km structure maps almost unattenuated throughout the E and F regions. At scale sizes below 5 km, however, significant attenuation occurs and can render the electric field essentially zero below 200 km for fields with scale sizes of 1 km or less that are applied at high altitudes.

The importance of electric field mapping in the ionosphere lies in the fact that electric fields created locally in a region of poor horizontal conductivity can map to an altitude where the horizontal conductivity is much larger. The presence of an electric field in a conducting region will drive a horizontal current that will tend to short the applied electric field. The electric field can then be maintained at the generator only if it can supply the current required in the conducting region. Let us consider the forces acting on a plasma concentration structure in an ionosphere with a vertical magnetic field. First, a pressure gradient force will exist on both the ions and the electron gases. But as shown above, owing to the different masses of these particles and their correspondingly different collision frequencies and gyrofrequencies, these two species will move at different rates. Any tendency for separation of the ions and electrons in this manner will produce an electric force.

The horizontal motion of the electrons and ions subject to these forces can be determined by manipulation of the steady-state equations of motion for each species. Rather complex expressions result if all the terms in these equations are retained. This is the usual practice in computer models, but in order to illustrate the dominant physics we consider only the horizontal ion motion and assume that electron–ion collisions have a negligible effect. At high latitudes we may assume $\mathbf{g} \times \mathbf{B} = 0$, in which case the horizontal motions are not affected by gravity. Then from (2.36b)

$$V_{ix} = (\sigma_{Pi}/en)E - (D_{i\perp}/n)\,\nabla_x n \qquad (8.16)$$

Here x again denotes the direction of the gradient and we have set $n_i = n_e = n$ due to the quasi-neutrality condition. The ion Pedersen conductivity, σ_{Pi}, and the ion perpendicular diffusion coefficient, $D_{i\perp}$, are given in general form by

$$\sigma_{Pi} = (e^2 n \nu_{in})/M_i(\Omega_i^2 + \nu_{in}^2)$$

$$D_{i\perp} = (k_B T_i \nu_{in})/M_i(\Omega_i^2 + \nu_{in}^2)$$

Similarly, the horizontal electron velocity can be expressed as

$$V_{ex} = -(\sigma_{Pe}/en)E - (D_{e\perp}/n)\,\nabla_x n + RV_{ix} \qquad (8.17)$$

where the factor R is given by

$$R = \frac{\Omega_e^2 \nu_{ie}/\nu_{in} + \nu_{ei}\nu_e}{\Omega_e^2 + \nu_e^2}$$

and arises because ion–electron collisions are not always negligible for the electrons. Finally, Eq. (8.16) and (8.17) can be combined to yield an expression for the local horizontal Pedersen current,

$$J_P = [(1 - R)\sigma_{Pi} + \sigma_{Pe}]E - e[(1 - R)D_{i\perp} - D_{e\perp}]\,\nabla_x n \qquad (8.18)$$

The magnitude of ionospheric structure and its time evolution are related by the equations of continuity and momentum, both of which contain the plasma velocity. If we consider an ionospheric structure not subject to any production, then the continuity equation for the ions may be written as

$$\partial n/\partial t = (-L) - \nabla \cdot (n\mathbf{V}_{i\perp}) - \nabla \cdot (n\mathbf{V}_{i\|})$$

Here the ion velocity has been expressed in terms of its components perpendicular ($\perp$) and parallel ($\|$) to the magnetic field. If we further simplify the problem by considering electric fields produced by the structure itself and neglect any parallel ion motion produced by the structure, then the ion velocity is given by (8.16) and the continuity equation can be rewritten in the form

$$\partial n/\partial t = -L - \nabla \cdot (\sigma_{\mathrm{P}i}/e)\mathbf{E} + \nabla \cdot (D_{i\perp}\nabla n)$$

At this point it is a common practice to "linearize" the equations by expressing each of the plasma properties as the sum of a background value and a small perturbation value due to the existence of the structure. Then all terms containing products of perturbation values are ignored, since they are much smaller than the other terms. If we denote all background values with the superscript "o", assume a horizontally stratified background ionosphere and ignore losses, then the continuity equation may be written finally as

$$\partial n/\partial t = -(\sigma_{\mathrm{P}i}^{o}/e)\,\nabla_x E + D_{i\perp}^{o}\,\nabla_x^2 n \qquad (8.19)$$

Now consider two extreme situations. First, suppose the local electric field associated with the plasma structure in the F region is completely shorted ($E = 0$) by mapping to a highly conducting E region. Investigation of (8.19) shows that for $E = 0$ the classical diffusion equation results with $D_\perp = D_{i\perp}$, the ion diffusion coefficient. This is the maximum possible value of $D_\perp$ under the circumstances considered. In the other limit, suppose that the structure cannot drive a Pedersen current because it maps to an E region that is an insulator. Then we may solve for the electric field obtained by setting the current in (8.18) to zero and substitute the result in (8.19). In the F region, where $R \ll 1$, the electric field from (8.18) becomes

$$E_x = \frac{e(D_{i\perp}^{o} - D_{e\perp}^{o})}{\sigma_{iP}^{o} + \sigma_{eP}^{o}}\,\nabla_x n$$

Substituting this result into (8.19) yields

$$\frac{\partial n}{\partial t} = \left[D_{i\perp}^{o} - \sigma_{iP}^{o}\left(\frac{D_{i\perp}^{o} + D_{e\perp}^{o}}{\sigma_{iP}^{o} + \sigma_{eP}^{o}} \right) \right] \nabla_x^2 n$$

Since in the F region $\sigma_{iP}^{o} \gg \sigma_{eP}^{o}$ this reduces to

$$\partial n/\partial t = D_{e\perp}(\nabla_x^2 n)$$

In this case therefore the velocity and associated decay rate of the structure will be the minimum possible under the circumstances considered and will be characterized by the electron diffusion coefficient $D_{e\perp}$.

From the above discussion it should be clear that the decay rate of an F-region irregularity depends not only on its scale size and the local ionospheric conditions but also on the mapping properties of the electric field and on the conductivity of the regions throughout which the field maps. In our description we considered two extreme examples where the electric field was easily determined. In practice, however, the electric field will be determined by applying the current continuity condition on an intermediate state. This yields slightly more complicated equations but with essentially the same characteristics.

One further point to consider is the effect that the current produced by a structure in the F region might have in the E region. This can be an important consideration because the relatively high collision frequencies in the E region mean that the plasma is compressible. If this compressional force can overcome the horizontal diffusion forces that oppose it in the E region, then a structure can form there that mimics the one in the F region. In order to be significant, this structure formation process in the E region must also overcome the high rate of dissociative recombination that exists at such altitudes. We now investigate the conditions under which such an "image" structure can be formed.

The image formation process can be understood from Fig. 8.19a for a pure decaying cylindrical irregularity in which the only electric field is due to the ambipolar effect. For a large enough structure the E field maps as shown and E-region ions are gathered together in the center region because of their Pedersen drifts. The requirement of charge neutrality is met by electrons flowing down **B** to the E region. The net result is that the E-region plasma density (ions and electrons) increases at the center. That is, an *image* high-density region forms in the center and two low-density regions form outside. The image amplitude will eventually be limited by recombination in the E region, which is proportional to n^2. Notice that the plasma-gathering process can be discussed in terms of compressibility of plasma in the E region in response to the applied electric field. The F-region plasma, on the other hand, is virtually incompressible perpendicular to **B**.

Although interesting in their own right, we have not explained the role images play in F-region diffusion. The E-region plasma density gradient due to an image drives an ion current opposite in direction to the mapped electric field. The net field-aligned current from the F region is therefore decreased and the F-region diffusion coefficient is lowered, tending more toward the low value of $D_\perp$ which pertains for an insulating E region. Now since the mapping process is itself scale size dependent, we conclude that the effective diffusion coefficient of high-latitude plasma irregularities may also be scale size dependent if the E-region density is neither too high nor too low. This important result is only now being included in models of irregularity formation and decay at high latitudes.

There is some experimental evidence that such image structures do occur, at least at low latitudes. The right-hand side of Fig. 8.21 shows the total plasma concentration observed by a near-equatorial rocket sounding. The high-altitude structure (just below the F peak) is due to equatorial spread F. The lower-altitude structure near 190 km shown with an arrow labeled "image region" is hard to explain by a plasma instability since it occurs in a region of zero vertical density

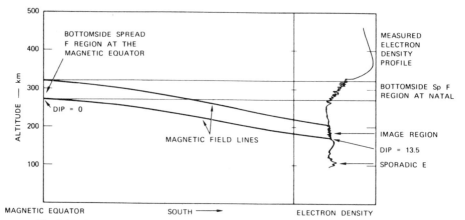

Fig. 8.21. Schematic diagram of the magnetic field mapping geometry used to explain the observed irregularities by the image mechanism. [After Vickrey *et al.* (1984). Reproduced with permission of the American Geophysical Union.]

gradient. Vickrey *et al.* (1984) suggested that equatorial spread F occurring at the magnetic equator (up the field line as shown in the figure) could have produced the observed structures by the image process since no gradient is needed to form images. Vickrey *et al.* (1984) calculated the expected image spectrum by considering an E-region system in field-aligned contact with an equatorial F-region plasma instability. Using (see Chapter 4)

$$\delta E/E_{\text{eff}} = \delta n/n$$

the $\delta n(k)/n$ spectrum measured at higher altitude on the same rocket was used to calculate $\delta E(k)$. As a first step they ignored electric field mapping effects and calculated the driven E-region image spectrum due to the applied $\delta E(k)$ assuming perfect mapping from the F-region source to the E region. Since the images themselves are damped as $k^2 D$, a power law in $\delta n(k)/n$ yields a *peaked* image spectrum in k space. The calculated spectrum is compared to an observed spectrum (from the data in Fig. 8.21) in Fig. 8.22. The solid line shows the calculated image spectrum for a "typical" bottomside spread F event. The dashed line shows the theory normalized to the peak in the observed image spectrum. The agreement is quite good in spectral form. The calculated amplitude was high by a factor of 3 (+ 10 dB), which is well within the spread of spread F fluctuation spectra. It is interesting to note that this driven preferred scale size is the same as that which is derivable from consideration of the electric field mapping process (LaBelle, 1985) so both effects may contribute.

To summarize, at high latitudes the electrical coupling of the E and F regions can have a significant effect on the lifetime and evolution of ionospheric structure, since the convection pattern will move such structure through regions with

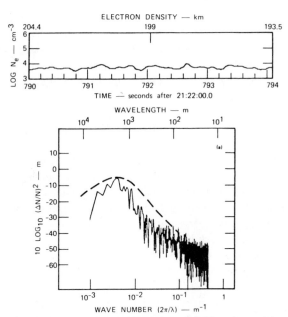

Fig. 8.22. Waveform and spectrum of "image" irregularities observed between 194- and 204-km altitude. The dashed line is the theoretical calculation normalized to the peak in the observed spectrum. The calculated curve was decreased by 10 dB, which corresponds to a factor of about 3 in irregularity amplitude. [After Vickrey *et al.* (1984). Reproduced with permission of the American Geophysical Union.]

greatly varying E-region conductivities. In the winter polar cap, for example, the E-region conductivity can be quite small and thus allow structure-producing fields to exist at most scale sizes. In the cusp region, the precipitation of electrons in the energy range 100 to 500 eV can itself be a mechanism for producing structure in the F-region plasma concentration. The scale size of such structure would be larger than about 50 km, reflecting the spatial extent of the cusp and the convection speed of the plasma through it. These large structures would easily survive across the entire polar cap and their amplitude would have a strong seasonal dependence. Such structures, convecting across the polar cap, would be subject to the $\mathbf{E} \times \mathbf{B}$ drift instability. Thus, many smaller-scale structures could be seen whose amplitudes increase in the nighttime polar cap. In the auroral zone itself the E region is generally a good conductor, and thus we would expect that the generation of intermediate-scale structure would be inhibited and that their magnitude might be generally lower than would be found in a region of lower E-region conductivity. However, the auroral zone can be the site of mixing by structured velocity fields that also lead to the generation of intermediate-scale structures. The combination of all possible structure-producing mechanisms with the convective history of the plasma with the goal of arriving at a predicted distribution of irregularities may well be an unrealistic task, particularly since the system changes on time scales shorter than the convective cycle. However, some steps establishing the validity of different structure-producing mechanisms can now be made and progress beyond the simple convection–diffusion model of Kelley et al. (1982a) is possible (e.g., Schunk and Sojka, 1987).

8.3   Small-Scale Waves in the High-Latitude F Region

For this purpose we define small-scale wavelengths to be less than 100 m. Research in this area has not progressed as rapidly as it has in the equatorial zone. One reason is that the observational data base using radars is not very large. In the equatorial case the Jicamarca and Altair radars can easily scatter from irregularities with wave numbers perpendicular to **B** since the geometry is favorable. Since these waves have much larger growth rates than waves with finite $k_\parallel$, they dominate the spectrum. At the high-latitude radar sites it is geometrically impossible to obtain the appropriate backscatter angle in the F region using VHF and UHF systems.

F-region radar observations are possible at HF frequencies since ionospheric refraction can be used to bend the radar signals and attain a scattering geometry nearly perpendicular to **B**. Successful measurements of this type have been made in Scandinavia (Villian et al., 1985) and from a transmitting station in Goose Bay, Labrador, Canada (Greenwald et al., 1983). The observation volume for the latter site is located over the Sondre Stromfjord, Greenland, incoherent scatter radar station. The preliminary results show that echoes are indeed received in the tens of meters wavelength range but a detailed explanation does not yet exist for the scattering structures.

Clues concerning the origin of these waves come from the Doppler shift of the returned HF signal. During one event simultaneous measurements were made with the incoherent scatter radar at Sondre Stromfjord. The latter can be used independently to determine the plasma drift velocity projected along the HF radar line of sight. The two data sets are superimposed in Fig. 8.23 and show quite good agreement. The implication is that the irregularities are frozen into the plasma flow, that is, that the phase velocity of the structures is small in the plasma reference frame.

The long-wavelength instability processes discussed in Section 8.2 do have this low-phase velocity property. However, it is easy to show that at a wavelength of 10 m, such local instabilities are stable. For example, we may let $E_{0\perp} = 20$ mV/m, $L = 10$ km, and $D_\perp = 1$ m²/s, which is on the small side, and solve the equation

$$\gamma = mE_{0\perp}/LB - k_m^2 D_\perp = 0$$

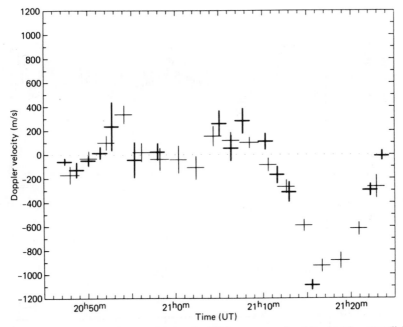

Fig. 8.23. Comparison of time series data for HF (heavy crosses) and incoherent scatter (light crosses) velocity estimates calculated for the returns received from the nearby volumes. [After Ruohoniemi *et al.* (1987). Reproduced with permission of the American Geophysical Union.]

for the marginally stable wave number k_m. This yields a wavelength $\lambda_m = 30$ m. A larger diffusion coefficient will only make the value of λ_m larger. For example, in equatorial spread F (see Chapter 4) anomalous diffusion sets in and cuts off the fluctuation spectrum near $\lambda = 100$ m.

Unless some additional instability arises at small scales, a possibility we address below, the only way waves with $\lambda \leq 10$ m can arise is through a cascade of structure from long to short λ. There is a curious difference between high- and low-latitude irregularities in this regard. As discussed in Chapter 4, equatorial spread F evolves in such a way that anomalous diffusion due to drift waves determines the spectrum. A very steep spectrum (k^{-5}) evolves for $\delta n^2(k)/n^2$ when $\lambda \leq 100$ m. This spectral range looks very much like a viscous subrange in neutral turbulence. To date, no evidence for such an anomalous diffusion process has been found at high latitudes and no analogy to the viscous subrange has been reported.

Measurements on satellites and rockets show that the spectrum of plasma density fluctuations varies as k^{-n} in the wavelength regime from tens of meters to tens of kilometers (Dyson *et al.*, 1974; Sagalyn *et al.*, 1974; Cerisier *et al.*,

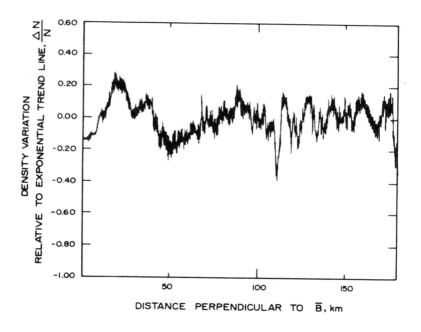

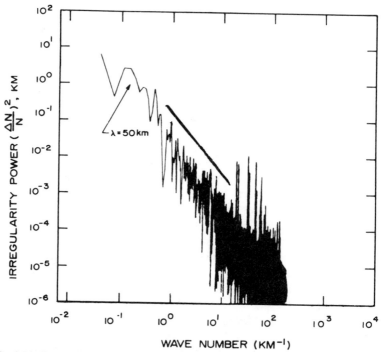

Fig. 8.24. Horizontal variations in electron density relative to an exponential dependence with altitude (top) and power spectrum of the data (bottom). [After Kelley *et al.* (1980). Reproduced with permission of the American Geophysical Union.]

1985) where typically $1.5 \le n \le 2.5$. As discussed in the section on equatorial spread F, this spectral form has also been reported for equatorial measurements and barium cloud striations. In both bottomside spread F and low-altitude barium striations such spectra were shown to be due to wave steepening and not to turbulence (Costa and Kelley, 1978; Kelley *et al.*, 1979). There has been no evidence to date that steepening occurs at high latitudes; rather, turbulent processes appear to dominate. An example from a rocket flight is reproduced in Fig. 8.24. The altitude dependence of the plasma density has been removed from the data by a detrending technique, and the quantity $\Delta n/n$ has been plotted as a function of distance perpendicular to **B.** The Fourier transform of these data is also shown and indicates a power spectral index n of about 1.6, which is consistent with a Kolmogorov ($n = \frac{5}{3}$) spectrum (the straight line has a $\frac{5}{3}$ slope). These auroral rocket data were obtained within minutes of the poleward surge of a magnetospheric substorm. The signal level was above the noise for $\lambda \ge 60$ m.

Information on even shorter-wavelength waves (≤ 10 m) comes almost entirely from rocket measurements, due to the high data rates required. Three distinct types of irregularities were observed during a rocket flight into the recovery phase of a magnetospheric substorm on April 2, 1970 (Bering *et al.*, 1975; Kelley *et al.*, 1975; Kelley and Carlson, 1977). At the edge of an auroral arc, as detected by onboard particle detectors, intense (5 mV/m) oxygen electrostatic ion cyclotron waves were observed in data from the electric field wave receiver. A spectrum from the rocket results is presented in Fig. 8.25. Existence of such waves is consistent with auroral backscatter measurements of cyclotron waves in the ionosphere. Both data sets indicate that the waves occur in a very localized area. The waves detected on the rocket were observed in conjunction with an intense upward field-aligned current. Hydrogen ion cyclotron waves are com-

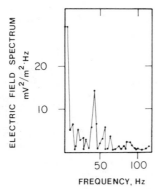

Fig. 8.25. Detection of a pure oxygen cyclotron wave near the edge of an auroral arc. [After Kelley *et al.* (1975). Reproduced with permission of the American Institute of Physics.]

monly observed above 5000 km (Kintner *et al.*, 1978) but observations of oxygen cyclotron waves such as these are more rare.

Kelley and Carlson (1977) also showed evidence for less intense and more broadband electrostatic waves in regions of intense velocity shear and field-aligned currents at the edge of the same auroral arc. The velocity shear measured by the dc electric field detector in the same region was 20 s^{-1}, which is very high and implies a change in the plasma flow velocity comparable to the sound speed in one ion gyroradius! Less intense irregularities in the same wavelength regime were observed equatorward of the arc but not above the arc itself or poleward of it. The dc electric field was such that the ionospheric plasma convection had an equatorward component in the whole region probed, which suggests that the structures may have been formed at the arc boundary and were transported equatorward. Similar spectra have been reported in the dayside auroral oval by Earle (1988).

The possible origins of ionospheric structure with scales in the range of a few meters ($\approx r_{gi}$) to a few tens of meters are quite numerous and no unifying mechanism is likely to explain all of the observations. It is very likely that some energy cascades into this wavelength range from larger scales in a manner similar to neutral turbulence. In a plasma the situation is made much more complicated by the possible generation of wave modes which compete with or modify the cascade concept. We have seen in Chapter 4 that in the equatorial spread F case, drift waves arise when the turbulence becomes strong *and* when the altitude is high enough that collisions are very weak. This wave mode has a very low growth rate, however, and requires a fairly well-behaved plasma to be important. In other words, the high-latitude ionosphere has additional free energy sources which may swamp the drift mode. At any rate, the existence of drift waves has not been shown in the high-latitude case although they may well be present.

Of the several possible free energy sources available, most theoretical effort has gone into calculations of the generation of electrostatic ion cyclotron waves by field-aligned currents and, as noted above, evidence does exist for their generation. This wave mode arises when

$$V_i^{th} \leq J_\parallel/ne \leq V_e^{th}$$

which corresponds to parallel current due to a differential drift between ions and electrons which falls between the ion and electron thermal speeds. In the lower ionosphere collisions must be taken into account, while higher up collisionless theory is adequate. The threshold parallel current density for O$^+$ and NO$^+$ cyclotron waves is plotted in Fig. 8.26 for a reference ionospheric profile and for a wavelength corresponding to $(kr_{gi}) = \sqrt{2}$. The calculation includes collisions. The required current densities are higher than the average observed field-aligned currents but not higher than some of the largest reported examples of $J_\parallel$ (e.g., Burke *et al.*, 1980). Thus, oxygen cyclotron waves should occur in the iono-

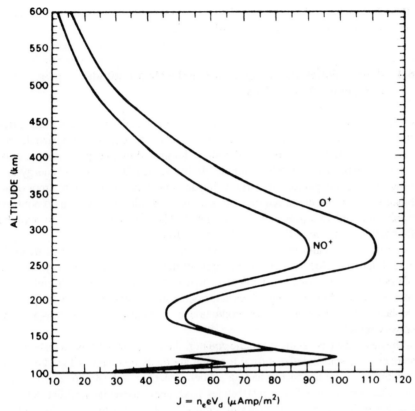

Fig. 8.26. Threshold currents required to excite O^+ and NO^+ EIC waves in O^+ and NO^+ plasmas, respectively. The parameters used are $k\rho_i = (2)^{1/2}$ and $k_\parallel/k_\perp = 0.06$. [After Satyanarayana and Chaturvedi (1985). Reproduced with permission of the American Geophysical Union.]

sphere but are rare and may be restricted to the edges of auroral arcs and/or regions of strong velocity turbulence.

Although most theoretical effort has involved current-driven modes, Ganguli *et al.* (1985) have pointed out that intense shears can also generate ion cyclotron waves. Since shears and intense field-aligned currents are often colocated [see Eq. (8.4)] it is not yet clear which free energy source is most important. Mixing and turbulent cascade will occur as well. Basu *et al.* (1986) have also shown a tendency for intense irregularities at large scales to arise in regions of velocity shear. An ionospheric Kelvin–Helmholtz instability may occur due to these shears (Keskinen *et al.*, 1988). Finally, it seems very likely that turbulence in the auroral acceleration zone will propagate via an Alfven wave mode to ionospheric altitudes and create considerable structure in the plasma flow velocity and the plasma density (Knudsen *et al.*, 1988).

8.4 Plasma Waves and Irregularities in the High-Latitude E Region—Observations

Due to the high neutral density at E-region heights, plasma production, diffusion, and recombination all proceed very rapidly. The large-scale horizontal organization of E-region plasma is therefore dominated by the spatial character of solar and particle precipitation sources. In the vertical direction, the energy distributions of the photon or particle impact ionization sources are important. Since we are more interested in dynamical processes in this text, we do not treat production and loss processes in any detail. The reader is referred to texts by Rishbeth and Garriott (1969) and Banks and Kockarts (1973).

Gravity waves are somewhat less important as a source of vertical structuring at high latitudes due to the large dip angle of the magnetic field. At any rate, the physics of the layering is identical to that discussed in Chapter 5 and is not repeated here. Any layering which does occur due to gravity waves tends to be washed out by particle impact production of plasma, which smooths out the density profile.

Just as is the case at the magnetic equator, the two primary plasma instabilities sources of structure in the auroral E region are the two-stream and gradient drift instabilities, with the corresponding echoes sometimes referred to as type 1 and type 2, respectively. These instabilities have been discussed in great detail in Chapter 4. There are some major differences in the character of the ionosphere at high latitudes, however, which affect the relative importance of these two processes. In addition, there is evidence for a third mode (type 3) and for a special type of two-stream event associated with anomalous electron heating by the electrojet waves (type 4). The electric field in the polar cap and auroral zone is applied from "above" and is appreciably larger and more widespread than in the equatorial region. The average auroral electric field, for example, has a strong diurnal component with an amplitude of 30 mV/m and maxima near 0500 and 1800 LT (Mozer and Lucht, 1974). Fields as high as 50 mV/m are common in both the oval and polar cap, and numerous measurements exceeding 100 mV/m have been reported. The latter corresponds to a drift velocity of 2000 m/s, which is a value five times higher than the threshold of the two-stream instability.

The plasma density structure is much more variable in the auroral zone than in the equatorial region. However, in the auroral region the magnetic field is inclined at about 10° to the vertical, and therefore a vertical electron density gradient has a density scale length perpendicular to the magnetic field which is appreciably larger than the same vertical gradient would have at the magnetic equator. This in turn means that the gradient drift instability tends to have a lower growth rate at high latitudes for the same size vertical gradients even though the plasma drifts are larger. In the auroral region there are also important horizontal density gradients due to spatial variations in the particle precipitation, which must be considered but which are difficult to measure.

The dynamic nature of both electric fields and density gradients makes analysis of auroral E-region instabilities even more challenging than their equatorial counterpart. On the other hand, a wealth of radar data does exist at E-region heights. These data have been supplemented by results from a number of rocket flights and a fairly complete experimental picture has emerged. In this chapter we concentrate on the plasma instabilities which operate in the auroral E region. In the next few sections some of the salient experimental data are presented. This is followed by a theoretical discussion in which additional data are introduced where appropriate.

Since auroral currents can be quite large, the term auroral electrojet is often used. The "jet" in this case refers to the relatively localized height range of the current rather than a narrow latitudinal range which characterizes the equatorial jet as well.

8.4.1 Radar Observations

In the equatorial electrojet most radar observations resolve into two classes: type 1 irregularities, which display a narrow Doppler spike that is offset from zero by the acoustic frequency $\omega_A = kC_s$; and type 2 irregularities, which display a broad Doppler spread centered on the frequency corresponding to the line-of-sight electron drift velocity. Both of these echo types have their counterparts in the auroral case. However, as summarized in a review by Fejer and Kelley (1980), there are various Doppler signatures associated with the "radar aurora." Since that review was written, considerable progress has been made in theoretical and experimental studies aimed at sorting out the origins of the various spectral signatures.

The four Doppler spectral types are shown in Fig. 8.27. The spectrum in the upper panel is very much like a narrow type 1 echo at the magnetic equator which locks onto the acoustic frequency. The second spectrum is very broad and is not unlike the equatorial type 2 case.

In the third panel a very narrow spectrum is shown that has a Doppler velocity which is much less than the acoustic speed and which corresponds to a Doppler shift of about 70 Hz. The signals in this mode are very strong and hence when the spectra are normalized the narrow spike becomes a dominant feature. These waves were first reported by Fejer et al. (1984a) and have inspired considerable theoretical interest. The implication of the narrow spike is that a very coherent

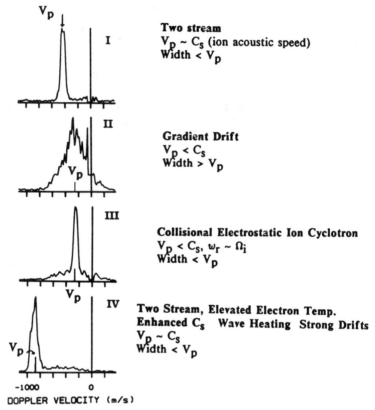

Fig. 8.27. The four types of radar spectra that are observed in the auroral electrojet. (Figure courtesy of J. Providakes.)

wave is present. The 70-Hz Doppler shift is about 40% above the gyrofrequency of O^+, which has led to a number of studies dealing with the excitation of oxygen cyclotron waves in the upper E region or lower F region. We return to this topic in the discussion section below.

We turn now to the fourth type of Doppler signature. High-latitude Doppler spectra often display such large "type 1" Doppler shifts that it was originally thought that in the high-latitude case, 3-m waves did not reach a limiting phase velocity at C_s. The Doppler drift was therefore assumed to be equal to the line-of-sight electron drift speed, a result in agreement with the linear theory of two-stream irregularities. This interpretation is very much in doubt now, and a considerable literature exists on anomalous electron heating due to electrojet

ITHACA, NY SEPTEMBER 6, 1982

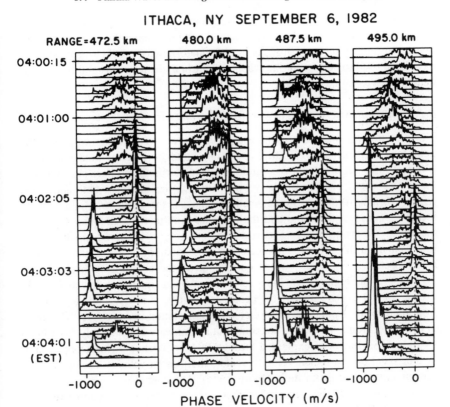

Fig. 8.28. Examples of the temporal and spatial variations of backscattered power and Doppler velocity at 50 MHz in the auroral oval. The spectra were normalized to the same value and the integration time was about 6 s. [After Providakes *et al.* (1985). Reproduced with permission of the American Geophysical Union.]

turbulence. This topic is taken up separately in some detail in Section 8.5.2. As we shall see, the result is that the type 4 waves might still saturate at a phase velocity equal to C_s but that the sound velocity is much higher than expected.

An example showing the complexity of the radar aurora is given in Fig. 8.28. Consecutive spectra separated by about 6 s are shown from four different ranges. Many examples of broad type 2 Doppler shifts were obtained during this time period, but in addition very strong sporadic echoes were seen which have very large Doppler shifts. The spectra look like type 1 events but have shifts much larger than expected for model values of the temperature and hence the acoustic speed at E-region altitudes. These are referred to as type 4 echoes.

Since the coherent scatter radar data are complicated to interpret, it is important to determine the zero-order plasma properties and the k spectrum of the waves independently in order to distinguish and understand the various possible plasma instability mechanisms. As in the equatorial case, rocket measurements have been used to add to our knowledge of auroral electrojet instabilities. These data are discussed next.

8.4.2   Rocket Observations of Auroral Electrojet Instabilities

The plot in Fig. 8.29 is a composite from six different rocket wave electric field measurements in very different background dc electric field conditions (Pfaff, 1985). The frequency–altitude–intensity format used here shows where the waves occur and their relative intensity. In the bottom right-hand panel ($E_{dc} = 10$ mV/m) there is no detectable signal. As the dc electric field strength increases the layer thickness and the intensity both increase dramatically. Near the top of the layer a discrete narrow band emission is observed in some cases (e.g., panels a through d). This feature is also apparent in the sequence of discrete spectra in Fig. 8.30, which correspond to the experiment in panel b of Fig. 8.29. At a height of 120 km a very narrow band signal was detected which, except for the exact value of the center frequency, is nearly indistinguishable from the topside *equatorial* electrojet spectrum plotted in Fig. 4.30. The value of the center frequency in this coherent feature can vary considerably from flight to flight. This has been explained through the dependence of the Doppler shift of the signal on the rocket velocity vector, the plasma velocity vector, and the acoustic speed in a given flight (Pfaff, 1985). The data seem to indicate a very turbulent "heart" to the electrojet with coherent structures in the upper regions. This description is very similar to that of the daytime equatorial electrojet rocket data discussed in Chapter 4.

Most rocket measurements, including those shown here, emphasize waves with $\lambda \leq$ tens of meters. This is also true in the equatorial electrojet. This limitation is due to the high velocity at which most rockets penetrate the E region. Like the Condor experiments (see Chapter 4), the Auroral-E rocket experiment was dedicated to low-altitude studies and hence the rocket spent more time in the electrojet (Pfaff, 1985). This allows longer-wavelength waves to be studied. The electric field data from the latter experiment are shown in Fig. 8.31e. The electric field signal is modulated by the rocket spin. The envelope of this modulation is smoothly varying above about 102 km. Below this height the spin signal is modulated by the geophysical phenomenon of interest here. Analysis indicates that these modulations correspond to roughly a 300-m horizontal wavelength for the electric field fluctuation. These waves occurred in a region of vertical plasma density gradient (see Fig. 8.31a) and of northward dc electric field. This, of course, is the proper orientation for instability of the gradient drift wave. Notice that the waves are localized in altitude even though the near-vertical magnetic

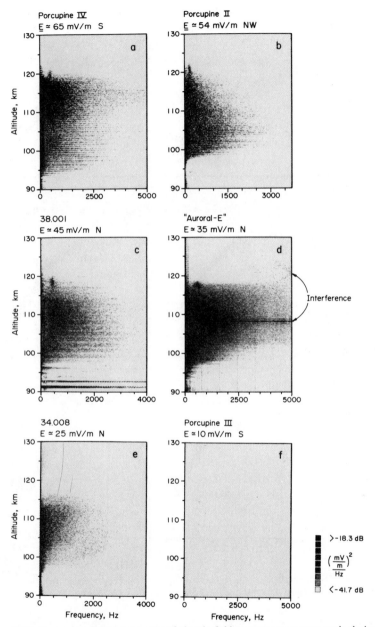

Fig. 8.29. Frequency–height sonograms of electric field wave measurements made during six different auroral zone experiments. The gray levels are identical for all six sonograms and the dc electric field is noted for each panel. (Figure courtesy of R. Pfaff.)

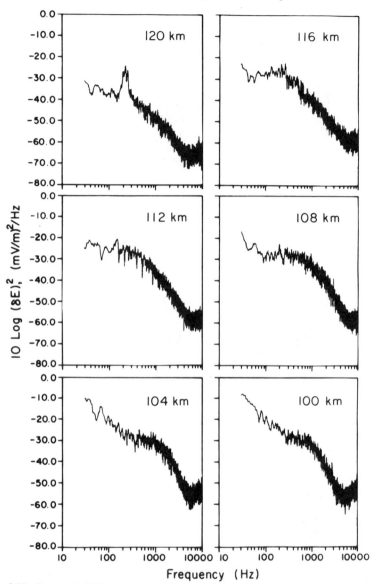

Fig. 8.30. Six electric field spectra spaced 4 km apart as a rocket descended through the electrojet. The spectra each represent 1.092 s of data. (Figure courtesy of R. Pfaff.)

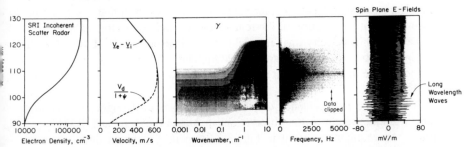

Auroral Electrojet — Poker Flat, Alaska
7 March 1981

Fig. 8.31. Growth rate modeling for the combined two-stream and gradient drift instability and comparison to the Auroral-E data. The model assumes a stable, vertical electron density gradient as measured by the SRI incoherent scatter radar and a 35-mV/m ambient electric field. The electron density is assumed to be uniform in horizontal extent. The *in situ* measured electric field data are shown in the two panels to the far right. (Figure courtesy of R. Pfaff.)

field allows very efficient coupling along **B.** We return to a discussion of this problem and these same data below.

8.4.3   Simultaneous Data Sets

As in the equatorial case, significant advances are possible when more than one measurement technique is applied in a given experiment. The data in Fig. 8.31 are of this type, since the incoherent scatter radar at Chatanika was used to provide the electron density profile shown in panel a as well as the ambient electric field. Armed with these data, the vertical profile of $V_e - V_i$ was calculated (panel b) and the linear theory (see Section 8.5) was used to calculate the growth rate as a function of wave number in panel c. When compared to the electric field wave data in panels d and e, it is clear at once that high-frequency (high k value) waves are observed in the upper electrojet, where the two-stream term dominates. However, the long-wavelength waves are present in the bottom-side gradient region, where the quantity $V_d/(1 + \psi)$ is below the threshold value (C_s) required for two-stream instability. It is clear from this study that both the gradient drift and the two-stream condition play a role in the auroral electrojet.

A controversy over interpretation of the phase velocity of meter-wavelength waves at high latitudes has been resolved by simultaneous coherent scatter radar and incoherent scatter radar observations. For years the STARE coherent scatter Doppler data were interpreted as representing the projection of $(\mathbf{E} \times \mathbf{B}/B^2)$ on the radar line of sight. This interpretation agrees with linear theory since (see Chapter 4) that theory predicts

$$\omega \simeq \mathbf{k} \cdot \mathbf{V}_D/(1 + \psi) \qquad\qquad (8.20)$$

but disagrees with the equatorial electrojet observations, which show that $\omega/k = C_s$ at any angle to the current. The equatorial result was not applied at high latitudes, since such large Doppler velocities were observed that it seemed unreasonable that the electron or ion temperatures could be as large as required if the Doppler shift velocity was set equal to C_s. On the other hand, $\mathbf{E} \times \mathbf{B}/B^2$ drifts as large as 2000 m/s are not uncommon at high latitudes, so it was not unreasonable to assume that the coherent scatter Doppler shift yields the projection of the $\mathbf{E} \times \mathbf{B}$ velocity.

Some of the first evidence that E-region temperatures might be very high came from incoherent scatter observations in Alaska (Schlegel and St.-Maurice, 1981). An example is shown in Fig. 8.32, where T_e is about three times the neutral gas temperature. The dc electric field in this event was measured by the same radar to be 85 mV/m, which clearly must have been driving very intense waves in the electrojet (e.g., see Fig. 8.29). For reference, the sound speed for a mean ion mass of 31 amu, $T_e = 1200$ K, and $T_i = 500$ K is 900 m/s. Simultaneous observations of the plasma drift velocity and the electron temperature were made at EISCAT in conjunction with the STARE coherent scatter radar (Nielsen and Schlegel, 1985). The data are compared in Fig. 8.33 for (a) the linear "fluid" model and (b) the "acoustic" model. The acoustic model is clearly in better agreement with the data.

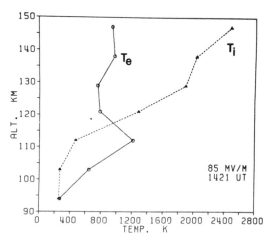

Fig. 8.32. Typical electron (solid line) and ion (dashed line) temperature profiles in the polar E region in the presence of a high electric field (southward, 85 mV/m in this example) measured at 1421 UT on November 13, 1979. [After Schlegel and St.-Maurice (1981). Reproduced with permission of the American Geophysical Union.]

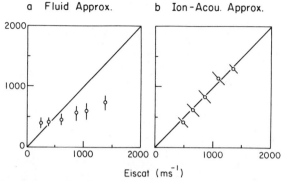

Fig. 8.33. Magnitude and direction of the electron drift velocity estimated from the STARE Doppler velocity measurements, applying (a) the cosine relationship and (b) the ion acoustic approach, as a function of the electron drift velocity determined by the incoherent scatter facility EISCAT. [After Neilsen and Schlegel (1985). Reproduced with permission of the American Geophysical Union.]

From these results it seems clear that the type 4 echo occurs when strong electron heating occurs in the electrojet due to the waves themselves. The anomalous heating mechanism is not entirely understood at present but there seems little doubt that it is occurring (Providakes *et al.*, 1988). This interesting nonlinear plasma problem is treated further below.

8.4.4   Laboratory Results

Considerable effort has gone into laboratory experiments involving the Farley–Buneman instability. The results are very interesting and not entirely consistent with the space results. For example, D'Angelo *et al.* (1974) found that the pure two-stream mode traveled with a phase velocity equal in magnitude to the $\mathbf{E} \times \mathbf{B}/B^2$ velocity. That is, they found no evidence for saturation for the phase velocity at C_s. Perhaps their experimental waves are similar to waves at the top of the equatorial electrojet, for which Pfaff *et al.* (1988) have presented evidence that horizontally propagating waves are such that $\omega/k > C_s$. These waves cannot be detected by ground-based radars. This would imply that fundamentally different physics pertains in the two-dimensionally turbulent electrojet as distinguished from the laboratory experiments. The coherent waves at the top of auroral electrojet may also be of this type.

In another set of laboratory experiments, Kustom *et al.* (1985) drove the plasma with drift velocities of several times C_s to study the effect on the wave spectrum. They found that the power level increases at high drift speeds but that the spectral form of fluctuations does not change. They found that the potential

$\delta\phi^2$ varied as $k^{-3.5}$ at all levels of turbulence. Since $\delta E = -i k \, \delta\phi$, this implies an electric field power spectrum varying as $k^{-1.5}$. These results are not dissimilar to the rocket δE^2 measurements presented in Fig. 8.30 at high frequencies. The latter show a $k^{-1.5}$ spectrum which breaks to a much steeper value near 1000 Hz (frequency measured in the rocket frame). The break could be due to a transition in the measurement when $\lambda \leq d$, the probe separation, or to some "viscous" cutoff. For such short-wavelength waves the probes respond to potential fluctuations rather than to electric fields.

Putting these details aside, Kustom *et al.* (1985) and Pfaff *et al.* (1984) agree that the fluctuation spectrum is monotonically decreasing and that the electron heating theories cannot be simply attributed to intense band-limited short-wavelength waves as originally suggested (St.-Maurice *et al.*, 1981).

8.5   Auroral Electrojet Theories

The local linear theory for the gradient drift and two-stream instabilities in the auroral electrojet does not differ very much from its counterpart in the equatorial case (see Chapter 4). In this section we delineate some of the additional factors which arise at high latitudes. By local in this context we mean that we ignore coupling along the magnetic field lines. We follow the presentation by Fejer and Providakes (1988) quite closely below.

In the lower E region (between about 90 and 100 km) the ions are unmagnetized ($\nu_i \gg \Omega_i$) and the electrons are collisionless to zero order ($\Omega_e \gg \nu_e$), so that $V_D = E \times B / B^2$. Linear theory then is identical to the result in Chapter 4 and the oscillation frequency and growth rate, in the reference frame of the neutral wind, are given by

$$\omega_r = \mathbf{k} \cdot (\mathbf{V}_D + \Psi \mathbf{V}_{Di})/(1 + \Psi) \tag{8.21}$$

$$\gamma = (1 + \Psi)^{-1}\{(\Psi/\nu_i)[(\omega_r - \mathbf{k} \cdot \mathbf{V}_{Di})^2 - k^2 C_s^2] + (1/Lk^2)(\omega_r - \mathbf{k} \cdot \mathbf{V}_{Di})(\nu_i/\Omega_i)k_y\} - 2\alpha n_0 \tag{8.22}$$

where

$$\Psi = \Psi_0[(k_\perp^2/k^2) + (\Omega_e^2/\nu_e^2)(k_\parallel^2/k^2)]$$

and

$$\Psi_0 = \nu_e \nu_i / \Omega_e \Omega_i$$

In this expression $\mathbf{V}_{Di}$ is the ion drift velocity, which could be due either to neutral winds or to the ambient electric field. The assumptions used are $\omega_r \ll \nu_i$, or in other words that the wavelengths are much larger than the ion mean free path, and $|\gamma| \ll \omega_r$. In addition, we consider $k \gg k_0$, where $k_0 = (\nu_i/\Omega_i)$ $[(1 + \Psi)L_N]^{-1}$, so that the linear waves are nondispersive. These approxima-

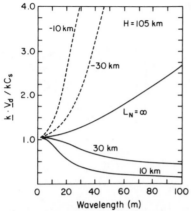

Fig. 8.34. Variation of the normalized threshold drift velocity with wavelength and electron density gradient length for waves perpendicular to the magnetic field. The solid (dashed) curves correspond to destabilizing (stabilizing) electron density gradients. [After Fejer *et al.* (1984b). Reproduced with permission of the American Geophysical Union.]

tions are valid for wavelengths between a few meters and a few hundred meters. For shorter wavelengths, kinetic theories are needed. The first term in the right-hand side of (8.22) is the two-stream term, which includes a diffusive damping term of the form k^2C_s. The second term describes the gradient drift instability, while the last term is due to recombinational damping. The two-stream term is dominant at short wavelengths (1 m $\leq \lambda \leq$ 20 m) and yields instability when $\mathbf{k} \cdot \mathbf{V}_D > kC_s(1 + \Psi)$. For reference a set of high-latitude parameters was chosen and the threshold drift velocity for instability evaluated as a function of wavelength. The results are shown in Fig. 8.34 for $k_\parallel = 0$ at a height of 105 km. The parameters used are $\nu_e = 4 \times 10^4 \text{ s}^{-1}$, $\nu_i = 2.5 \times 10^3 \text{ s}^{-1}$, $\Omega_e = 10^7 \text{ s}^{-1}$, $\Omega_i = 180 \text{ s}^{-1}$, $C_s = 360$ m/s, and $2\alpha n_0 = 0.06 \text{ s}^{-1}$. Several different gradient scale lengths were used. Large-scale waves are easily excited only if the electron density gradient is destabilizing (L_N is negative). Note also that the two-stream threshold drift ($L_N = \infty$) increases rapidly with wavelength for $\lambda \geq 20$ m. Therefore, the two-stream instability mechanism is essentially restricted to short wavelengths.

8.5.1   The Gradient Drift Instability

There are several ways to obtain density gradients at high latitudes. Figure 8.35 illustrates the case when a *vertical* density gradient exists in the presence of a dip angle which is large but is not equal to 90°. This situation applies when solar illumination produces the E-region plasma or when diffuse auroral precipitation is the source. Then a vertically upward gradient is unstable for a northward

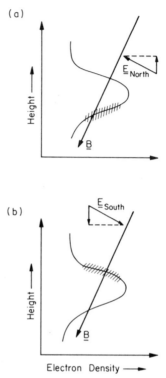

Fig. 8.35. Sketch showing (a) unstable positive gradient in the vertical electron density profile in the presence of a poleward (northward in the northern hemisphere) electric field and (b) unstable negative gradient in the presence of an equatorial (southward in the northern hemisphere) electric field. (Figure courtesy of R. Pfaff.)

electric field while a downward gradient is unstable for a southward field. The gradient scale length perpendicular to B is $L_\perp = L_N (\cos I)^{-1}$, where I is the dip angle. For the same value of L_N, $L_\perp$ is much larger at high latitudes than in the mid-latitude or equatorial case. Even when classical sporadic E layers form at high latitudes, the cos I term reduces their importance. Steep horizontal gradients may arise due to very intense localized auroral arcs. However, such arcs have a high electron density, which drastically increases the recombinational damping. Hence, the gradient drift process is less important at high latitudes than elsewhere in the ionosphere.

That said, we now proceed to show that the gradients still cannot be totally ignored. In Fig. 8.36 the electron density profile and intermediate- to short-wavelength wave data are plotted side by side for two rocket flights through the auroral electrojet. In the upper case **E** had a northward component, while in the lower case **E** was southward. In each case the solid trace on the left shows

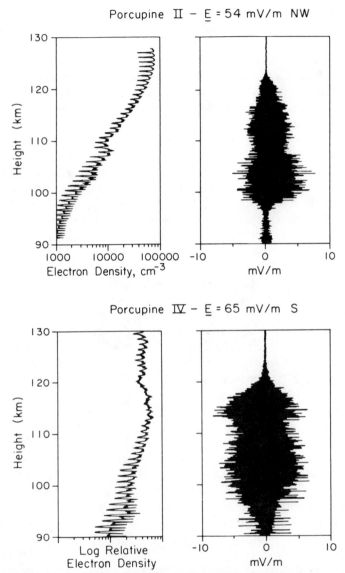

Fig. 8.36. Wave electric field and fixed-bias Langmuir probe data from the traversal of the Porcupine II and Porcupine IV payloads through the electrojet region. The Porcupine II experiment was conducted during a northward electric field, whereas the Porcupine IV experiment was flown in the presence of a southward electric field. The solid line in the left-hand panel shows the density profile. The density decreases are due to probe end effects. (Figure courtesy of R. Pfaff.)

the true electron density variation with altitude. In the upper plot the wave activity shows a local minimum exactly where the density gradient changes sign. In the lower example the wave activity increases when the vertical density gradient changes sign. The implication is that the electron density gradient controls or at least affects the intensity of electrojet turbulence but that the two-stream process is the dominant source of wave activity. In both of these events the electron drift speed was well over 1000 m/s, much higher than the typical sound speed.

One example which clearly seems to show an auroral event in which a dominant role is played by the gradient drift instability was shown earlier in Fig. 8.31. Here large-wavelength electric field fluctuations were observed on the bottomside. The waves are quite intense ($\delta E \approx 25$ mV/m) and are polarized in the direction of the $\mathbf{E} \times \mathbf{B}$ drift (Pfaff, 1986). Of course, this low-apogee flight was one of the few rocket experiments capable of detecting large-scale waves on the bottomside and hence they may well exist much of the time when the electric field is northward and when the electron density n_0 is not so high that recombination dominates.

This example is potentially quite interesting since it illustrates the altitude dependence of the wave amplitude. A detailed study of the linear theory for the event has been carried out and the results were plotted in panel c of Fig. 8.31. Indeed, the long-wavelength waves are observed at the same altitude as those where the growth rate in panel c shows positive values at long wavelengths. The observed longest-wavelength waves (panel e) seem to cut off just above 100-km altitude, although the intermediate- to short-wavelength waves (panel d) remain strong between 100 and 110 km.

We now consider the extent to which these long-wavelength waves map to higher altitudes along the magnetic field lines. The theory of the mapping of electrostatic fields (which can include the fields associated with the long-wavelength waves) along magnetic field lines was reviewed in Chapter 2. Although the magnetic field lines may sometimes be considered equipotentials, in practice a nonzero attenuation will occur for wave electric fields. The parallel attenuation length resulting from the mapping of an electric field perpendicular to the magnetic field may be approximated by

$$\lambda_\parallel \simeq (\sigma_0/\sigma_P)^{+1/2}\lambda_\perp \tag{8.23}$$

where σ_0 and σ_P are the parallel and Pedersen conductivities. In Fig. 8.37 $(\sigma_0\sigma_P)^{1/2}$ is plotted as a function of altitude using a model neutral atmosphere and the geomagnetic field at Poker Flat, Alaska. The computation of this ratio was very sensitive to the particular expression used for the collision frequencies and may vary quite a bit at these low altitudes. In addition, since these collision frequency values depend critically on the electron temperature, which also varies rapidly here, we have plotted two curves corresponding to $T_e = 500$ K and $T_e =$

Poker Flat, Alaska

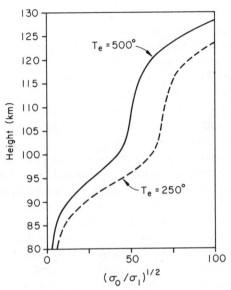

Fig. 8.37. Curves representing $(\sigma_0/\sigma_P)^{1/2}$ in the auroral zone for constant temperature profiles corresponding to $T_e = 500$ K and $T_e = 250$ K. (Figure courtesy of R. Pfaff.)

250 K. Notice that the conductivity ratio changes considerably in the region we are considering. For the sake of discussion, consider the value of this ratio at 96 km to be approximately 35. Thus, using (8.23), a 300-m wave would map through a distance of 10.5 km as it decays along the field lines. The calculated distance is somewhat larger than that given by the observations but the basic idea seems correct. In any event, this calculation is consistent with the hypothesis that the source region of the waves was very narrow in height. In turn, this supports the claim that the waves were produced by a locally unstable electron density gradient and that the observed dominant wavelength may have corresponded to the peak in the linear growth rate for the gradient drift instability. The perturbation electric field mapped to a region where the waves were stable. Electrojet theory has not yet come to grips with such mapping effects in any detail. This set of measurements could, in fact, be a good starting point for a nonlocal study of electrojet instabilities including such electric field mapping.

8.5.2 *The Two-Stream Instability and Type 4 Radar Echoes*

Of course the two-stream instability is very easily excited in the auroral case and it is probably responsible for most radar echoes. The possibility that such waves heat E-region electrons to temperatures far in excess of that of the neutral gas makes their study very interesting, since such wave-induced processes are very important in space plasma physics in general.

The type 4 observations discussed earlier give evidence for local increases in the ion acoustic speed of up to 900 m/s. For a mean ion mass of 31 amu and an ion temperature of about 500 K this corresponds to an electron temperature of about 2500 K for isothermal ions and electrons. For different specific heat ratios the temperature would be reduced slightly. Electron temperatures of this magnitude have been measured with incoherent radars during highly active periods (see Fig. 8.32). Cosmic noise absorption levels measured during events of this type are also consistent with large increases in the effective electron temperature in the locally heated region (Stauning, 1984).

Schlegel and St.-Maurice (1981) argued that the large electron temperature enhancements in the unstable electrojet layer cannot be explained by either particle precipitation or classical Joule heating. These enhancements maximize at the height (about 110 km) where two-stream waves are strongest. This led St.-Maurice et al. (1981) to suggest that these waves are responsible for the electron heating throughout the unstable region. Several theories were developed to explain the possible heating of the electron gas by strongly driven two-stream waves. Originally, St.-Maurice et al. (1981) considered a quasi-linear modified two-stream instability theory with the assumption that the wave energy is concentrated in a narrow band of short-wavelength ($\lambda - 20$ cm) waves where the linear growth rate is a maximum. However, the wave amplitudes necessary to generate the observed electron temperatures are much larger than measured by *in situ* probes in this wavelength range (e.g., Pfaff *et al.*, 1984). St.-Maurice and Laher (1985) then suggested that heating of the electron gas is caused by parallel electric fields associated with long-wavelength gradient drift waves, while Primdahl and Bahnsen (1985) emphasized the role of an anomalous collision frequency, ν_*. Robinson (1986) also considered the heating of the electrons by the perpendicular gradient drift wave component in the presence of anomalous electron collisions. In this formulation, the electron heating rate can also be written

$$dW_e/dt \Big|_{\text{waves}} = n_0 m \nu_* (V_D - C_s k_\parallel/k)^2$$

where ν_* is the anomalous electron collision frequency. Self-consistent calculations of anomalous electron collision frequencies necessary for the saturation of the two-stream waves and of the corresponding electron temperatures were in excellent agreement with the experimental results. (As discussed below in Section 8.5.5, the concept of a ν_* was first proposed by Sudan (1983) who used it to explain limitation of the wave phase velocity.)

A formula useful for studying electron heating by both collisional and wave-related phenomena has been given by St.-Maurice *et al.* (1986).

$$Q = N_e m_e (v_e + v_*) \left[\left(\frac{E}{B} \right)^2 + \left(\frac{\delta E}{B} \right)^2 \right] + N_e m_e v_e \left(\frac{\Omega_e k_\parallel}{v_e k_\perp} \right)^2 \left(\frac{\delta E}{B} \right)^2$$

In this expression $v_e + v_*$ is the sum of the collisional and anomalous electron collisions. The first term includes both the Joule heating due to the zero order dc field and the Joule heating due to the wave field. Since $\delta E \lesssim E$ the first term is anomalous to the extent that $v_* > v_e$. The second term is entirely due to classical collisions, but the heating electric field in this case is due to the component of δE parallel to **B**. This term shows that even a small $k_\parallel$ can lead to very efficient heating due to waves.

Providakes *et al.* (1988) have argued that since radar experiments show that $k_\parallel$ is large, the latter term is the most important. Furthermore, they point out that anomalous collisions could be included as well in this part of the expression which would increase the heating even more if $v_* > v_e$. Explaining the large value of $k_\parallel$ is probably the single most important theoretical problem at this time.

8.5.3 Type 3 Radar Echoes: Are They Due to Ion Cyclotron Waves?

The type 3 observations seem to require additional flexibility in the theoretical analysis. Since there is evidence that such echoes come from the upper E region and are related to field-aligned currents, a general linear dispersion relation has been developed by Fejer *et al.* (1984b) for primary (directly excited) nearly field-aligned ($k_\perp \gg k_\parallel$) waves in the high-latitude E region, including cross-field and field-aligned drifts, ion inertia and magnetization, electron density gradients, and recombination. The general dispersion relation they derived describes the two-stream and gradient drift instability driven by Hall currents in the electrojet region, the magnetized two-stream ion cyclotron instability, which can be driven by either Hall or field-aligned currents between about 120 and 130 km, and the collisional electrostatic ion cyclotron instability driven by field-aligned currents above about 130 km.

In addition to the pulsed radar results shown in Fig. 8.27, observations with the University of Saskatchewan CW radar have shown type 3 spectra centered around $+30$ and $+55$ Hz during several periods of intense magnetic activity in the premidnight sector (Haldoupis *et al.*, 1985). These resonant spectral peaks suggest the existence of cyclotron waves with frequencies the same as gyro-frequencies of the E-region ion constituents (i.e., O^+, O_2^+, and NO^+). The different spectral components are probably generated in separate regions within the scattering volume. For field-aligned current-driven waves a positive Doppler shift corresponds to an upward current. These echoes are usually localized in space and time and are associated with substantial increases of the echo power but show very little variation in their Doppler shifts, implying highly coherent wave structures. One very significant feature of the observations is the large deviation of the scattering wave vector from the plane perpendicular to **B**. In other words, the waves seem to propagate at angles in the range of 82° to 85° from **B**. At such large angles the linear theory described above indicated that the plasma should be stable. HF radar measurements show that the real part of ω is in excellent agreement with ion cyclotron theory (Villain *et al.* (1987)).

Interferometer observations generally show the region of type 3 echoes to be associated with horizontal shears in the cross-field plasma flow and with highly localized and irregular structures within the scattering volume. Providakes *et al.* (1985) reported type 3 echoes along the northern edge of a discrete visible auroral arc. These correspond to the darkened spectral peaks shown in Fig. 8.38. The type 3 echoes were associated with a negative velocity shear $W_z = -0.5 \text{ s}^{-1}$, which as noted earlier in this chapter suggests the presence of large downward field-aligned currents. Rocket and satellite data show that the edges of auroral arcs are characterized by very large shears, which in turn are associated with intense field-aligned currents of many tens of microamperes per square meter.

The theory of electrostatic ion cyclotron (EIC) waves in the auroral ionosphere has been discussed by many investigators. Kindel and Kennel (1971) derived a kinetic linear dispersion relation for electrostatic waves in a weakly collisional ($\nu_e < k_\parallel V_e$) uniformly magnetized plasma. We are interested here, however, in the excitation of EIC waves in the upper E region at wavelengths for which the electrons are strongly collisional, and so the weak collision theory is not appropriate.

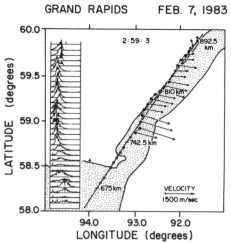

Fig. 8.38. Fifty-megahertz Doppler spectra and radar velocity–position map during a discrete visual arc event in the morning sector. [After Providakes *et al.* (1985). Reproduced with permission of the American Geophysical Union.]

Kinetic studies of collisional EIC waves ($\nu_e > k_\| V_e$) in the lower auroral iono-sphere were pursued by Providakes *et al.* (1985) and Satyanarayana and Chatur-vedi (1985). These authors have shown that the destabilizing effect of electron collisions is necessary for the excitation of EIC waves in the upper E region, where ion–neutral collisions are stabilizing.

Satyanarayana and Chaturvedi (1985) showed that the maximum growth rates for collisional EIC waves in the bottomside ionosphere are comparable to those of collisionless EIC waves above the F region, where ion–neutral collisions are negligible and ion–ion collisions become an important damping mechanism. The critical electron drift velocity to excite EIC waves depends on electron density, on T_e/T_i, and on the ion mass. Figure 8.39 shows three typical ionospheric elec-tron density profiles and the corresponding critical drift velocities for generation of EIC waves. The critical drift velocity has a minimum between 150 and 200 km. This height range is consistent with recent VHF coherent radar obser-vations. At 150 km the electron thermal speed (V_e) is about 100 km/s, which gives a threshold parallel drift velocity $V_D \simeq 20$ km/s.

The theories discussed above indicate that field-aligned drifts of about 30 km/s can excite collisional EIC waves at wavelengths of about 20 m in the upper E region. For an electron density of 5×10^4 cm^{-3} this corresponds to a field-aligned current density of about 100 μA/m^2. As noted above, field-aligned currents within a factor of three of this magnitude are sometimes observed in highly localized regions by rockets and satellites. These results provide consid-erable evidence that auroral field-aligned currents are large enough to generate collisional EIC waves in the upper E region at wavelengths of 10–20 m. The

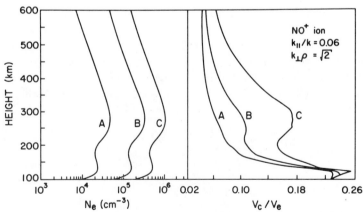

Fig. 8.39. Three typical electron density profiles and the corresponding plot of the critical drift velocity for the generation of EIC. V_e is the electron thermal speed. (Figure courtesy of J. Providakes.)

most favorable altitude predicted by theory (about 150 km) is consistent with the observations, and the measured frequency range of 50–80 Hz is just above the gyrofrequencies of the E-region ion constituents. However, 3-m ion cyclotron waves can be directly excited only by unreasonably large field-aligned drifts (greater than the electron thermal speed). In addition, the theory predicts that for minimum threshold, $k_\parallel/k = 0.007$ (aspect angle of 89.6°), whereas radar measurements indicate that aspect angles are as small as 80° ($k_\parallel/k \simeq 0.17$). The threshold velocity is much larger at these small aspect angles. It is possible that some nonlinear mechanism could generate coherent (resonant) 3-m waves from linearly unstable 20-m waves. In fact, Satyanarayana and Chaturvedi (1985) suggested that ion resonance broadening could provide such a cascade to short wavelengths. However, ion resonance broadening would generate broad spectra (spectral width much greater than Ω_i) at 3 m and, therefore, cannot explain the radar data. Considerable work thus remains before we finally understand the type 3 echoes.

8.5.4 Nonlinear Theories

The analytic (turbulence-based) gradient drift instability theory of Sudan and co-workers has already been discussed in Chapter 4, as have the simulation results of McDonald and co-workers. The theory should be applicable to the auroral case, although pure primary gradient drift waves are not as important at high altitudes as they are at the equator and at mid-latitudes. In this approach the 3-m waves responsible for 50-MHz backscatter are due to a cascade of energy from unstable long-wavelength waves to stable short-wavelength modes. Because of the interest in electron heating at high latitudes we pursue the case of the nonlinear two-stream wave here.

Sudan (1983a, b) has suggested that momentum transfer between electrons and short-wavelength waves leads to an anomalous cross-field diffusion coefficient D_*. Since D_* is proportional to the collision frequency there is a corresponding anomalous electron collision frequency ν_* which increases with the wave amplitude. The nonlinear dispersion relation can be obtained by replacing ν_e with ν_*. The nonlinear two-stream oscillation frequency and growth rate are then given by

$$\omega_r^{NL} = \frac{\mathbf{k} \cdot \mathbf{V}_D}{1 + \Psi + \Psi_*} = kC_s \tag{8.24}$$

$$\gamma^{NL} = \frac{\Psi + \Psi_*}{\nu_i(1 + \Psi + \Psi_*)}[(\omega_r^{NL})^2 - k^2 C_s^2] \tag{8.25}$$

where ω_r^{NL} is the nonlinear oscillation frequency and $\Psi/\Psi_* = \nu/\nu_*$. In this theory the increasing value of Ψ_* in the denominator of (8.24) is what limits the phase velocity to values near marginal stability, $\omega_r^{NL}/k = C_s$, even though $\hat{k} \cdot \mathbf{V}_D$ is much larger than C_s.

Large values of ν_*/ν_e are generally required to keep the two-stream phase velocity equal to C_s in the high-latitude electrojet region during periods of large plasma drifts, even though both C_s and ν_e increase on these occasions. For example, for an altitude of 105 km, using $V_D = 1000$ m/s, $C_s = 420$ m/s, $\nu_e = 8.7 \times 10^4$ s^{-1}, $\Omega_e = 10^7$ s^{-1}, and $\Omega_i = 180$ s^{-1}, we must have $\nu_* \simeq 18\nu_e$ $\simeq 1.5 \times 10^6$ s^{-1}. Even so Primdahl and Bahnsen (1985) have applied this theory to rocket observations with promising results.

Whether or not these large anomalous electron collision frequencies can be attained is still an open question, but there is considerable experimental evidence

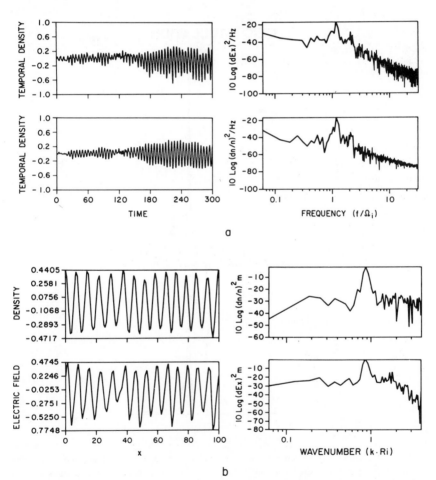

Fig. 8.40. (a) Density and electric field time histories at a fixed spatial point and their corresponding temporal power spectra for a collisional ion cyclotron wave. (b) Density and electric field waveforms at a fixed time and their spatial power spectra for a collisional ion cyclotron wave. [After Seyler and Providakes (1987). Reproduced with permission of the American Institute of Physics.]

that the effective electron collision frequency is indeed enhanced during disturbed conditions. There is, of course, the classical observation that the radar backscatter at equatorial latitudes "saturates" at C_s at all angles to the electrojet (see Chapter 4). We have also mentioned the success of Robinson's analysis, which relies heavily on an effective ν_*. In addition, the comparison in Fig. 8.33 of high-latitude 1-m wave phase velocity, plasma drift velocity, and ion acoustic velocity seems to show clearly that the phase velocity locks onto the acoustic speed rather than the drift velocity. However, the acoustic speed does increase with drift velocity, presumably due to wave-related heating.

Turning to nonlinear theories for type 3 waves, one-dimensional simulations

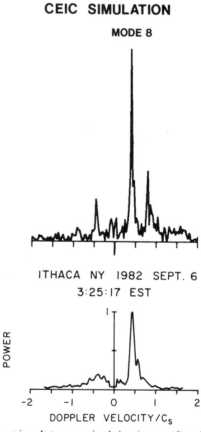

Fig. 8.41. Direct comparison between a simulation (top panel) and actual radar data (lower panel). In each panel the Doppler spread is plotted at a fixed wave number corresponding to a 50-MHz backscatter radar. [After Seyler and Providakes (1987). Reproduced with permission of the American Institute of Physics.]

have been carried out by Seyler and Providakes (1987). For current-driven collisional ion cyclotron waves they found that an extremely coherent wave was generated with the $\delta n/n$ and δE characteristics in both time and space shown in Fig. 8.40. In a collisionless run using similar parameters, higher harmonics were generated. Less energy was found in the fundamental than is displayed in the collisional case shown here. To directly compare the simulation with radar data, they investigated the phase velocity characteristics of the waves at the radar wave number. The simulation result and the radar data are shown in Fig. 8.41. The agreement is quite good and is another illustration of an important new trend in simulation diagnostics wherein the actual quantity measured is reproduced in the simulation.

8.6 Summary

In this chapter we have only scratched the surface in the plasma physics of high-latitude phenomena. We have in some cases been able to export results from other latitudes but, in general, the high-latitude regime has added features which complicate the physics. These problems are likely to keep ionospheric plasma physics an exciting discipline for the foreseeable future.

References

Banks, P. M., and Kockarts, G. (1973). "Aeronomy," Parts A and B. Academic Press, New York.

Basu, Su., Basu, Sa., Senior, C., Weimer, D., Neilsen, E., and Fougere, P. F. (1986). Velocity shears and sub-km scale irregularities in the nighttime auroral F-region. *Geophys. Res. Lett.* **13,** 101.

Bering, E. A., Kelley, M. C., and Mozer, F. S. (1975). Observations of an intense field-aligned thermal ion flow and associated intense narrow band electric field oscillations. *JGR, J. Geophys. Res.* **80,** 4612.

Burke, W. J., Kelley, M. C., Sagalyn, R. C., Smiddy, M., and Lai, S. T. (1979). Polar cap electric field structures with a northward interplanetary magnetic field. *Geophys. Res. Lett.* **6,** 21.

Burke, W. J., Hardy, D. A., Rich, F. J., Kelley, M. C., Smiddy, M., Shuman, B., Sagalyn, R. C., Vancour, R. P., Wildman, P. J. L., and Lai, S. T. (1980). Electrodynamic structure of the late evening sector of the auroral zone. *JGR, J. Geophys. Res.* **85,** 1179.

Burke, W. J., Gussenhoven, M. S., Kelley, M. C., Hardy, D. A., and Rich, F. J. (1982). Electric and magnetic field characteristics and discrete arcs in the polar cap. *JGR, J. Geophys. Res.* **87,** 2431.

Cerisier, J. C., Berthelier, J. J., and Beghin, C. (1985). Unstable density gradients in the high-latitude ionosphere. *Radio Sci.* **20,** 755.

Cole, K. D. (1965). Stable auroral red arcs, sinks for energy of DST main phase. *J. Geophys Res.* **70,** 1689.

Costa, E., and Kelley, M. C. (1978). On the role of steepened structures and drift waves in equatorial spread F. *JGR, J. Geophys. Res.* **83,** 4359.

D'Angelo, N., Pecseli, H. L., and Petersen, P. I. (1974). The Farley instability: A laboratory test. *JGR, J. Geophys. Res.* **79**, 4747.

Dyson, P. L., and Winningham, J. D. (1974). Topside ionospheric spread F and particle precipitation in the dayside magnetospheric clefts. *JGR, J. Geophys. Res.* **79**, 5219.

Dyson, P. L., McClure, J. P., and Hanson, W. B. (1974). In situ measurements of the spectral characteristics of F region ionospheric irregularities. *JGR, J. Geophys. Res.* **79**, 1497.

Earle, G. D. (1988). Electrostatic plasma waves and turbulence near auroral arcs, Ph.D. Thesis. Cornell University, Ithaca, New York.

Farley, D. T. (1959). A theory of electrostatic fields in a horizontally stratified ionosphere subject to a vertical magnetic field. *JGR, J. Geophys. Res.* **64**, 1225.

Fejer, B. G., and Providakes, J. (1988). High latitude E region irregularities: New results. *Phys. Scr.* **T18**, 167.

Fejer, B. G., and Kelley, M. C. (1980). Ionosphere irregularities. *Rev. Geophys. Space Phys.* **18**, 401.

Fejer, B. G., Reed, R. W., Farley, D. T., Swartz, W. E., and Kelley, M. C. (1984a). Ion cyclotron waves as a possible source of resonant auroral radar echoes. *JGR, J. Geophys. Res.* **89**, 187.

Fejer, B. G., Providakes, J. and Farley, D. T. (1984b). Theory of plasma waves in the auroral E region. *JGR, J. Geophys. Res.* **89**, 7487.

Ganguli, G., Palmadesso, P., and Lee, C. Y. (1985). Electrostatic ion cyclotron instability caused by a nonuniform electric field perpendicular to the external magnetic field. *Phys. Fluids* **28**, 761.

Goldman, S. R., Baker, L., Ossakow, S. L., and Scannapieco, A. J. (1976). Striation formation associated with barium clouds in an inhomogeneous ionosphere. *JGR, J. Geophys. Res.* **81**, 5097.

Greenwald, R. A., Baker, K. D., and Villain, J. P. (1983). Initial studies of small-scale F-region irregularities at very high latitudes. *Radio Sci.* **18**, 1122.

Haldoupis, C. I., Prikryl, P., Sojko, G. J., and Koehler, D. J. (1985). Evidence for 50 MHz bistatic radio observations of electrostatic ion cyclotron waves in the auroral plasma. *JGR, J. Geophys. Res.* **90**, 10983.

Heelis, R. A., Vickrey, J. F., and Walker, N. B. (1985). Electrical coupling effects on the temporal evolution of F layer plasma structure. *JGR, J. Geophys. Res.* **90**, 437.

Kelley, M. C. (1986). Intense sheared flow as the origin of large-scale undulations of the edge of diffuse aurora. *JGR, J. Geophys. Res.* **91**, 3225.

Kelley, M. C., and Carlson, C. W. (1977). Observation of intense velocity shear and associated electrostatic waves near an auroral arc. *JGR, J. Geophys. Res.* **82**, 2343.

Kelley, M. C., Bering, E. A., and Mozer, F. S. (1975). Evidence that the ion cyclotron instability is saturated by ion heating. *Phys. Fluids* **18**, 1590.

Kelley, M. C., Baker, K. D., and Ulwick, J. C. (1979). Late time barium cloud striations and their relationships to equatorial spread F. *JGR, J. Geophys. Res.* **84**, 1898.

Kelley, M. C., Baker, K. D., Ulwick, J. C., Rino, C. L., and Baron, M. J. (1980). Simultaneous rocket probe scintillation and incoherent scatter radar observations of irregularities in the auroral zone ionosphere. *Radio Sci.* **15**, 491.

Kelley, M. C., Vickrey, J. F., Carlson, C. W., and Torbert, R. (1982a). On the origin and spatial extent of high-latitude F region irregularities. *JGR, J. Geophys. Res.* **87**, 4469.

Kelley, M. C., Livingston, R. C., Rino, C. L., and Tsunoda, R. T. (1982b). The vertical wave number spectrum of topside equatorial spread F: Estimates of backscatter levels and implications for a unified theory. *JGR, J. Geophys. Res.* **87**, 5217.

Keskinen, M. J., Mitchell, H. G., Fedder, J. A., Satyanarayama, P., Zalesak, S. T., and Huba, J. D. (1988). Nonlinear evolution of the Kelvin–Helmholtz instability in the high latitude ionosphere. *JGR, J. Geophys. Res.* **93**, 137.

Kindel, J. M., and Kennel, C. F. (1971). Topside current instabilities. *J. Geophys. Res.* **76**, 3055.

Kintner, P. M., and Seyler, C. E. (1985). The status of observations and theory of high latitude ionospheric and magnetospheric plasma turbulence. *Space Sci. Rev.* **41**, 91.

Kintner, P. M., Kelley, M. C., and Mozer, F. S. (1978). Electrostatic hydrogen cyclotron waves near one earth radius in the polar magnetosphere. *Geophys. Res. Lett.* **5**, 139.

Knudsen, D. J., Seyler, C. E., Kelley, M. C., and Vickrey, J. F. (1988). 2D simulation of ionosphere/magnetosphere coupling at small scales. *EOS Trans. AGU*, p. 1369.

Kustom, B., D'Angelo, N., and Merlino, R. L. (1985). A laboratory investigation of the high-frequency Farley–Buneman instability. *JGR, J. Geophys. Res.* **90**, 1698.

LaBelle, J. (1985). The mapping of electric field structures from the equatorial F region to the underlying E region. *JGR, J. Geophys. Res.* **90**, 4341.

Lui, A. T. Y., Meng, C.-I., and Ismail, S. (1982). Large amplitude undulations on the equatorward boundary of the diffuse aurora. *JGR, J. Geophys. Res.* **87**, 2385.

Mozer, F. S., and Lucht, P. (1974). The average auroral zone electric field. *JGR, J. Geophys. Res.* **79**, 1001.

Nielsen, E., and Schlegel, K. (1985). Coherent radar Doppler measurements and their relationship to the ionospheric electron drift velocity. *JGR, J. Geophys. Res.* **90**, 3498.

Ossakow, S. L., and Chaturvedi, P. K. (1979). Current convective instability in the diffuse aurora. *Geophys. Res. Lett.* **6**, 332.

Pfaff, R. F. (1985). Rocket measurements of plasma turbulence in the equatorial and auroral elements, Ph.D. Thesis. Cornell University, Ithaca, New York.

Pfaff, R. F., Kelley, M. C., Fejer, B. G., Kudeki, E., Carlson, C. W., Pedersen, A., and Haüsler, B. (1984). Electric field and plasma density measurements in the auroral electrojet. *JGR, J. Geophys. Res.* **89**, 236.

Pfaff, R. F., Kelley, M. C., Kudeki, E., Fejer, B. G., and Baker, K. D. (1988). Electric field and plasma density measurements in the strongly-driven daytime equatorial electrojet. 2. Two-stream waves. *JGR, J. Geophys. Res.* **92**, 13597.

Primdahl, F., and Bahnsen, A. (1985). Auroral E-region diagnosis by means of nonlinearly stabilized plasma waves. *Ann. Geophys.* **3**, 57.

Providakes, J. F., Farley, D. T., Swartz, W. E., and Riggin, D. (1985). Plasma irregularities associated with a morning discrete auroral arc: Radar interferometer observations and theory. *JGR, J. Geophys. Res.* **90**, 7513.

Providakes, J. F., Farley, D. T., Fejer, B. G., Sahr, J., Swartz, W. E., Häggström, I., Hedberg, Å., and Nordling, J. A. (1988). Observations of auroral E-region plasma waves and electron heating with EISCAT and a VHF radar interferometer. *J. Atmos. Terr. Phys.* **50**, 339.

Providakes, Jam. F., Kelley, M. C., Swartz, W. E., Mendillo, M., and Holt, J. (1989). Radar and optical measurements of ionospheric processes associated with large electric fields near the plasmapause. *JGR, J. Geophys. Res.* **94** (in press).

Rich, F. J., Burke, W. J., Kelley, M. C., and Smiddy, M. (1980). Observations of field-aligned currents in association with strong convection electric fields at subauroral latitudes. *JGR, J. Geophys. Res.* **85**, 2335.

Rishbeth, H., and Garriott, O. K. (1969). "Introduction to Ionospheric Physics." Int. Geophys. Ser. Vol. 14. Academic Press, New York.

Robinson, R. M., Evans, D. S., Potemra, T. A., and Kelly, J. D. (1984). Radar and satellite measurements of an F-region ionization enhancement in the post-noon sector. *Geophys. Res. Lett.* **1**, 899.

Robinson, R. M., Tsunoda, R. T., Vickrey, J. F., and Guérin, L. (1985). Sources of F region ionization enhancements in the nighttime auroral zone. *JGR, J. Geophys. Res.* **90**, 7533.

Robinson, T. R. (1986). Towards a self-consistent nonlinear theory of radar auroral backscatter. *J. Atmos. Terr. Phys.* **48**, 417.

Roble, R. G., and Rees, M. H. (1977). Time-dependent studies of the aurora: Effects of particle precipitation on the dynamic morphology of ionospheric and atmospheric properties. *Planet. Space Sci.* **25**, 991.

Ruohoniemi, J. M., Greenwald, A., Baker, K. D., Villain, J. P., and McCready, M. A. (1987). Drift motions of small scale irregularities in the high-latitude F region: An experimental comparison with plasma drift motions. *JGR, J. Geophys. Res.* **92**, 4553.

Sagalyn, R. C., Smiddy, M., and Ahmed, M. (1974). High-latitude irregularities in the topside ionosphere based on Isis 1 thermal probe. *JGR, J. Geophys. Res.* **79**, 4252.

St.-Maurice, J.-P., and Laher, R. (1985). Are observed broadband plasma wave amplitudes large enough to explain the enhanced electron temperatures of the high-latitude E region. *JGR, J. Geophys. Res.* **90**, 2843.

St.-Maurice, J.-P., Schlegel, K., and Banks, P. M. (1981). Anomalous heating of the polar F region by unstable plasma waves. 2. Theory. *JGR, J. Geophys. Res.* **86**, 1453.

St.-Maurice, J.-P., Hanuise, C., and Kudeki, E. (1986). Anomalous heating in the auroral electrojet. *JGR, J. Geophys. Res.* **91**, 13493.

Satyanarayana, P., and Chaturvedi, P. K. (1985). Theory of the current-driven ion cyclotron instability in the bottomside ionosphere. *JGR, J. Geophys. Res.* **90**, 12209.

Schlegel, K., and St.-Maurice, J. P. (1981). Anomalous heating of the polar E region by unstable plasma waves. 1. Observations. *JGR, J. Geophys. Res.* **86**, 1447.

Schunk, R. W., and Sojka, J. J. (1987). A theoretical study of the lifetime and transport of large ionospheric density structures. *JGR, J. Geophys. Res.* **92**, 12343.

Schunk, R. W., Banks, P. M., and Raitt, W. J. (1976). Effects of electric fields and other processes upon the nighttime high latitude F layer. *JGR, J. Geophys. Res.* **81**, 3271.

Seyler, C. E., and Providakes, J. (1987). Particle and fluid simulations of resistive current-driven electrostatic ion cyclotron waves. *Phys. Fluids* **30**, 3113.

Smiddy, M., Kelley, M. C., Burke, W., Rich, R., Sagalyn, R., Shuman, B., Hays, R., and Lai, S. (1977). Intense poleward-directed electric fields near the ionospheric projection of the plasmapause. *Geophys. Res. Lett.* **4**, 543.

Stauning, P. (1984). Absorption of cosmic noise in the E-region during electron heating events. A new class of riometer absorption events. *Geophys. Res. Lett.* **11**, 1184.

Sudan, R. N. (1983a). Unified theory of type I and type II irregularities in the equatorial electrojet. *JGR, J. Geophys. Res.* **88**, 4853.

Sudan, R. N. (1983b). Nonlinear theory of type I irregularities in the equatorial electrojet. *Geophys. Res. Lett.* **10**, 983.

Tanskanen, P. J., Hardy, D. A., and Burke, W. J. (1981). Spectral characteristics of precipitation electrons associated with visible aurora in the premidnight oval during periods of substorm activity. *JGR, J. Geophys. Res.* **86**, 1379.

Tsunoda, R. T. (1989). High latitude F region irregularities: A review and synthesis. *Rev. Geophys. Space Phys.* (in press).

Vickrey, J. F., and Kelley, M. C. (1982). The effects of a conducting E layer on classical F region cross-field plasma diffusion. *JGR, J. Geophys. Res.* **87**, 4461.

Vickrey, J. F., Rino, C. L., and Potemra, T. A. (1980). Chatanika/TRIAD observations of unstable ionization enhancements in the auroral F-region. *Geophys. Res., Lett.* **7**, 789.

Vickrey, J. F., Kelley, M. C., Pfaff, R., and Goldman, S. R. (1984). Low-altitude image striations associated with bottomside equatorial spread F: Observations and theory. *JGR, J. Geophys. Res.* **89**, 2955.

Vickrey, J. F., Livingston, R. C., Walker, N. B., Potemra, T. A., Heelis, R. A., Kelley, M. C., and Rich, F. J. (1986). On the current–voltage relationship of the magnetospheric generator at intermediate spatial scales. *Geophys. Res. Lett.* **13**, 495.

Villain, J. P., Caudal, G., and Hanuise, C. (1985). A SAFRI–EISCAT comparison between the velocity of F region small-scale irregularities and the ion drift. *JGR, J. Geophys. Res.* **90**, 8433.

Villain, J. P., Greenwald, R. A., Baker, K. B., and Ronhoniemi, J. M. (1987). HF radar observations of E-region plasma irregularities produced by oblique electron streaming. *JGR, J. Geophys. Res.* **92,** 12327.

Weber, E. J., Tsunoda, R. T., Buchau, J., Sheehan, R. E., Strickland, D. J., Whiting, W., and Moore, J. G. (1985). Coordinated measurements of auroral zone plasma enhancements. *JGR, J. Geophys. Res.* **90,** 6497.

Weber, E. J., Klobuchar, J. A., Buchau, J., Carlson, H. C., Jr., Livingston, R. C., De La Beaujardiere, O., McCready, M., Moore, J. G., and Bishop, G. J. (1986). Polar cap F layer patches: Structure and dynamics. *JGR, J. Geophys. Res.* **91,** 12121–12129.

Weber, E. J., Kelley, M. C., Ballenthin, J. D., Basu, S., Carlson, H. C., Fleischman, J. R., Hardy, D. A., Maynard, N. C., Pfaff, R. F., Rodriguez, P., Sheehan, R. E., and Smiddy, M. (1989). Rocket measurements within a polar cap arc: Plasma, particle and electric circuit parameters. *JGR, J. Geophys. Res.* **94** (in press).

Appendix A | Ionospheric Measurement Techniques*

A.1 Radio Wave Techniques in Ionospheric Physics

Rishbeth and Garriott (1969) describe the operation of ionosondes in some detail in their book and we refer the reader to that source for information on that device. Ionosondes were the workhorse in ionospheric research for many decades and still are very useful instruments. Digital techniques have greatly improved their versatility and a new generation of "digisondes" is now in operation. We refer the reader to the paper by Wright and Pitteway (1979) for more information. In this section we discuss incoherent and coherent scatter radars and the radio wave scintillation technique. Further material may be found in articles by Fejer and Kelley (1980) and Farley (1979).

A.1.1. Incoherent Scatter Radars

The ground was broken for the first incoherent scatter radar (ISR) site in Arecibo, Puerto Rico, in 1959. At the time of writing, there are at least seven such sites in regular operation with one additional one used intermittently. The locations of those eight are given in Table A.1 along with other relevant information. Data from every one of these sites are used somewhere in the text, as are data from a ninth site (Chatanika, Alaska), which is also listed in the table. The Chatanika radar was moved to Sondre Stromfjord, Greenland, in 1983.

Like an ionosonde, an ISR transmits a radio wave signal and receives a returned "echo" sometime later. Ionosondes operate over frequencies typically in

---

*Optical techniques are not described in this appendix. See Hernandez (1987) for details concerning the Fabry–Pirot method.

Table A.1

Relevant Parameters and Coordinates

Observatory	Frequency (MHz)	Geographic latitude	Geographic longitude	L	UT − LT
Jicamarca, Peru	49.9	11.9°S	76°W	1.05 (300 km)	5
Millstone Hill, Massachusetts	440	42.6°N	71.5°W	3.12	5
Chatanika, Alaska	1290	64.9°N	147.7°W	5.51	10
Arecibo, Puerto Rico	430	18.3°N	66.75°W	1.43	4
St. Santin, France	935	44.6°N	2.2°E	1.76	0
EISCAT (Tromso, Norway)	224/933	69.6°N	19.2°E	6.3	−2
Sondre Stromfjord, Greenland	1290	67.0°N	50.95°W	15+	3
Altair, Kwajalein	155.5	8.8°N	167.5°E	1.05 (300 km)	−11
MU, Japan	46.5	34.85°N	136.10°E	1.33	−9

the range 1–20 MHz and rely on the fact that such a signal is reflected when its frequency f is equal to the local plasma frequency f_p, where

$$f_p = (2\pi)^{-1}(ne^2/m_e\varepsilon_0)^{1/2}$$

To a good approximation,

$$f_p = 9000\sqrt{n}\quad \text{Hz}$$

where n is the plasma density in reciprocal cubic centimeters. Since the peak plasma density in the ionosphere is a few times 10^6 cm^{-3}, $f_p \leq 12$ MHz. Vertically transmitted frequencies above the peak plasma frequency in the overhead ionosphere go right through. Consequently, ground-based ionosondes yield no information above the height of the peak on the F-region plasma density. If there is a dense E layer it can block the F layer entirely. Occasionally a thin low-altitude layer, such as the sporadic E layers discussed in Chapter 5, will allow some energy to tunnel through, revealing the F layer above.

The ISR technique circumvents both the reflection and topside problems by using frequencies well above any reasonable plasma frequency in the natural ionosphere. The lowest frequency listed in Table A.1 is about 50 MHz. Such waves are virtually unattenuated by the ionosphere and pass through almost unaffected into space. "Almost" is the key word here, since it is the small amount of energy scattered by the ionospheric electrons which is used by the ISR method. To gain some perspective, the amount of energy scattered back to a typical ISR antenna from 300 km is roughly comparable to the target represented by a small copper coin at the same range. It should be no surprise, then, that the largest of the ISR sites measure their effective power–aperture product in tens of megawatt-acres or in more reasonable units a few times 10^{11} W·m^2.

As with most radar systems, the signal is transmitted in pulses with range from the radar to the echoing region determined by dividing half the delay time by the speed of light. Since the ionosphere is an extended target, various ranges can be interrogated using the same set of pulses by analyzing the returned signal at appropriate time delays.

The information in the returned signal is remarkably rich in physical content. The power in the returned echo is proportional to the number density of electrons in the volume irradiated. This stems from the fact that each electron incoherently radiates back a small amount of the incident energy. The electric field in the transmitted wave causes any and all electrons encountered during transit of the pulse to oscillate, resulting in radiation of a signal at almost the same frequency. This fact alone was sufficient to generate the enthusiasm necessary to build the first site at Arecibo, since, for the first time, information on the ionosphere would be available above the F peak. The richness in information content stems as much from the spectrum of the returned signal as from the power returned, however. The electrons are consistently in thermal motion and hence the radiation is Doppler shifted from the incident frequency. The result is a spread in the returned radio wave spectrum which contains considerable information about the velocities present in the medium and hence about its dynamics.

The original idea (Gordon, 1958) hypothesized quite naturally that the Doppler width of the returned spectrum would be of order

$$\delta f/f \cong \langle V \rangle_e / c$$

where $\langle V \rangle_e$ is the mean thermal electron velocity. Surprisingly, the first experiments, by Bowles (1958), showed the Doppler width to be of order $\langle V \rangle_i / c$, where $\langle V \rangle_i$ is the ion thermal speed. This is explained by the fact that, when the probing wavelength is greater than the Debye length, the Debye cloud around each electron and the ions inside that cloud all contribute to the scattering.

A schematic diagram of the Doppler spread due to scattering from thermal fluctuations in the so-called ion line is presented in Fig. A.1. The power under the curve is proportional to the number density, while the width δf can be used

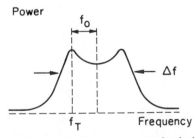

Fig. A.1. Schematic diagram of the Doppler spectrum associated with backscatter from thermal fluctuation in the upper atmosphere. Here f_T is the transmitted frequency, f_0 is the mean Doppler shift, which yields the line-of-sight velocity, and Δf is a measure of the width of the spectrum. Only the ion line is shown here.

to determine the ion temperature. The relative intensity of the "wings" in the spectrum yields the electron temperature, and the overall shift of the spectrum, f_D, from the transmitted frequency f_T gives the line-of-sight component of the mean ion velocity V_i. If several positions are used, the complete vector flow velocity can be found. In the F region the ion flow velocity yields the perpendicular electric field components via the relationship

$$\mathbf{E}_\perp = \mathbf{B} \times \mathbf{V}_i$$

The ion flow velocity parallel to $\mathbf{B}$ is much more complicated since many factors contribute, such as the component of the neutral wind along $\mathbf{B}$, the ion pressure gradient, and gravitational forces (see Chapter 2).

   The theory of scattering from thermal fluctuations is very well understood. In fact, as pointed out by Farley (1979), when viewed as a test of the crucial Landau damping method for carrying out Vlasov plasma theory, measurements of the incoherent scatter spectrum have verified that method to better than 1% accuracy. The radar scatters from irregularities in the medium, $\mathbf{k}_m$, according to the relationship

$$\mathbf{k}_T = \mathbf{k}_s + \mathbf{k}_m \tag{A.1}$$

where $\mathbf{k}_T$ is the transmitted wave and $\mathbf{k}_s$ the scattered wave. Since $\mathbf{k}_s = -\mathbf{k}_T$ for backscatter,

$$\mathbf{k}_m = 2\mathbf{k}_T$$

Thus, the scattering wavelength is one-half the transmitted wavelength. Equation (A.1) represents conservation of momentum since each photon carries a momentum equal to $\hbar \mathbf{k}$. Conservation of energy (each photon energy equal to $\hbar\omega$) requires

$$\omega_T = \omega_s + \omega_m \tag{A.2}$$

Thus, the Doppler width of the returned spectrum ($\omega_s - \omega_T$) is related to the frequencies of waves in the medium, ω_m. Since thermal fluctuations can be considered as a superposition of damped sound waves, the spread in ω_m is of the order

$$\omega_m \approx |\mathbf{k}_m| C_s$$

where C_s is the sound speed in a plasma,

$$C_s^2 = (kT_e + kT_i)/M_i$$

The width of the ion line for a 50-MHz radar in a plasma with $C_s \approx 1000$ m/s is thus of order $[(4\pi/6)\text{ m}](1000\text{ m/s}) \approx 4000$ rad/s ≈ 600 Hz. For a medium of the same temperature the Doppler width will just scale with the radar frequency, so at Arecibo (430 MHz) the width of the "ion line" is about eight times larger.

   The peaks or wings in the ion line spectrum correspond to the frequency of the normal modes in the medium, that is, the frequency at which a finite-amplitude sinusoidal sound wave of wave number $\pm \mathbf{k}_m$ would exist. In thermal equilibrium such waves are damped. The time dependence associated with this damping fills in the remainder of the spectrum above and below $|\omega_s| = |\mathbf{k}_m| C_s$. When $T_e > T_i$, Vlasov theory predicts that the sound wave damping will be less. Indeed, as mentioned above, these wings increase in amplitude relative to the rest of the spectrum when T_e/T_i increases.

In ion sound waves, both electrons and ions participate. The electron gas can itself support thermal fluctuations, which at the scattering wave number occur at frequencies of order $k_m \langle V \rangle_e$. This frequency is larger by a factor M_i/m_e than the frequency of the wings in the ion line. In the ionosphere for O^+ this ratio is about 3×10^4, so the pure electron waves are Doppler shifted considerably more. The first simultaneous observations of both ion and electron scattering lines are reproduced in Fig. A.2. The data were taken in the altitude range 1460–2486 km over Arecibo, where H^+ is the dominant ion. The electron line has nearly the same power as the ion line but is spread over a much wider bandwidth. This makes the signal-to-noise ratio much smaller and far fewer measurements of this scattering source exist.

In practice, most ISR systems measure the temporal autocorrelation function of the signals scattered from a fixed point in space by sending up a series of

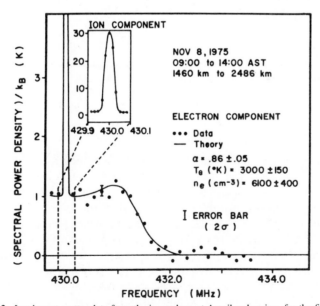

Fig. A.2. Incoherent scatter data from the ionosphere at Arecibo showing, for the first time, the electronic as well as ionic portions of the spectrum. Due to the wide bandwidth and resulting low signal-to-noise ratio, a great deal of averaging in both time and space is required. [After Hagen and Behnke (1976). Reproduced with permission of the American Geophysical Union.]

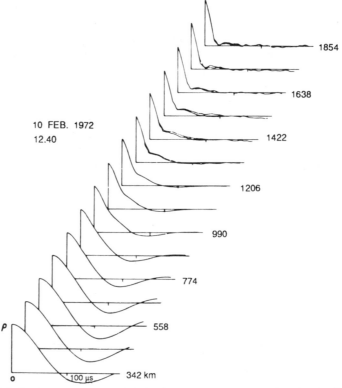

10 FEB. 1972
12.40

Fig. A.3. Incoherent scatter autocorrelation function measurements made at Arecibo. The pulse length used was 2 ms and the integration time was 20 min. The experimental and fitted theoretical curves are plotted together, but the agreement is so good that the two often cannot be distinguished. [After Hagen and Hsu (1974). Reproduced with permission of the American Geophysical Union.]

pulses. Since the spectrum is the Fourier transform of the autocorrelation function (ACF), it can be generated at will on the ground using the latter. The temporal resolution is then determined by how many such measurements must be averaged to generate a reasonable ACF. Since the theory is exact, least-squares fitting methods can be used on the ACF waveform which include second-order ion species, differential drifts of the various ions, and so forth. This further increases the information on the ionosphere available via the ISR method. Examples of ACF measurements are presented in Fig. A.3 as a function of altitude over Arecibo. The fitted theoretical ACFs are superposed on the measurements and can hardly be distinguished. Notice that as the altitude increases the ACF narrows in time, corresponding to the widening of the Doppler spectrum as the plasma changes from O^+ to H^+ with increasing altitude.

A.1.2   Coherent Scatter Radars

When plasma instabilities are present in the ionosphere, the amplitude of fluctuations in the medium can grow to values much greater than the thermal level. If the wave number of these fluctuations matches the requirement in Eq. (A.1), then smaller radar systems can be used to detect these large fluctuations. The Doppler spectrum in such a case is then representative of the phase velocity of the nonthermal waves rather than of the bulk motion and temperature of the plasma. Of course, a large ISR system will also detect such waves if they are present. The spread F maps made using the Jicamarca ISR system and presented in Chapter 4 are of this type. Signal strengths more than 60 dB above the thermal fluctuation level have been recorded there.

Most of the plasma instabilities detected by coherent scatter radars (CSRs) produce waves with **k** vectors nearly perpendicular to the magnetic field. At the magnetic equator the plane perpendicular to **B** includes the vertical direction. Thus, coherent scattering can (and does) occur at all elevations in that plane. The large antenna at Jicamarca can be oriented only a few degrees off vertical but smaller steerable antennas are used to look obliquely east or west to study electrojet and spread F instabilities. It is quite easy to keep the beam perpendicular to **B** in the equatorial case.

At higher latitudes the perpendicularity requirement places a severe constraint on the location of the radar system and its field of view. To study instabilities in the auroral electrojet at, say, 110 km, the radar must be located many hundreds of kilometers south of the region of interest. An example of the "aspect" angle contours for a CSR radar located in Ithaca, New York (42.3°N, 75.4°E) is shown in Fig. A.4. The L shell is also plotted in that figure and in Fig. A.5, which shows the geographic location. The angle α is near 90° over a broad range of L shells and geographic distance. Linear theory predicts that echoes should occur only for aspect angles within a few degrees of perpendicularity. However, experiments show that during conditions of strong auroral activity, echoes are seen up to 8° off perpendicular. Although not completely understood theoretically, this is very fortunate from the experimental viewpoint since (a) the total range over which echoes are received increases and (b) locations thought to be inaccessible to the technique are actually feasible. For example, although the nearly vertical magnetic field conditions in Greenland yield poor aspect angles, strong echoes were received over Sondre Stromfjord, Greenland, by a 50-MHz radar located at the southern tip of Greenland.

Most CSR success has been at VHF and UHF frequencies. In equatorial spread F coherent echoes have been found at wavelengths of 3 m, 0.95 m, and even 0.11 m (scattering) during extreme conditions. Equatorial electrojet measurements at lower frequencies (larger scattering wavelengths) have been made occasionally at Jicamarca and quite extensively in Africa (Hanuise and Crochet, 1978). These so-called HF radars suffer more refraction and the echoing mecha-

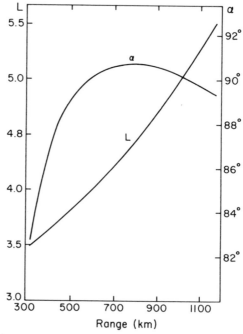

Fig. A.4. L values and aspect angle for altitude 110 km along Ithaca radar beam.

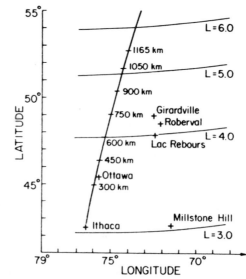

Fig. A.5. Location of the Ithaca auroral radar and its bearing.

nism is sometimes ambiguous due to the proximity of the frequency to ambient plasma frequencies. That is, it is not always clear whether the system is operating as a radar or more like a classical ionosonde. At very high latitudes the refractive properties at HF can actually be an advantage since, as noted above, it is not easy to obtain the line-of-sight aspect angles perpendicular to **B** needed by VHF systems. A rotating HF system was used near the magnetic dip pole, for example, to yield nearly uniform coverage in azimuth since the refraction relative to a vertical magnetic field line yielded identical aspect angles at all azimuths (Tsunoda *et al.*, 1976).

The latter HF technique is particularly useful in the summer, when the polar cap is illuminated by sunlight and refraction occurs virtually all day long. At this point it is worth reiterating that all radar scattering is proportional to $\delta n(\mathbf{k}_s)$, not to $\delta n(\mathbf{k}_s)/n_0$. That is, electrostatic plasma waves which have the same $(\delta n/n_0)$ amplitude will scatter more strongly at high background density levels than at low ones. For this reason the daytime equatorial electrojet can be detected with low-power systems, e.g., some tens of kilowatts, but nighttime echoes are much harder to get even though the plasma waves are just as intense.

The Doppler shift of the returned signal is not nearly so straightforward to interpret as is that from an incoherently scattered signal. In fact, even today, 20 years after their first detection, the Doppler shift of echoes from the equatorial electrojet remains somewhat enigmatic. The linear theory for production of a wave within a supersonic current suggests that the wave phase velocity satisfy the condition

$$V_P \propto (V_d \cos \theta) \tag{A.3}$$

where V_d is the differential drift velocity between electrons and ions and θ is the angle between the **k** vector and the current. Experiment does not verify this relationship during conditions when the plasma is two-stream unstable. In fact, the Doppler shift equals the acoustic speed no matter what the angle is between the current and the radar line of sight. On the other hand, when the plasma drift speed is below threshold, Eq. (A.3) seems to hold. The intersecting nonlinear plasma physics behind these results is discussed in the text at some length. Here we just reiterate this as an example of the complexity of coherent scatter radar measurements.

To pursue this one step further, analysis of data from the STARE auroral zone coherent scatter radar system has sometimes used the assumption that (A.3) holds in converting from the line-of-sight Doppler to an electron drift and then to an ambient electric field. That is, two independent radars at different locations were used to yield the vector perpendicular phase velocity $\mathbf{V}_p$ from the same scattering volume and then the ionospheric electric field calculated from

$$\mathbf{E}_\perp = \mathbf{B} \times \mathbf{V}_p$$

As discussed in Chapter 8, this assumption now is somewhat suspect and more theoretical and experimental work is clearly needed if $\mathbf{V}_p$ is to be fully understood.

Recently a method has been developed by Farley and co-workers to track the location of strong scattering regions across the coherent scatter radar field of view. Two antennas are used, separated by a known distance in, say, the east (E)–west (W) direction, and the complex cross spectrum is determined:

$$S_{EW}(\omega) = \frac{\langle F_E(\omega)F_W^*(\omega)\rangle}{\langle|F_E(\omega)|^2\rangle^{1/2}\langle|F_W(\omega)|^2\rangle^{1/2}}$$

where $F(\omega)$ is the Fourier transform of the digitized signals from the respective antennas. For each line-of-sight Doppler frequency the magnitude of $S_{EW}(\omega)$ is the normalized coherence of the scatterer. When the coherence is high in some frequency band it means that the scatterer is localized within the beamwidth. In such a case, the phase angle ϕ of the complex number $S_{EW}(\omega)$ determines the position of the scatterer within the beam. As ϕ changes in time, the scatterer can be tracked across the field of view and its cross velocity therefore determined. In some cases this velocity is very likely equal to the plasma flow velocity [e.g., the equatorial F-region case studied by Kudeki *et al.* (1981)]. In the E region the method yields the phase velocity of the large-scale waves which dominate the physics. In some sense the two situations are equivalent; the large-scale F-region waves move with the plasma flow, so the interferometer in both cases measures the "phase velocity" of the large features which are creating conditions conducive to generation of the 3-m waves that scatter the 50-MHz waves. The technique has also been used in the auroral zone (Providakes *et al.*, 1983). Note that if the two antennas are separated north–south, the vertical position of the scattering centers may be found. This has considerable potential for auroral physics since it is not yet entirely clear at what height the scattering takes place.

A.1.3 Scintillation Techniques

The scintillation method is used to study the ionosphere by measuring the fluctuations of a radio signal due to its traversal through an irregular medium. In this method, the medium is considered to be equivalent to a diffracting screen with random density irregularities which are frozen in the uniform background and move with a fixed velocity. It is assumed that absorption by the medium is negligible. If the diffracting region is thin, the variations in the emerging wave front are present only in the phase and not in the amplitude of the signal. As the wave propagates beyond the screen, fluctuations in amplitude begin to develop, due to interference effects. This approach was initially used in a number of limiting cases by Hewish (1952) and Wagner (1962), who considered a one-dimensional thin screen; by Bowhill (1961) and Mercier (1962), who considered a two-dimensional thin screen; and by Tatarskii (1961) and Budden (1965), who generalized these studies by assuming a two-dimensional thick screen. These limiting cases were reviewed by Salpeter (1967), who extended the theory to important regimes which had not been previously considered, derived sufficient conditions for the validity of the "thin phase screen" approximation, and investigated the effect of a finite angular source size on the scintillation spectrum for any regime. It was found that, under the weak-scattering, thin-screen approximation, the power spectrum of the density fluctuations must be multiplied by a function that depends on the height of the irregularity layer and on the frequency of the incident wave to yield the power spectrum of the observed scintillation. That is, the power spectrum of the scintillation is a linearly filtered version of the power spectrum of the density fluctuations, under these approximations. The multiplying function, known as the Fresnel filtering factor, acts as a high-pass filter and has deep minima at points which are proportional to the square root of integer numbers (Bowhill, 1961; Budden, 1965; Salpeter, 1967). These minima are smeared when the irregularity layer is thick (Tatarskii, 1961; Budden, 1965) but, when they are present in the scintillation power spectrum, they provide information which can be used to estimate the velocity of the medium (Lovelace *et al.*, 1970; Rufenach, 1972).

A crucial step in the theory of scintillation is the choice of a model for the power spectrum of the density irregularities. Prior to about 1970, all the authors assumed a Gaussian spectrum for both interplanetary and ionospheric irregularities. A Gaussian description is mathematically convenient and in the absence of further information was thought to be a reasonable representation. Many scintillation observations were then used to infer "dominant" density scales, both for the interplanetary medium and for the F-region ionosphere, which were several orders of magnitude smaller than the directly observed outer scale dimensions. In addition, calculated scintillation levels based on Gaussian irregularities could not explain the unexpectedly large scintillation levels reported at UHF and higher frequencies.

Since the results of *in situ* measurements of density fluctuations in both the interplanetary space and the ionosphere can be interpreted in terms of a power law spectrum, this shape has been assumed by several authors in the study of interplanetary (Cronyn, 1970; Lovelace *et al.*, 1970) and ionospheric scintillation (Rufenach, 1971, 1972, 1975). Rufenach (1975) derived approximate expressions, which were subsequently determined exactly by Costa and Kelley (1977), which show that the scintillation levels based on power law irregularities are larger than those based on Gaussian irregularities for similar rms density fluctuation levels.

In the weak-scattering, thin-screen approximation, the medium is replaced by a plane surface along which the phase of the radio wave fluctuates but the amplitude remains constant. As the wave propagates beyond the screen, fluctuations in amplitude begin to develop, due to interference effects. Let $P_\phi(k_x, k_y)$ be the power spectrum of the phase fluctuations in the wave emerging from an irregularity layer and $P_1(k_x, k_y)$ be the power spectrum of the intensity fluctuations in the received signal, expressed in a reference frame with the k_z axis aligned with the direction of the incident wave. It has been shown (Bowhill, 1967; Salpeter, 1967; Rufenach, 1975) that the following relations apply in the weak-scattering limit:

$$P_\phi(k_x, k_y) = 2\pi(r_e\lambda)^2(L \sec \psi)P_N(k_x, k_y, k_z = 0) \tag{A.4}$$

and

$$P_1(k_x, k_y) = 4 \sin^2[(k_x^2 + k_y^2)/k_f^2]P_\phi(k_x, k_y) \tag{A.5}$$

where $k_f^2 = 4\pi/Lz$, λ is the wavelength of the incident wave, z is the distance (measured along the ray path) from the source to the observation point, ψ is the angle between the ray path and the vertical at the point where the ray path intercepts the phase screen, r_e is the classical electron radius ($r_e = 2.82 \times 10^{-15}$ m), L is the thickness of the irregularity layer, and $P_N(k_x, k_y, k_z = 0)$ is the two-dimensional wave number power spectrum in the electron density fluctuations in the plane perpendicular to the **k** vector of the incident radio wave.

Scintillation measurements of P_ϕ and P_1 thus yield information on the two-dimensional wave number spectrum of electron density fluctuations in the ionospheric plasma. These in turn can be compared to one-dimensional measurements of electron density fluctuations made by probes flown on rockets or satellites, which are related to $P_N(\mathbf{k})$ by

$$P_{1D}(k_x) = \int\int P_N(\mathbf{k}) \, dk_y \, dk_z \tag{A.6}$$

where the x axis is taken along the direction of travel. In the case of power law irregularities, one important effect of the integration performed by the spacecraft measurement is to decrease the slope of the power law by one unit relative to the phase spectrum P_ϕ. That is, a phase spectrum which varies as k^{-3} corresponds to a one-dimensional *in situ* spectrum which varies as k^{-2}. It is also clear from (A.4) and (A.5) that phase scintillation measurements are more closely related mathematically to the *in situ* fluctuations since the Fresnel factor, $\sin^2[(k_x^2 + k_y^2)/k_f^2]$, does not appear. Most recent measurements use P_ϕ, and considerable effort has gone into dedicated beacon experiments on satellites which radiate a number of frequencies all in phase with a high-frequency carrier (e.g., the Wideband and HILAT satellites).

Scintillations of signals from a rapidly moving low-altitude satellite beacon yield a rapid cut across an irregular region. The phase scintillation level will then vary in time as the satellite beam traverses different plasma regions. The *frequency* spectrum of the measured phase scintillations thus yields a measure of the *in situ wave number* power spectrum. Likewise, an airplane-based measurement system can use signals from a high-altitude satellite (with a very slow angular velocity across the region of interest) using the aircraft velocity to map out an *in situ* irregularity spectrum. A fixed ground site will also receive time-varying phase scintillations from a high-altitude satellite as a structured medium passes overhead. If that velocity is known the *in situ* spectrum can be determined. If several "spaced" receivers are used, correlation methods can in principle be used to determine the velocity of the medium as it passes over the site. Such spaced receiver drift measurements are potentially of great interest, but further development and comparison with other techniques are still needed at this time.

A.2 *In Situ* Measurements

Satellite- and rocket-borne instrumentation of all kinds has greatly added to our knowledge of the ionosphere. Here we concentrate on techniques which are used to determine ionospheric parameters such as density, temperature, drift velocity, and electric fields. The treatment is not exhaustive but is representative of some of the instrumentation referred to in the text. Other reviews of interest are by Bauer and Nagy (1975) and Mozer (1973).

A.2.1 *Langmuir Probes, Retarding Potential Analyses, and Drift Meters*

In this section we deal with a variety of instruments used to measure the temperature, concentration, and drift velocity of either the ambient thermal electrons or the thermal ions. These instruments are mounted on satellites and rockets that are moving through the plasma at velocities between 1 and 9 km/s. In such cases any conducting surface will collect an electron current and an ion current that can be calculated by assuming that the plasma has a drifting Maxwellian distribution function. If the conductor is held at some potential P, the current is calculated by integrating the distribution function over the surface area of the collector for all energies greater than P.

In the earth's ionosphere the bulk velocity of the electrons with respect to a satellite or rocket is much smaller than the thermal velocity of the electrons. Electron current is therefore collected by all exposed conducting surfaces at a potential that will allow it. The bulk velocity of the ions with respect to a satellite or rocket is, however, comparable to or greater than their thermal velocity. Thus, the ion current to a probe can depend on the orientation of the collecting surface with respect to the relative velocity vector. If the spacecraft were maintained at zero potential with respect to the plasma, then it can be seen that it would collect more electron current due to the higher mobility and larger available collection area than ion current. The spacecraft does not, however, draw a current from the plasma, and therefore the spacecraft will assume a negative potential with respect to the plasma such that the net current is zero. The spacecraft potential, which is also the ground potential of any probe connected to it, causes a region of positive charge to build up around it from which the electrons are repelled. This region is known as the sheath.

A.2.1.a   Electron Temperature Measurements—the Langmuir Probe

Mott-Smith and Langmuir (1926) published a classical paper on the current collection properties of a probe in a plasma from which the modern Langmuir probe derives its name. Most Langmuir probes consist of small conducting surfaces with cylindrical or spherical geometries. Typical sensor dimensions are a 2-cm-diameter sphere or a 20-cm-long, 0.2-cm-diameter cylinder. The probes are usually mounted on short booms of about 20-cm length to project beyond the spacecraft sheath. As mentioned before, the current collection properties of these probes depend on the shape and area of the collector. As we shall see, however, they do not have a large effect on the determination of electron temperature. The instruments function by applying a varying voltage to the probe that covers the range of energies of interest. Typical probe voltages may vary between $+5$ and -5 V. Figure A.6 shows a characteristic curve of probe current versus voltage that one might obtain from a cylindrical Langmuir probe. In fact, all Langmuir characteristic curves have the same regions. The ion saturation region occurs where the probe potential is sufficiently negative to repel all the thermal electrons and begin to attract the ions. The electron retardation region exists where the ion current is not affected greatly by the potential on the probe, but some electrons are repelled. Finally, the electron saturation region exists where all the thermal electrons are attracted to the probe and where the probe potential repels the ion current. In the electron retardation region the probe current is given by

$$I_e = N_e q A (k_B T_e / 2\pi m_e)^{1/2} \exp(-q\Phi/k_B T_e)$$

where A is the collector area, N_e the electron concentration, q the electron

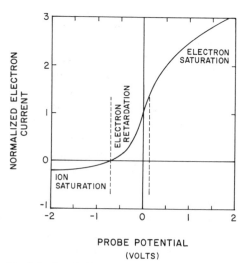

Fig. A.6. A typical Langmuir probe I–V characteristic.

charge, k_B the Boltzmann constant, T_e the electron temperature, m_e the electron mass, and Φ the retarding potential applied to the probe.

A variety of techniques can be employed to extract the electron temperature from such a characteristic curve. If a least-squares fit to the original data can be performed, then either knowledge of the electron concentration is required or this parameter must also be a least-squares variable. An alternative lies in considering the logarithm of the electron current rather than the current itself. Taking the logarithm of both sides of the previous equation,

$$\ln(I_e) = -q\Phi/k_B T_e + \ln(\text{const})$$

Thus, the logarithm of the output from this device in the electron retardation region should be a straight line for which the slope is proportional to the electron temperature. Using this result, it is possible to use on-board microprocessors to derive the electron temperature directly.

In the two saturation regions the ion current and the electron current can be obtained by integrating the appropriate drifting Maxwellian energy distribution function over the collector surface for all energies less than the probe potential. These currents depend on the sensor geometry and, as can be seen in Fig. A.6, they need not saturate at potentials less than 5 V. We will deal with the ion saturation of a planar probe in the discussion of retarding potential analyzers. In the so-called electron saturation region the electron current to a cylindrical collector is given approximately by

$$I_e = 2AN_e q \left(\frac{k_B T_e}{2\pi m_e}\right)^{1/2} \frac{1}{\pi^{1/2}} \left(1 + \frac{q\Phi}{k_B T_e}\right)^{1/2}$$

This expression is derived from the assumption of an infinitely long cylinder and therefore has an "end effect" correction that is usually small for typical cylindrical probe dimensions (Szuszczewicz and Takacs, 1979). Having derived the electron temperature from the electron retardation region, it can be seen that the electron saturation portion of the curve can be simply used to derive the electron concentration. In fact, the ion saturation portion of the curve can be similarly used to calculate the total ion concentration. The latter has the advantage that when the probe velocity is much greater than the thermal velocity of the ions, the expression for the total ion concentration is essentially independent of the ion temperature and a very small end effect correction. Since the ionospheric plasma is charge neutral, this approach is frequently employed to derive the total plasma concentration from a Langmuir probe.

A.2.1.b Ion Temperature and Density Measurements—the Retarding Potential Analyzer

Retarding potential analyzers (RPAs) measure the ambient ion current to a collector as a function of an applied retarding potential. In a manner similar to a

Langmuir probe, a curve of ion current versus retarding potential is obtained from which the thermal ion temperature can be determined.

The velocity of an earth-orbiting satellite can be as high as 8 km/s in the ionosphere. At this velocity the kinetic energy of the ions ($\frac{1}{2} mV^2$) is equivalent to $\frac{1}{3}$ eV/amu. The presence of molecular and metallic ions (Fe^+) in the ionosphere therefore requires that retarding potentials between 20 and 30 V be applied in the sensor. An RPA sensor typically has a planar or spherical geometry. A spherical geometry has the advantage that it can be mounted on a spacecraft in such a way that it has access to the flowing ion gas for most orientations of the spacecraft. The plasma sensor must, however, be mounted to view along the velocity vector of the satellite, thus precluding measurements more than once per spin period if the vehicle is spinning. The planar sensor has the advantage that its known orientation with respect to the spacecraft velocity allows additional information to be extracted from the curve of ion current versus retarding potential. In addition, the construction of a planar geometry leads to more uniform electric potentials in the sensor.

The sensor consists of a collector to which the current-measuring electronics is attached and which is usually grounded to the spacecraft. Placed between the collector and incoming plasma stream is a series of grids. We will describe the operation of a planar sensor as shown schematically in cross section in Fig. A.7. In this configuration the collector S7 is grounded through the current-measuring electronics and the grid S6 is grounded to ensure that no induced currents are seen by the collector. The grid S5 is biased negatively (typically -25 V) to prevent ambient electrons from reaching the collector and to suppress any electrons liberated from the collector by solar ultraviolet radiation. The grids S4 and S3 have the retarding potential applied to them. This potential may take a variety of waveforms in the range 0 to $+30$ V. Finally, the entrance grids S2 and S1 are grounded to the external plane of the instrument to ensure that an electrically uniform, uncontaminated environment is presented to the plasma.

The ion current at a given retarding potential depends on the ion flux reaching the collector for all ion species with energy greater than the applied potential.

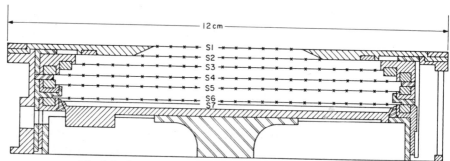

Fig. A.7. Grid arrangement for a plasma retarding potential analyzer.

This quantity can be obtained by integrating the ion velocity distribution function over the collector surface. The simplest expression is obtained for a planar surface and is given by

$$I = \tfrac{1}{2}\alpha V A q \sum_i N_i[1 + \mathrm{erf}(\beta_i f_i) + \frac{1}{\pi^{1/2}\beta_i V} \exp(-\beta_i^2 f_i^2)]$$

where

$$I = \text{collector current}$$
$$\alpha = \text{grid transmission factor}$$
$$A = \text{aperture area}$$
$$q = \text{ionic charge}$$
$$N_i = \text{concentration of } i\text{th ion}$$
$$V = \mathbf{v} \cdot \hat{n}$$
$$\hat{n} = \text{unit vector normal to sensor face}$$
$$\mathbf{v} = \mathbf{v_d} - \mathbf{v_p}$$
$$\mathbf{v_p} = \text{sensor velocity}$$
$$\mathbf{v_d} = \text{ambient ion drift velocity}$$
$$f_i = V - [2q(\Phi + \Phi_0)/m_i]^{1/2}$$
$$\beta_i = (m_i/2k_B T_i)^{1/2}$$
$$m_i = \text{mass of } i\text{th ion}$$
$$T_i = \text{ion temperature}$$
$$k_B = \text{Boltzmann constant}$$
$$\Phi = \text{retarding grid potential}$$
$$\Phi_0 = \text{spacecraft potential}$$

Figure A.8 shows some representative curves of the ion current versus retarding potential that would result from such a sensor operated on an earth-orbiting satellite. The light ions H^+ and He^+ have very little ram energy and thus appear at low retarding voltages. Successively heavier masses appear at higher retarding potentials due to their higher ram energy. The effect of larger ion temperatures is also apparent when comparing frames 4 and 5 of Fig. A.8. Higher temperature also supplies the ions with higher energies, and thus the curve appears "fatter" than a curve with lower ion temperature. Given the number of mass species and the mass numbers, a nonlinear least-squares fitting of the data to the expression above will yield a common ion temperature, the constituent ion concentrations, and the bulk ambient ion drift velocity.

A.2.1.c   *Ion Drift Velocity Measurements—the Ion Drift Meter*

We have seen that a planar RPA is capable of measuring the component of the ion drift velocity along the look direction of the sensor. In the earth's ionosphere, where an orbiting spacecraft moves at a velocity much larger than the ion thermal speed, a rather simple device called an ion drift meter (IDM) can be used to measure the other two mutually perpendicular ion drift velocity components. The

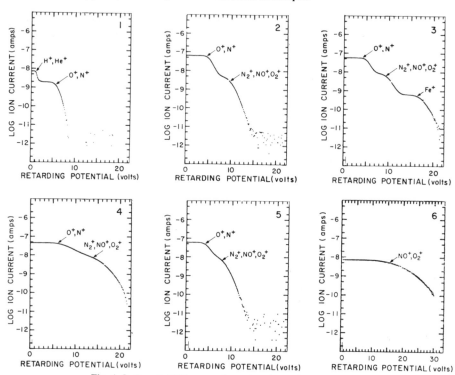

Fig. A.8. Sample RPA curves for various ionospheric conditions.

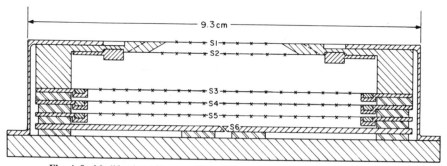

Fig. A.9. Modification of a plasma ion sensor to measure the plasma drift velocity.

IDM has a planar geometry similar to the RPA but has a square entrance aperture and a segmented collector as shown schematically in Fig. A.9.

The grid S5 is biased negatively to prevent ambient electrons from reaching the collector and to suppress any photoelectrons liberated from the collector surface. All other grids are grounded to ensure a field-free region between S2 and S3 through which the ambient ions can drift. Since the ions are moving supersonically with respect to the sensor, they form a collimated beam in the manner shown in Fig. A.10. The collector area which they illuminate will therefore depend on the angle at which they arrive at the sensor. Since the current collected on each collector segment is proportional to the area struck by the ion beam, it can be shown quite easily that the ratio of the currents of two segments is proportional to the tangent of the ion arrival angle. If the entrance aperture has straight edges and the currents to each collector pair are denoted by I_1 and I_2, then

$$\log I_1 - \log I_2 = \log(I_1/I_2) = \log[(H + D \tan \alpha)/(H - D \tan \alpha)]$$

If we let $I_1 - I_2 = \Delta I$ and let $I = I_1 + I_2$, then to first order in $\Delta I/I$ we can show that

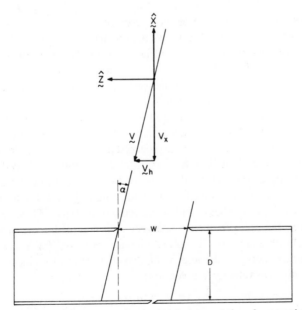

Fig. A.10. Illustration of the geometry involved in the interpretation of measured particle arrival angles.

$$\log I_1 - \log I_2 = (2H/D) \tan \alpha$$

Thus, by using a combination of logarithmic amplifiers to measure the collector current and a linear difference amplifier to supply the difference in the signals, it is possible to obtain the ion arrival angle directly. A similar geometry in the plane perpendicular to the paper can be obtained by considering two appropriate pairs from quandrant segments of the collector. Thus, two mutually perpendicular ion arrival angles can be obtained from a single sensor. These arrival angles can easily be converted to velocities relative to the spacecraft from knowledge of the ram drift obtained from the RPA. Without this information usually only a small error is introduced by using the spacecraft velocity along the look direction. To convert all these velocities into ambient ion drifts it is necessary to subtract the components of the spacecraft velocity from each direction. This requires knowledge of the orientation (attitude) in inertial space, which is usually obtained from a combination of star sensors, sun sensors, and horizon sensors.

A.2.1.d   Ion Composition Measurements—the Mass Spectrometer

The thermal ion devices we have described so far require the detection of the current produced by an ion striking a conducting surface. These devices are therefore incapable of detecting ions if the conducting surface is too small or the number of ions striking the conductor is too small. The RPA we have described distinguishes only a lower limit to the energy of the ions striking the collector and it therefore cannot easily distinguish a small number of ions with small mass immersed in a larger number of ions of larger mass. This is frequently the situation between, say, 500 and 1000 km, where the O^+ concentration exceeds the H^+ and He^+ concentrations. Mass spectrometers are utilized for the detection of very low concentrations of constituent species.

Most mass spectrometers employ high-sensitivity detectors for measuring the constituent ion concentrations. These detectors are called electron multipliers, and they detect the presence of a single charged particle by a cascade of electrons that is produced when it strikes the surface of the multiplier. Preceding the detector is an analyzer that selects the mass of the particle to be detected. Many different techniques are employed in distinguishing ions of different masses, but in common use for ionospheric studies are three basic types. These are the magnetic analyzer, the radio frequency analyzer, and the quadrupole analyzer. The magnetic and drift analyzers use a preacceleration potential P that is much larger than the thermal energy of the ions being detected. Thus, all ions enter the analyzer with a discrete velocity V given by

$$V^2 = 2Z\Phi/M$$

where M is the ion mass and Z denotes the charge state of the ion.

In a magnetic analyzer the application of a uniform magnetic field B perpendicular to the initial velocity vector will cause the ion to begin a gyro motion. Application of the simple equations of motion for a charged particle in a magnetic field will show that the radius of curvature of this motion is given by

$$R = (1/B)(2MV/Z)^{1/2}$$

Thus, ions of larger mass will have a larger radius of curvature than ions of smaller mass. Alternatively, one may arrange for ions of different masses to describe the same radius of curvature in the magnetic field by adjusting their entrance velocity, that is, by adjusting the potential Φ. Figure A.11 shows schematically the trajectories of ions in a 180° magnetic sector. At the image plane it is possible to use a single detector and scan the potential Φ to observe all masses. Alternatively, multiple-detector arrays, known as microchannel plates, can be used to detect simultaneously a portion of the mass spectrum that is directed to the image plane. Finally, we should note that an ambiguity can exist in distinguishing between mass number and charge state. For example, doubly charged helium ions of mass 4 and deuterium ions of mass 2 will appear at the same location for a given potential Φ. It is necessary to resolve the momentum differences between such ions in order to separate them in a magnetic analyzer. Magnetic spectrometers in use today have a variety of sector lengths from about 60° to 180°.

The radio frequency mass spectrometer was introduced by Bennett and is sometimes called a Bennett tube. It functions somewhat like a linear accelerator and is shown schematically in Fig. A.12. The spectrometer has several "stages" consisting of groups of rf modulating grids. The ions are injected into the instrument so that they all have approximately the same energy, as in the case of a

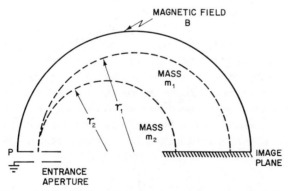

Fig. A.11. The trajectories of ions with different masses in a magnetic field may be used to distinguish them.

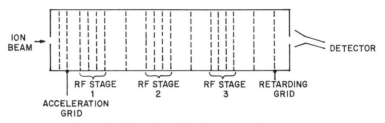

Fig. A.12. Illustration of the rf stages in a Bennett tube.

magnetic spectrometer. The rf grid system then selectively accelerates and de-celerates ions of different mass because they will have entered the instrument with different velocities. The incremental velocity added or subtracted from an ion will depend on the phase of the grid cycle, and maximum velocity addition will occur for unique values of the ion velocity (i.e., ion mass) and phase angle. At the exit of the detector a retarding grid is used to select only those ions receiving the maximum acceleration. Electron multiplier detectors are again util-ized and mass scanning is achieved by appropriately adjusting the analyzer draw-in potential, the frequency of the rf oscillator, and the bias of the retarding grid.

In addition to the magnetic and rf spectrometers, the quadrupole spectrometer is in common use for detecting neutral and ion species in the ionosphere. This spectrometer is so called because it consists of four hyperbolically shaped elec-trodes of length l to which a combination of dc and radio frequency voltages is applied. This is shown schematically in Fig. A.13. The basic concept is to pro-vide a potential field within the instrument which is periodic in time and sym-metrical about the transmission axis. The frequency of the periodic potential is such that the transmission time along the axis length is long compared to the rf period. This spectrometer has the property that the draw-in potential is much less than that used in the magnetic or Bennett spectrometer, so the thermal motion of the ions is important. By adjusting the applied potentials it is possible to allow an ion of a particular mass-to-charge ratio to pass through the system and to ensure that all other ions undergo oscillating trajectories of increasing amplitude so that they are trapped on one of the electrodes before moving the distance l. Again, electron multiplier detectors are used at the instrument exit to achieve extremely high sensitivities. This device and the Bennett tube are not amenable to the use of microchannel plates but are frequently used when concerns about weight preclude the use of a magnetic analyzer or when neutral mass spectrome-try not requiring high time resolution is performed.

A.2.2   Electric Current Measurements—the Fluxgate Magnetometer

The presence of electric currents, both on the ground and in space, is detected by the variations from the background magnetic field that they produce. A mul-

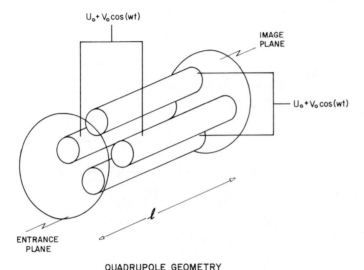

$U_o + V_o \cos(wt)$

IMAGE
PLANE

$U_o + V_o \cos(wt)$

ℓ

ENTRANCE
PLANE

QUADRUPOLE GEOMETRY

Fig. A.13. Geometry of a quadrupole ion mass spectrometer.

titude of such measurements made by instruments in low earth orbit has allowed us to specify the earth's field with great accuracy. Spherical harmonic expansions are generally used with coefficients that are periodically updated to reflect variations in the main field. If we let ΔB be a measured vector difference between the measured magnetic field and that calculated from a model, then Maxwell's equations allow us to specify this magnetic perturbation in terms of a current J by the expression

$$\mathrm{curl}(\Delta B) \;=\; \mu_0 J$$

Devices for measuring these magnetic perturbation vectors are called magnetometers. Various techniques are employed in these instruments, but the most commonly used both on the ground and in space is a fluxgate magnetometer. This device behaves like a transformer with a core material that is magnetically saturable. Figure A.14 shows the operating principle of such a device. The input signal to the primary winding is a sinusoidal waveform of amplitude H_0 and period T. This is approximated by a triangular waveform in the figure. The amplitude of this primary signal is much larger than the ambient magnetic field component ΔH along the core. The ambient field therefore produces a small offset in the drive field. This small offset will cause the core flux to spend more time in one saturated state than in the other. In the case illustrated the time of positive saturation T_s^+ is greater than the time of negative saturation T_s^-. A voltage is induced on the secondary winding of the core only when the core flux density is changing. Thus, the output voltage will consist of a series of pulses

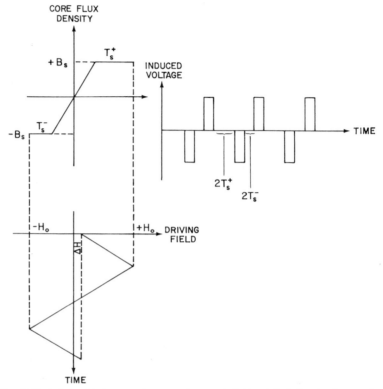

Fig. A.14. Relationship between the external magnetic field and the induced voltage at the output of a fluxgate magnetometer.

that are unevenly spaced due to the different times spent in the saturated state. This is illustrated by the plot of induced voltage versus time shown in Fig. A.14. A Fourier analysis of the output signal shows that second harmonics of the primary driving frequency are present only if the ambient field ΔH is nonzero. Further, the amplitude of the second harmonic is directly proportional to the sign and magnitude of $\Delta H/H_0$. In practice, the secondary harmonic output is obtained by appropriate electronic filtering of the output voltage. The so-called secondary harmonic type of fluxgate magnetometer has found the most extensive use in ground-based and spaceflight applications. It is important to note that J is not directly measured by such a system and that $\nabla \times \mathbf{B}$ is impossible to determine from a single spacecraft.

A.2.2.a   Other Current Measurement Technology

Attempts to measure all the charged particle fluxes and hence deduce the current have proved to be very difficult. In some auroral zone experiments a considerable fraction of intense upward currents has been accounted for by detection of the precipitating electron flux between, say, 10 eV and 20 keV. However, one is never quite sure how much current is carried by the lowest-energy electrons or thermal ions.

Faraday ring devices are now being developed to measure current directly (Torbert, personal communication, 1988). A laser is used to illuminate a loop of fiber optic cable. The change in polarization which results is a direct measure of the current threading the coil. The effect is small and technical problems abound but the reward would be well worth it if the development is successful.

A.2.3 Double-Probe Electric Field Detectors

The double-probe technique has been used successfully to measure electric fields on balloons, rockets, and satellites. In essence, at dc the technique makes a resistive contact to the plasma at two separated positions. If the two electrodes and their local interactions with the medium are sufficiently similar, then the difference of potential between the two electrodes equals the difference of potential between the two points in space. Dividing by the magnitude of the vector distance **d** between them yields the component of the electric field linking the two sensors.

The most symmetrical element is a sphere and many *in situ* electric field measurements use spherical electrodes mounted on insulating booms which are made as long as financial and mechanical constraints allow. The length is maximized since the voltage signal V_s is proportional to $V_s = -\mathbf{E}' \cdot \mathbf{d}$, whereas the sources of error are either independent of the separation from the spacecraft or decrease drastically with separation distance. Typical boom lengths on sounding rockets range from 1 to 15 m, and on satellites 100-m tip-to-tip separations are commonly used.

Some of these error sources can be understood from the interaction between a single electrode and the local plasma. Consider first a floating probe, that is, one from which no current is drawn by the attached electronics. Due to their higher velocity, the flux of electrons to a given surface, nv_e, exceeds the flux of ions to the same surface, nv_i. An electrode will thus charge up negatively, repelling just enough electrons that the electron and ion fluxes are equal. For a spherical probe at rest with area A in a plasma of temperature T, with electron and ion masses m and M, the current I to the probe is given by

$$I = (Ane/4)(8k_BT/\pi M)^{1/2} - (Ane/4)(8k_BT/\pi m)^{1/2} \exp(eV_F/k_BT) \quad (A.7)$$

(Fahleson, 1967). Setting $I = 0$ for a floating probe and solving (A.7) for V_F yields

$$V_F = (k_BT/e) \ln(m/M)^{1/2}$$

For a 1000 K, O^+ plasma, $V_F = -5.1k_BT/e \approx -0.44$ volts. Any asymmetries between the probes will make V_F differ and create an error signal. Sources of errors include asymmetric ion collection due to the spacecraft motion, which is usually at a velocity higher than the ion thermal speed, asymmetric photoemission, which behaves like an ion current to each probe, and asymmetric electron collection. The latter may arise if the main spacecraft emits photoelectrons which are collected by probes or if the magnetic field lines which thread one of the probes pass near the main spacecraft. An additional error signal comes from any possible difference in the average work function of the two surfaces.

A schematic diagram (Mozer, 1973) showing the potentials involved when two separated electrodes, with different work functions WF_1 and WF_2, different floating potentials V_1 and V_2, and separated by a distance **d** in an external electric field E' are connected to a differential electrometer with input resistance R, is given in Fig. A.15. To make the measurement a current I must be drawn by the electronics. This creates an additional potential difference at the two spheres denoted by $R_1 I$ and $R_2 I$, where

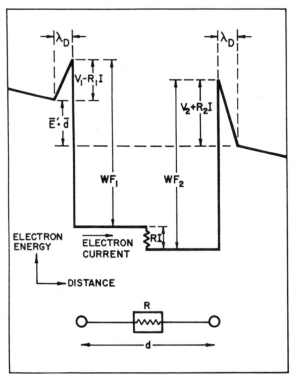

Fig. A.15. Electron potential energy as a function of position near a Langmuir double probe. (Figure courtesy of F. Mozer.)

$$R_j = (\partial V/\partial I)_{V_F}$$

is the dynamic resistance of the electrode–plasma contact evaluated at the floating potential. Differentiating Eq. (A.7) and evaluating R at the floating potential yields

$$R = (k_B T/e)/I_i$$

That is, the electron temperature measured in electron volts divided by the ion current to the probe yields the dynamic resistance for a floating probe. For a sphere of 100-cm² projected collection area, moving through a plasma of temperature 0.16 eV at a velocity of 1 km/s with a density of 10^4 cm^{-3}, $R \approx 10^7$ Ω. Analysis of the potential diagram shows that the electrometer measures

$$RI = [\mathbf{E'} \cdot \mathbf{d} + (V_1 - V_2) + (WF_2 - WF_1)]/[1 + (R_1/R) + (R_2 R)]$$

(Notice that the potential plotted is the electron potential energy, which is the negative of the normal potential in circuit analysis.) Clearly, to detect the full $\mathbf{E'} \cdot \mathbf{d}$ potential we need to have $R \gg R_j$, as well as to keep the error terms in the numerator small. This is quite easy with modern electronics, where input resistances of 10^{12} Ω are easily available.

For systems with $|\mathbf{d}|$ small it is essential to control the asymmetries and surface properties by using spherical electrodes which are coated with a colloidal suspension of carbon (Kelley, 1970) or are gold plated. Longer electrode baselines can employ cylindrical sensors, which are easier to manufacture and interface with the mechanical structure. These systems overcome the signal-to-noise problem by increasing the signal.

Concerning recent ionospheric satellites, the very successful S3-2 and S3-3 and Viking satellites used spherical sensors mounted on triaxial or multiconductor electronic cables held outward by centrifugal force. Separations of 40–100 m tip to tip were obtained. The DE satellites used quasi-rigid cylindrical booms or single conductor wire booms with similar separation distances. In the latter case, the last few meters of the insulation covering the metallic boom was removed to make electrical contact with the plasma. Ten or more ionospheric rockets per year have been flown in the past decade by various groups using either spherical or cylindrical sensors.

Motion of the spacecraft across the earth's magnetic field at velocity $\mathbf{V}_s$ creates an additional vector electric field contribution since in the moving frame the field $\mathbf{E'}$ is given by

$$\mathbf{E'} = \mathbf{E}_A + \mathbf{V}_s \times \mathbf{B}$$

where $\mathbf{E}_A$ is the desired ambient electric field. Subtraction of two vector requires accurate knowledge of $\mathbf{V}_s$ and the vehicle attitude. The magnetic field itself is usually known quite well at ionospheric heights and is not a problem in carrying out the required vector subtraction.

The balloon-borne double-probe method is essentially identical except that the electronics and the system in general are much more sensitive to the requirement ($R \gg R_j$). This is due to the very low charged particle density in the atmosphere and the correspondingly high value of R_j. What is surprising is that the horizontal electric field components at balloon heights (30 km) have anything at all to do with the ionosphere. In fact, a number of theoretical studies as well as the data itself show that ionospheric fields of large horizontal scale (≥ 100 km) map nearly unattenuated down to 30 km (Mozer and Serlin, 1969; Kelley and Mozer, 1975). Superpressure balloons have been used in the southern hemisphere, where very long duration flights are possible, several months long in fact. Communication with these balloons is done via satellite links (Holzworth et al., 1984).

A.2.4   Electrostatic Wave Measurements

Almost all the waves discussed in this text are electrostatic in nature; that is, there are no associated magnetic field fluctuations. From Maxwell's equation,

$$\nabla \times \delta\mathbf{E} = -\partial(\delta B)/\partial t = 0$$

so $\delta\mathbf{E}$ may be derived from a potential

$$\delta\mathbf{E} = -\nabla(\delta\phi) \tag{A.8}$$

and hence the term electrostatic is used. This term does not imply a static phenomenon, however, since these waves can travel at quite high velocities and display high-frequency fluctuations. The linearized version of (A.8) is

$$\delta\mathbf{E} = -i\mathbf{k}(\delta\phi) \tag{A.9}$$

which shows that $\delta\mathbf{E}$ is parallel to the propagation vector.

For frequencies below either the ion plasma frequency or the lower hybrid frequency, whichever is lower, both ions and electrons participate in the response of the medium to wave electric fields. This implies further that there are density fluctuations $\delta n/n$ and a velocity fluctuation δv associated with the wave. Electrostatic waves may thus be detected by the density probes and by the drift meters described above.

Because of the motion of the spacecraft relative to the plasma, even pure spatial variations appear as temporal fluctuations in the vehicle frame. If the wave phase velocity in the plasma is much less than the relative velocity, then the frequency ω in the spacecraft frame is just equal to $\mathbf{k} \cdot \mathbf{V}_R$. This assumption is almost always valid in the case of satellite measurements because of the large velocities (~ 7000 m/s) of the spacecraft and is often, but not always, valid for rockets as well. In the auroral zone, plasma drifts can easily exceed the component of rocket velocity of the vehicle and plasma perpendicular to $\mathbf{B}$. In this case, it is important to measure the quasi-dc electric field. In fact, the relative velocity of the vehicle and plasma perpendicular to $\mathbf{B}$ is $\mathbf{E_R} \times \mathbf{B}/B^2$, where $\mathbf{E_R}$ is the electric field measured directly with the rocket instruments in the rocket reference frame. If the wave frequency in the plasma frame is of the order of $\mathbf{k} \cdot \mathbf{V_R}$, the analysis is more complicated. Low-apogee rockets flown to study the equatorial and auroral electrojets fall in this category.

Although probes and radar systems both respond to irregularities in the plasma density, there is a fundamental difference which must be kept in mind in comparing results. As discussed above, the backscatter radar responds to irregularities with a unique wave number $\mathbf{k}$ which is nearly a three-dimensional delta function. A probe responds to any wave number for which $\mathbf{k} \cdot \mathbf{V_R} \neq 0$. Even at a given frequency ω' the wave number is not unique, since there are infinitely many wave numbers $\mathbf{k}$ such that $\omega' = \mathbf{k} \cdot \mathbf{V_R}$. The frequency response $S(\omega')$ of a density detector to a spectrum of pure irregularities $P_N(\mathbf{k})$ (zero frequency in the plasma frame) is given by

$$S(\omega') = (1/V_R) \iint P_N(k_1, k_2, \omega'/V_R)\, dk_1\, dk_2 \qquad \text{(A.10)}$$

where axis 3 is parallel to $\mathbf{V_R}$ and $\omega' = k_3 V_R$. As an example, consider the response of a detector to the isotropic power law spectrum of irregularities often discussed in relation to scintillations

$$P_N(\mathbf{k}) = (2^{1/2}\alpha N_0^2)/(2\pi^2 k_0^2)\{1 + [k_0^{-4}(k_x^2 + k_y^2 + \alpha^2 k_z^2)]^2\}^{-1} \qquad \text{(A.11)}$$

For a velocity perpendicular to the symmetry axis the double integration (Costa and Kelley, 1977) yields a spectrum which, for $\omega' > V_R k_0$, varies as ω^{-2}. Thus, a k^{-4} spectrum reduces to ω^{-2} for a one-dimensional measurement made by a spacecraft. More detailed discussions of this effect can be found in Costa (1978) and Fredericks and Coroniti (1976). The latter reference shows that in some cases a single measurement cannot be unambiguously interpreted in terms of ω and k if the Doppler shift frequency is comparable to ω. Comparison with theory is in principle most straightforward with computer simulations, since the one-dimensional spectrum can be found simply by sampling the computed density along a desired direction in the simulation. Such a comparison has been made by Kelley *et al.* (1987), as discussed in Chapter 4.

Since the density is a scalar quantity, no information is available on the angular distribution of irregularities. Thus, either theory or some other measurement techniques must be employed to relate the measured density fluctuation spectra to ambient spatial variations.

Although relative density detectors can be used to study irregularities, absolute measurements are important, since they pertain to zero-order conditions in the ionosphere and therefore to conditions which may be unstable to the production of irregularities. In cases where the plasma is stable at the operating wavelength, an incoherent scatter radar can be used to measure the background density structure and, in principle, long-wavelength irregularities. A notable exception is the Jicamarca radar, since during interesting events the system is usually dominated by 3-m irregularities due to instabilities.

Information on the propagation direction of the waves can be found from electric field measurements, since the electric field is a vector quantity. In an electrostatic wave, $\mathbf{k} \cdot \delta\mathbf{E} = 0$, and hence the direction of $\delta\mathbf{E}$ can be used to determine $\mathbf{k}$. In the case of a single plane wave the solution is straightforward, although there may still be an ambiguity in the sign of the phase velocity. In practice, a spectrum of irregularities exists which depends not only on $\mathbf{k}$ but also on the direction of propagation. Furthermore, the discussion surrounding (A.10) applies just as well to one-dimensional measurements of fluctuating electric fields (Fredericks and Coroniti, 1976). Thus, an inspired blending of theoretical understanding and other measurement results must be combined with electric field measurements to yield definitive interpretation of the angular dependence and the frequency spectra. Nonetheless, such measurements contain essential information and, in conjunction with radar or scintillation measurements, can be extremely valuable in understanding ionospheric instabilities.

If the wavelength of an electrostatic wave is shorter than the separation distance $\mathbf{d}$ between electrodes, then the full electric field potential $-\mathbf{E} \cdot \mathbf{d}$ is not measured. In fact, if an integral number of half-wavelengths separates the sensors, the voltage difference is zero. For a plane wave the voltage response is

$$V_S = [-\mathbf{E} \cdot \mathbf{d} \cos(\omega t - \theta)] \sin(\mathbf{k} \cdot \mathbf{d}/2)]/(\mathbf{k} \cdot \mathbf{d}/2) \qquad \text{(A.12)}$$

If $\mathbf{k} \cdot \mathbf{d} \ll 1$, this reduces to $\mathbf{E} \cdot \mathbf{d}$. For $\mathbf{k} \cdot \mathbf{d}/2$ large and a finite spread of wave numbers about some k_0, the sensor response is proportional to the wave potential E_0/k_0 and not the wave electric field. Under certain conditions, these nulls can be used to determine the wavelength of the electrostatic wave since $\mathbf{d}$ is known (Temerin, 1978). Such information is invaluable for sorting out which wave mode causes a given fluctuation.

Simultaneous measurements of $\delta\mathbf{E}(\omega)$ or $\delta V(\omega)$ and the density fluctuation spectrum $\delta n(\omega)/n$ can sometimes be used to determine the wave mode, since theory can be used to predict the relationship between the various components of the wave. This method has been used, for example, to identify drift waves in equatorial spread F (e.g., Chapter 4; Kelley *et al.*, 1982).

A.2.5 Barium Ion Cloud Measurements

Since the electric field is of crucial importance to the understanding of ionospheric dynamics, tracer techniques were developed in the 1960s to measure the flow velocity of the ionospheric plasma. The basic idea is to inject a small number of atoms into the medium under conditions such that the cloud is fully sunlit but observers on the ground are in darkness. For certain materials the sunlight both ionizes the material and makes it visible via a resonant scattering process. Barium vapor has proved to be the best material for this experiment and well over 100 releases have been carried out over the years at altitudes from 150 to 60,000 km.

To vaporize the barium metal, very high temperatures are needed, and a thermite reaction is used to attain the required heat of vaporization. As discussed in the text, small clouds are needed if the tracer aspect is important. "Small" in this case means that the height-integrated conductivity of the cloud must be less than that of the ionosphere. Following Haerendel *et al.* (1967), at F-region altitudes the electric field is given by

$$\mathbf{E}_\perp = \frac{1 + \lambda^*}{2} B[\hat{B} \times \mathbf{V}_\perp + \frac{1}{\kappa_i} (\mathbf{V}_\perp - \mathbf{U}_{n\perp}) + \frac{\lambda^* - 1}{\lambda^* + 1} \mathbf{U}_{n\perp} \times \mathbf{B}] \quad (A.13)$$

where $\mathbf{V}_\perp$ is the velocity of the ion cloud perpendicular to $\mathbf{B}$, λ^* is the ratio of height-integrated Pedersen conductivities in the presence and in the absence of the cloud, κ_i is the ratio of the gyration frequency to the collision frequencies of a barium ion, $\mathbf{U}_{n\perp}$ is the neutral wind velocity in the reference frame fixed to the earth, $\hat{B}$ is a unit vector parallel to the local magnetic field direction, and $\mathbf{E}$ and $\mathbf{E}_\perp$ represent the ionospheric electric field in a frame of reference fixed to the earth in the plane perpendicular to the local magnetic field direction. For $\lambda^* = 1$ and κ_i large (F-region altitudes) $\mathbf{V}_\perp$ is equal to $(\mathbf{E} \times \mathbf{B})/B^2$. The method has proved to be extremely valuable for measuring electric fields in the ionosphere. This is particularly true at low latitudes, where probe techniques and drift meters must contend with satellite velocities much in excess of the plasma velocity.

References

Bauer, S. J., and Nagy, A. F. (1975). Ionospheric direct measurement techniques. *Proc. IEEE* **63**, 1.

Bowhill, S. A. (1961). Statistics of a radio wave diffracted by a random ionosphere. *J. Res. Natl. Bur. Stand., Sect. D* **65**, 275.

Bowles, K. L. (1958). Observations of vertical incidence scatter from the ionosphere at 41 Mc/sec. *Phys. Rev. Lett* **1**, 454.

Budden, K. G. (1965). The amplitude fluctuations of the radio wave scattered from a thick ionospheric layer with weak irregularities. *J. Atmos. Terr. Phys.* **27**, 155.

Costa, E. (1978). Aspects of the linear and nonlinear development of equatorial spread F with application to ionospheric scintillation. Ph.D. Thesis, Cornell University, Ithaca, New York.

Costa, E., and Kelley, M. C. (1977). Ionospheric scintillation calculations based on *in-situ* irregularity spectra. *Radio Sci.* **12**, 797.

Cronyn, W. M. (1970). The analysis of radio scattering and space probe observations of small scale structure in the interplanetary medium. *Astrophys. J.* **161**, 755.

Fahleson, U. V. (1967). Theory of electric field measurements conducted in the magnetosphere with electric probes. *Space Sci. Rev.* **7**, 238.

Farley, D. T. (1979). The ionospheric plasma. *Solar Syst. Plasma Phys.* **3**.

Fejer, B., and Kelley, M. C. (1980). Ionospheric irregularities. *Rev. Geophys. Space Res.* **18**, 401.

Fredericks, R. W., and Coroniti, F. V. (1976). Ambiguities in the deduction of rest frame fluctuation spectra from spectra computed in moving frames. *JGR, J. Geophys. Res.* **81,** 5591.

Gordon, W. E. (1958). Incoherent scattering of radio waves by free electrons with application to space exploration by radar. *Proc. TRE* **46,** 1824.

Haerendel, G., Lüst, R., and Rieger, E. (1967). Motion of artificial ion clouds in the upper atmosphere. *Planet. Space Sci.* **15,** 1.

Hagen, J. B., and Behnke, R. A. (1976). Detection of the electron component of the spectrum in incoherent scatter of radio waves by the ionosphere. *JGR, J. Geophys. Res.* **81,** 3441.

Hagen, J. B., and Hsu, P. Y. (1974). The structure of the protonosphere above Arecibo. *JGR, J. Geophys. Res.* **79,** 4269.

Hanuise, C., and Crochet, M. (1978). Oblique HF radar studies of plasma instabilities in the equatorial electrojet in Africa. *J. Atmos. Terr. Phys.* **40,** 49.

Hernandez, G. (1987). "Fabry–Perot Interferometers." Cambridge Univ. Press, Cambridge, England.

Hewish, A. (1952). The diffraction of galactic radio waves as a method of investigating the irregular structure of the ionosphere. *Proc. R. Soc. London, Ser. A.* **214,** 494.

Holzworth, R. H., Onsager, T., and Powell, S. (1984). Planetary-scale variability of the fair-weather vertical electric field in the stratosphere. *Phys. Rev. Lett.* **53,** 1398.

Kelley, M. C. (1970). Auroral zone electric field measurements on sounding rockets. Ph.D. Thesis, Physics Department, University of California at Berkeley.

Kelley, M. C., and Mozer, F. S. (1975). Simultaneous measurement of the horizontal components of the earth's electric field in the atmosphere and in the ionosphere. *JGR, J. Geophys. Res.* **80,** 3275.

Kelley, M. C., Pfaff, R., Baker, K. D., Ulwick, J. C., Livingston, R., Rino, C., and Tsunoda, R. (1982). Simultaneous rocket probe and radar measurements of equatorial spread F—transitional and short wavelength results. *JGR, J. Geophys. Res.* **87,** 1575.

Kelley, M. C., Seyler, C. E., and Zargham, S. (1987). Collisional interchange instability. 2. A comparison of the numerical simulations with the in situ experimental data. *JGR, J. Geophys. Res.* **92,** 10089.

Kudeki, E., Fejer, B. G., Farley, D. T., and Ierkic, H. M. (1981). Interferometer studies of equatorial F region irregularities and drifts. *Geophys. Res. Lett.* **8,** 377.

Lovelace, R. E., Salpeter, E. E., Sharp, L. E., and Harris, D. E. (1970). Analysis of observations of interplanetary scintillations. *Astrophys. J.* **159,** 1047.

Mercier, R. P. (1962). Diffraction by a screen causing large random fluctuations. *Proc. Cambridge Philos. Soc.* **58,** 382.

Mott-Smith, H., and Langmuir, I. (1926). The theory of collections in gaseous discharges. *Phys. Rev.* **28,** 727.

Mozer, F. S. (1973). Analysis of techniques for measuring DC and AC electric fields in the magnetosphere. *Space Sci. Rev.* **14,** 272.

Mozer, F. S., and Serlin, R. (1969). Magnetospheric electric field measurements with balloons. *J. Geophys. Res.* **74,** 4739.

Providakes, J. F., Swartz, W. E., Farley, D. T., and Fejer, B. G. (1983). First VHF auroral radar interferometer observations. *Geophys. Res. Lett.* **10,** 401.

Rishbeth, H., and Garriott, O. K. (1969). "Introduction to Ionospheric Physics." Academic Press, New York.

Rufenach, C. L. (1971). A radio scintillation method of estimating the small-scale structure in the ionosphere. *J. Atmos. Terr. Phys.* **33,** 1941.

Rufenach, C. L. (1972). Power law wave number spectrum deduced from ionospheric scintillation observations. *J. Geophys. Res.* **77,** 4761.

Rufenach, C. L. (1975). Ionospheric scintillation by a random phase screen: Spectral approach. *Radio Sci.* **10,** 155.

Salpeter, E. E. (1967). Interplanetary scintillation. I. Theory. *Astrophys. J.* **147,** 433.

Szuszczewicz, E. P., and Takacs, P. Z. (1979). Magnetosheath effects on cylindrical Langmuir probes. *Phys. Fluids* **22,** 2424.

Tatarskii, V. I. (1961). "Wave Propagation in a Turbulent Medium." Dover, New York.

Temerin, M. (1978). The polarization, frequency, and wavelengths of high-latitude turbulence. *JGR, J. Geophys. Res.* **83,** 2609.

Tsunoda, R. T., Perreault, P. D., and Hodges, J. C. (1976). Azimuthal distribution of HF slant E region echoes and its relationship to the polar cap electric field. *JGR, J. Geophys. Res.* **81,** 3834.

Wagner, L. S. (1962). Diffraction by a thin phase-changing ionospheric layer with applications to radio star scintillation. *J. Geophys. Res.* **67,** 4195.

Wright, J. W., and Pitteway, M. L. V. (1979). Real-time data acquisition and interpretation capabilities of the dynasonde. 2. Determination and magnetoionic mode and echo location using a small spaced receiving array. *Radio Sci.* **14,** 827.

Appendix B | Reference Material and Equations

In this appendix we gather a number of useful parameters and relationships for quick reference. We also give a brief description of the various magnetic indices used in ionospheric research as well as definitions of some of the esoteric terms which creep into a discipline over the years.

B.1 Atmospheric and Ionospheric Structure

In this section we reproduce some tables and curves from the "Satellite Environment Handbook" (Johnson, 1961)[1] which offer a quick set of reference values for a number of ionospheric and atmospheric parameters, including such derived quantities as conductivity and collision frequency. The reader is cautioned that considerable information on the atmosphere has been accumulated since these curves were generated and that any detailed study should use newer sets of data. Nonetheless, these curves are internally consistent and have proved very useful over the years for quick estimates of the quantities displayed.

We start with the four reference ionospheres shown in Fig. B.1. These are representative of mid-latitude conditions for daytime and nighttime for two extreme portions of the solar cycle, solar maximum and solar minimum. To derive collision frequencies and conductivities from these plots an atmospheric model is necessary, due to the importance of the collisions with the neutral gas. The model used is given in Table B.1, which corresponds to solar maximum, and Table B.2, which corresponds to solar minimum.

[1] Material from Johnson (1961) is reproduced with the permission of Stanford Univ. Press. Copyright © 1961 by the Board of Trustees of the Leland Stanford Junior University.

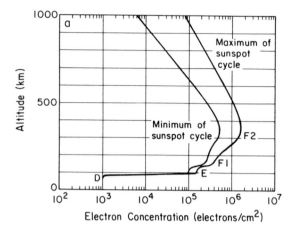

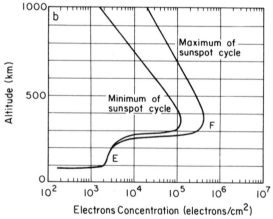

Fig. B.1. Typical mid-latitude distributions at the extremes of the sunspot cycle for daytime (a) and nighttime (b) conditions. [From Johnson (1961).]

The first derived quantity is the ion collision frequency (Chapman, 1956), given by

$$\nu_{in} = (2.6 \times 10^{-9})(n_n + n_i)A^{-1/2} \quad \text{s}^{-1}$$

where n_i is the ion concentration and n_n the neutral density in reciprocal cubic centimeters, and A is the mean molecular weight of the neutrals and ions, which are taken to be the same. The electrons are so light that they have little effect on the ions at ionospheric altitudes. The total collision frequency $\nu_i = \nu_{in} + \nu_{ie}$ then equals ν_{in}. In the magnetosphere, where n_n is essentially zero, ion–electron

Table B.1

Atmospheric Parameters as a Function of Altitude Near Sunspot Maximum[a]

h (km)	T (K)	M	n (particles/cm^3)	ρ (g/cm^3)	p (dyne/cm^2)	H (km)
0	288	29.0	2.5×10^{19}	1.22×10^{-3}	1.01×10^{6}	8.43
10	223	29.0	8.6×10^{18}	4.1×10^{-4}	2.65×10^{5}	6.56
20	217	29.0	1.85×10^{18}	8.9×10^{-5}	5.5×10^{4}	6.38
30	231	29.0	3.7×10^{17}	1.79×10^{-5}	1.19×10^{4}	6.83
40	261	29.0	8.3×10^{16}	4.0×10^{-6}	3.0×10^{3}	8.40
50	283	29.0	2.3×10^{16}	1.08×10^{-6}	9.0×10^{2}	8.11
60	245	29.0	7.53×10^{15}	3.7×10^{-7}	2.55×10^{2}	7.35
70	173	29.0	1.96×10^{15}	9.4×10^{-8}	4.7×10^{1}	5.21
80	168	29.0	2.84×10^{14}	1.36×10^{-8}	6.6×10^{0}	5.08
90	176	28.8	3.9×10^{13}	1.88×10^{-9}	9.5×10^{-1}	5.35
100	208	27.8	6.0×10^{12}	2.8×10^{-10}	1.74×10^{-1}	6.54
120	390	26.1	6.3×10^{11}	2.9×10^{-11}	3.4×10^{-2}	13.1
140	662	24.5	1.07×10^{11}	4.7×10^{-12}	1.04×10^{-2}	24.0
160	926	23.7	4.0×10^{10}	1.52×10^{-12}	5.1×10^{-3}	34.8
180	1115	22.8	2.0×10^{10}	7.7×10^{-13}	3.1×10^{-3}	43.6
200	1230	22.0	1.07×10^{10}	4.2×10^{-13}	1.95×10^{-3}	50.2
220	1305	21.2	6.6×10^{9}	2.7×10^{-13}	1.20×10^{-3}	55.3
240	1356	20.6	4.6×10^{9}	1.70×10^{-13}	8.5×10^{-4}	60.0
260	1400	20.0	3.3×10^{9}	1.12×10^{-13}	6.4×10^{-4}	63.8
280	1430	19.5	2.35×10^{9}	7.9×10^{-14}	4.7×10^{-6}	67.0
300	1455	19.1	1.82×10^{9}	5.7×10^{-14}	3.6×10^{-4}	70.6
320	1472	18.7	1.32×10^{9}	4.3×10^{-14}	2.7×10^{-4}	73.0
340	1485	18.4	1.00×10^{9}	3.1×10^{-14}	2.04×10^{-4}	75.6
360	1491	18.0	7.6×10^{8}	2.3×10^{-14}	1.54×10^{-4}	77.8
380	1496	17.8	5.9×10^{8}	1.78×10^{-14}	1.23×10^{-4}	79.9
400	1500	17.5	4.7×10^{8}	1.38×10^{-14}	9.8×10^{-5}	81.8
450	1500	17.0	2.5×10^{8}	7.2×10^{-15}	5.2×10^{-5}	85.7
500	1500	16.6	1.44×10^{8}	4.1×10^{-15}	2.9×10^{-5}	88.6
600	1500	16.3	4.8×10^{7}	1.32×10^{-15}	1.00×10^{-5}	93.1
700	1500	16.1	1.70×10^{7}	4.6×10^{-16}	3.5×10^{-6}	97.0
800	1500	16.0	6.3×10^{6}	1.66×10^{-16}	1.32×10^{-6}	101
900	1500	15.8	2.35×10^{6}	6.0×10^{-17}	4.90×10^{-7}	105
1000	1500	15.7	9.1×10^{5}	2.4×10^{-17}	1.90×10^{-7}	108
1200	1500	15.2	1.52×10^{5}	3.8×10^{-18}	3.2×10^{-8}	118
1400	1500	13.0	3.2×10^{4}	6.6×10^{-19}	6.7×10^{-9}	145
1600	1500	8.3	1.00×10^{4}	1.35×10^{-19}	2.1×10^{-9}	239
1800	1500	3.5	5.7×10^{3}	3.5×10^{-20}	1.14×10^{-9}	836
2000	1500	1.8	4.6×10^{3}	1.44×10^{-20}	9.5×10^{-10}	1167
2500	1500	1.0	3.47×10^{3}	6.0×10^{-21}	7.2×10^{-10}	2095

Source: Johnson (1961).

[a] The temperature T, molecular weight M, number concentration n, density ρ, and scale height H are given as functions of altitude h.

Table B.2

Atmospheric Parameters as a Function of Altitude Near Sunspot Minimum[a]

h (km)	T (K)	M	n (particles/cm^3)	ρ (g/cm^3)	p (dyne/cm^2)	H (km)
100	208	27.8	6.0×10^{12}	2.8×10^{-10}	1.74×10^{-1}	6.54
120	340	26.1	4.5×10^{11}	1.94×10^{-11}	2.1×10^{-2}	11.38
140	500	24.3	6.6×10^{10}	2.9×10^{-12}	4.6×10^{-3}	18.1
160	628	22.9	2.15×10^{10}	7.7×10^{-13}	1.86×10^{-3}	23.8
180	732	21.5	9.1×10^{9}	3.0×10^{-13}	9.1×10^{-4}	29.5
200	807	20.5	4.5×10^{9}	1.48×10^{-13}	5.0×10^{-4}	35.2
220	865	19.5	2.35×10^{9}	7.7×10^{-14}	2.8×10^{-4}	39.6
240	906	18.9	1.42×10^{9}	4.3×10^{-14}	1.77×10^{-4}	44.1
260	937	18.3	8.9×10^{8}	2.6×10^{-14}	1.14×10^{-4}	47.2
280	959	17.9	5.8×10^{8}	1.62×10^{-14}	7.6×10^{-5}	49.9
300	973	17.5	3.8×10^{8}	1.04×10^{-14}	5.1×10^{-5}	51.9
320	984	17.2	2.6×10^{8}	6.9×10^{-15}	3.5×10^{-5}	53.3
340	991	16.9	1.70×10^{8}	4.8×10^{-15}	2.34×10^{-5}	54.9
360	996	16.7	1.20×10^{8}	3.3×10^{-15}	1.66×10^{-5}	56.0
380	998	16.5	8.3×10^{7}	2.2×10^{-15}	1.14×10^{-5}	57.1
400	1000	16.3	6.0×10^{7}	1.58×10^{-15}	8.3×10^{-6}	58.4
450	1000	16.3	2.6×10^{7}	6.8×10^{-16}	3.6×10^{-6}	60.1
500	1000	15.9	1.20×10^{7}	3.1×10^{-16}	1.66×10^{-6}	61.6
600	1000	15.6	2.45×10^{6}	6.3×10^{-17}	3.4×10^{-7}	66.0
700	1000	14.7	5.8×10^{5}	1.28×10^{-17}	7.9×10^{-8}	76.0
800	1000	10.2	1.74×10^{5}	2.9×10^{-18}	2.4×10^{-8}	157
900	1000	5.1	9.1×10^{4}	7.21×10^{-19}	1.26×10^{-8}	312
1000	1000	2.3	7.1×10^{4}	2.7×10^{-19}	9.8×10^{-9}	487
1200	1000	1.3	5.3×10^{4}	1.07×10^{-19}	7.2×10^{-9}	915
1400	1000	1.0	4.5×10^{4}	7.6×10^{-20}	6.2×10^{-9}	1220
1600	1000	1.0	3.7×10^{4}	6.3×10^{-20}	5.1×10^{-9}	1319
1800	1000	1.0	3.1×10^{4}	5.6×10^{-20}	4.3×10^{-9}	1390
2000	1000	1.0	2.7×10^{4}	4.8×10^{-20}	3.8×10^{-9}	1456
2500	1000	1.0	2.0×10^{4}	3.4×10^{-20}	2.8×10^{-9}	1634

Source: Johnson (1961).

[a] The temperature T, molecular weight M, number concentration n, density ρ, and scale height H are given as functions of altitude h.

collisions must be considered. Curves of ν_i are given in Fig. B.2 for the two sunspot conditions. Since $n_i \ll n_n$ we need not distinguish between daytime and nighttime ionospheres.

The electron collision frequency ν_e depends on both neutral and ion terms. Nicolet (1953) has given the following expression:

$$\nu_{en} = (5.4 \times 10^{-10}) n_n T_e^{1/2} \quad s^{-1}$$

where n_n is again in reciprocal cubic centimeters and T_e is expressed in kelvins. Similarly,

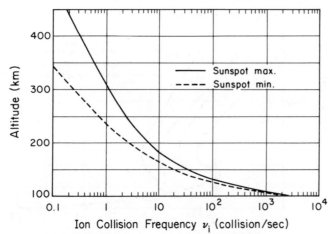

Fig. B.2. Ion collision frequency versus altitude. [From Johnson (1961).]

$$\nu_{ei} = [34 + 4.18 \ln(T_e^3/n_e)]n_e T_e^{-3/2} \quad s^{-1}$$

yields the Coulomb collision frequency. The sum of the two collision frequencies is presented in Fig. B.3 for the four possible combinations of conditions we are considering. In these plots $T_e \approx T_n$ at night and at low altitudes but has been taken to be somewhat larger during the daytime due to heat transfer from photo-electron flux.

The parallel or specific conductivity σ_0 may now be calculated from the expression

$$\sigma_0 = ne^2 (1/m\nu_e + 1/M\nu_i)$$

Notice that since ν_e is proportional to n, at high altitude σ_0 becomes independent of n and, in fact, is simply proportional to $T_e^{3/2}$. The results are presented in Fig. B.4.

The Pedersen conductivity is given by

$$\sigma_p = ne^2 \left(\frac{\nu_e}{m(\nu_e^2 + \Omega_e^2)} + \frac{\nu_i}{M(\nu_i^2 + \Omega_i^2)} \right)$$

and is plotted in Fig. B.5. The magnetic field was taken to be 0.5 gauss = 5×10^{-5} tesla. The Hall conductivity is given by

$$\sigma_H = ne^2 \left(\frac{\Omega_i}{M(\nu_i^2 + \Omega_i^2)} - \frac{\Omega_e}{m(\nu_e^2 + \Omega_e^2)} \right)$$

and is shown in Fig. B.6. Although not shown in Fig. B.6, the Hall conductivity

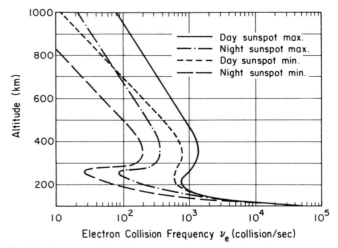

Fig. B.3. Electron collision frequency versus altitude. [From Johnson (1961).]

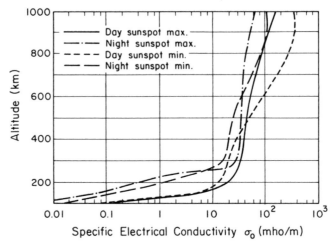

Fig. B.4. Parallel electrical conductivity σ_0 (zero-field conductivity) versus altitude. [From Johnson (1961).]

usually cuts off below 100-km altitude, since it is directly proportional to the electron density. This holds since the electrons $\mathbf{E} \times \mathbf{B}$ drift even as low as 80 km since $\Omega_e > \nu_e$, while the ions are locked into the neutral gas. In the altitude range 80–100 km, then, a Hall current flows, given by

$$\mathbf{J}_H = -ne(\mathbf{E} \times \mathbf{B})/B^2$$

In extremely energetic particle precipitation events, Hall currents can flow at altitudes considerably lower than 105 km.

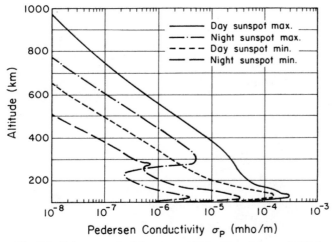

Fig. B.5. Pedersen conductivity versus altitude. [From Johnson (1961).]

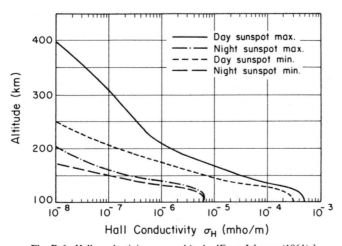

Fig. B.6. Hall conductivity versus altitude. [From Johnson (1961).]

B.2 Miscellaneous Formulas

In these expressions B is in gauss, n is in reciprocal cubic centimeters, $R_e = 6371$ km = one earth radius, temperature is expressed in electron volts, electric field is expressed in millivolts per meter, and A is in atomic mass units. Singly charged species are assumed. Many of the formulas were adapted from the "NRL Plasma Formulary" (Publication 0084-4040, Naval Research Laboratory, Washington, D.C. 20375-5000).

Earth's dipole magnetic field

$$|\mathbf{B}(r, \theta)| = \frac{0.3R_e^3}{r^3} (1 + 3 \sin^2 \theta)^{1/2}$$

$$\mathbf{B} = \frac{-0.6R_e^3 \sin \theta}{r^3} \hat{a}_r + \frac{0.3R_e^3}{r^3} \cos \theta \; \hat{a}_\theta$$

where $\hat{a}_r$ is a unit vector radially outward, $\hat{a}_\theta$ is a unit vector in the direction of increasing θ, and θ is the latitude.

Equation for dipole field line

$$r = L \cos^2 \theta$$

Magnitude of dipole field as a function of latitude on the same field line

$$B(\theta) = [B_e/\cos^6 \theta](1 + 3 \sin^2 \theta)^{1/2}$$

where B_e is the magnetic field value where the field line crosses the equator.

Electron plasma frequency

$$f_p = 8980\sqrt{n} \quad \text{Hz}$$

Ion plasma frequency

$$f_{pi} = 210\sqrt{n/A} \quad \text{Hz}$$

$$f_{pi}(\text{oxygen}) = 52.5\sqrt{n} \; \text{Hz}$$

Electron gyrofrequency

$$f_e = (\Omega_e/2\pi) = 2.8 \times 10^6 (B) \quad \text{Hz}$$

Ion gyrofrequency

$$f_i = (\Omega_i/2\pi) = 1.52 \times 10^3 (B/A) \; \text{Hz}$$

$$f_i(\text{oxygen}) = 95(B) \quad \text{Hz}$$

Lower hybrid frequency

  (a) High-density limit $f_p > (\Omega_e/2\pi)$

$$f_{LH} = 6.52 \times 10^4 \, (B/A^{1/2}) \quad \text{Hz}$$

$$f_{LH}(\text{oxygen}) = 16.3 \times 10^4 (B) \quad \text{Hz}$$

  (b) Low-density limit

$$f_{LH} = \text{ion plasma frequency}$$

Electron gyroradius

$$r_e = 2.38 T_e^{1/2}/B \quad \text{cm}$$

Ion gyroradius

$$r_i = 1.02 \, A^{1/2} T_i^{1/2}/B \quad \text{m}$$

Debye length

$$\lambda_d = 7.43 \times 10^2 \, T^{1/2}/n^{1/2} \quad \text{cm}$$

Electron thermal velocity

$$V_{th}^e = (k_B T_e/m)^{1/2} = 4.19 \times 10^2 T^{1/2} \quad \text{km/s}$$

Ion thermal velocity

$$V_{th}^i = (k_B T_i/M)^{1/2} = 9.79(T_i/A)^{1/2} \quad \text{km/s}$$

Alfvén speed

$$V_A = 2.18 \times 10^6 \, B/(An)^{1/2} \quad \text{km/s}$$

Beta

$$\beta = \text{thermal energy density/magnetic energy density}$$

$$\beta = 4.03 \times 10^{-11} nT/B^2$$

B.3 Surface Magnetic Field Measurements and Magnetic Activity Indices

Equations (6.8) and (6.12) represent the fundamental relationships between horizontal and field-aligned currents and the electric field and conductivity in the ionosphere. The discussion of the dependence of the ionospheric conductivity on altitude given in Chapter 2 shows that the region over which substantial currents flow perpendicular to the magnetic field lines is restricted to altitudes from about 90 to about 130 km in the sunlit ionosphere and may extend up to 300 km when no appreciable local ionization source is present. These horizontal currents are most easily detected from their magnetic field signature on the ground.

Ground-based magnetometers function somewhat differently from those on satellites, but their output—three mutually perpendicular components of the magnetic perturbation from a normal steady baseline—is the same. These magnetic field perturbations are usually resolved along the geographically north–south (positive north), east–west (positive east), and vertical (positive down) directions and are denoted by H, D, and Z components, respectively. Sometimes a geomagnetic coordinate system is used, in which case the symbols X, Y, and Z denote the magnetic perturbations in the geomagnetic north, geomagnetic east, and parallel to **B** directions (in the northern hemisphere). It is impossible to derive the true horizontal ionospheric current distribution uniquely from ground magnetic perturbations, since they are a superposition of contributions from the horizontal ionospheric currents, field-aligned currents, distant currents in the magnetosphere, and currents induced in the earth's surface. For these reasons the ground magnetic perturbations are usually expressed in terms of "equivalent" ionospheric currents. The study of magnetic perturbations and their interpretation as current systems in the earth and in space is extremely complex and we do not discuss this topic in detail. However, magnetic perturbations are used to describe phenomena such as magnetic storms and substorms and to derive indices such as DST, K_p, and AE that describe the magnetic activity in the earth's environment. It is therefore necessary to discuss briefly the nature of the measurements.

In the auroral zone this can be most simply done by considering the latitude profile of the magnetic perturbation produced by a current wedge of 3° latitudinal extent flowing westward through a 20° longitudinal extent of the ionosphere and closed by field-aligned currents at its edges. The resulting latitude profile for a meridian displaced 4° east of the center of the current wedge is shown schematically in Fig. B.7. Applying the right-hand rule for the magnetic perturbation, it can easily be seen that a maximum southward perturbation will exist directly under the current and it will diminish as we move to higher and lower latitudes. On either side of the current there will exist a vertical perturbation directed upward to the equatorward side and downward on the poleward side. Recall that the coordinate system we use is positive directed downward. The field-aligned currents will produce both north–south and east–west perturbations that have a nonzero effect because the current wedge has a finite longitudinal extent. The overhead horizontal current system will in general produce north–south and east–west perturbations, depending on the orientation of the current flow with respect to the magnetometer axes. In this case the contribution is purely in the north–south direction. It should be recognized that from any given magnetometer station the magnetic field perturbation is dependent on the strength of current system and its location with respect to the station.

The AE or auroral electrojet index is obtained from a number (usually greater than 10) of stations distributed in local time in the latitude region that is typical of the northern hemisphere auroral zone. For each of these stations the north–south magnetic perturbation H is recorded as a function of universal time. A superposition of these data from all the stations enables a lower bound or maximum negative excursion of the H component to be determined; this is called the AL index. Similarly, an upper bound or maximum positive excursion in H is determined; this is called the AU index. The difference between these two indices, AU − AL, is called the AE index. Notice that negative H perturbations occur when stations are under an eastward-flowing current. Thus the indices

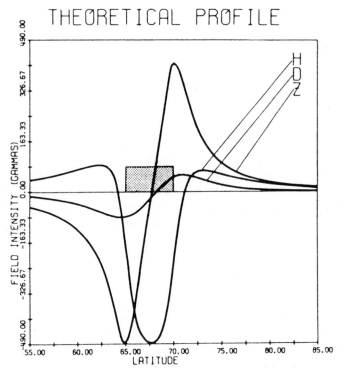

THEORETICAL PROFILE

Fig. B.7. Surface magnetic field perturbation for a current jet in the ionosphere. The curves show latitudinal profiles of *H*, *D*, and *Z* at a meridian displaced 4° east of the symmetry point. The conductivity of the earth and the associated image currents have been ignored. [After Kisabeth and Rostoker (1971). Reproduced with permission of the American Geophysical Union.]

AU and AL give some measure of the individual strengths of eastward and westward electrojets, while AE provides a measure of the overall horizontal current strength. Excursions in the AE index from a nominal daily baseline are called magnetospheric substorms and may have durations of tens of minutes to several hours.

The K_p index is obtained from a number of magnetometer stations at midlatitudes. When these stations are not greatly influenced by the auroral electrojet currents, conditions are termed magnetically quiet. If the auroral zone expands equatorward, however, these stations can record the effects of the auroral electrojet current system and of the magnetospheric ring current and field-aligned currents that can connect it to the ionosphere. This occurs during so-called magnetically disturbed periods. The mid-latitude stations are rarely directly under an intense horizontal current system and thus magnetic perturbations can be dominant in either the *H* or *D* component. The K_p index utilizes both these perturbations by taking the logarithm of the largest excursion in H or D over a 3-h period and placing it on a scale from 0 to 9.

The DST index is obtained from magnetometer stations near the equator but not so close that the E-region equatorial electrojet dominates the magnetic perturbations seen on the ground. At such latitudes the H component of the magnetic perturbation is dominated by the intensity of the magnetospheric ring current. Recall that the ring current is directed westward at all local times and therefore produces a negative H perturbation at low-latitude magnetometer stations. The DST index is a direct measure of the hourly average of this perturbation. Large negative perturbations are indicative of an increase in the intensity of the ring current and typically appear on time scales of about an hour. The decrease in intensity may take much longer, on the order of several hours. The entire period is called a magnetic storm with a relatively short main phase and a

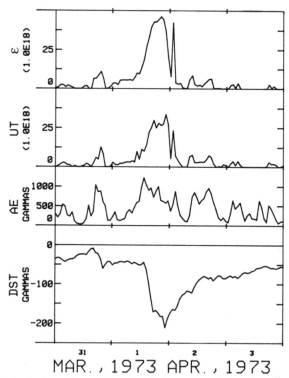

Fig. B.8. Example of the relationship between the power ε of the solar wind–magnetosphere dynamo and the total energy dissipation rate U_T of the magnetosphere for the storm of March 31 to April 3, 1973. The AE and DST indices are also shown. (Figure courtesy of S.-I. Akasofu.)

longer recovery phase. During a magnetic storm it is usual to observe several isolated or one prolonged substorm signature in the AE index.

The DST and AE indices for a typical magnetic storm are shown in the lower two panels of Fig. B.8. Initially DST increases at about 2000 UT on March 31. This is called sudden commencement and is due to increased values of the current at the magnetopause, which shields the magnetosphere from the solar wind. In another viewpoint, the magnetosphere becomes compressed and the field strength increases at the surface of the earth. Polar magnetic activity then begins, as seen in the AE index. Ten or more magnetic substorms occur in the next 72 h. These substorms create auroras and drive horizontal current systems, which dissipate energy in the ionosphere at the rate shown by UT in the second panel [units are (ergs/s) $\times$ 10^{-18}].

As the substorms occur they energize plasma. Some of this precipitates into the atmosphere and some is driven deep into the inner magnetosphere. The latter causes the ring current, which makes a loop around the earth with a net positive current to the west. Such a current loop makes a magnetic perturbation in the southward direction at the earth which is roughly independent of longitude and which causes a net *decrease* in the earth's magnetic field. The DST index therefore has a negative excursion which is the classical signature of a magnetic storm.

The top panel in Fig. B.8 is a measure of the energy flux in the solar wind times an area factor at the front of the magnetosphere (Akasofu, 1981). The agreement between the top two panels is quite remarkable and shows clearly that the system is solar wind driven.

References

Akasofu, S. I. (1981). Energy coupling between the solar wind and the magnetosphere. *Space Sci. Rev.* **28,** 121.

Chapman, S. (1956). The electric conductivity of the ionosphere: A review. *Nuovo Cimento* **5,** Suppl., 1385–1412.

Johnson, F. S., ed. (1961). "Satellite Environment Handbook." Stanford Univ. Press, Stanford, California.

Kisabeth, J. L., and Rostoker, G. (1971). Development of the polar electrojet during polar magnetic substorms. *J. Geophys. Res.* **76,** 6815–6828.

Index

International Geophysics Series

EDITED BY

J. VAN MIEGHEM
(1959–1976)

ANTON L. HALES
(1972–1979)

WILLIAM L. DONN
Lamont-Doherty Geological Observatory
Columbia University
Palisades, New York
(1980–1986)

Current Editors

RENATA DMOWSKA
Division of Applied Science
Harvard University

JAMES R. HOLTON
Department of Atmospheric Sciences
University of Washington
Seattle, Washington

*Out of print.

*Out of print.